Somatosensory Processing

Somatosensory Processing

From Single Neuron to Brain Imaging

Edited by

Mark J. Rowe

*Professor of Physiology, University of New South Wales,
Sydney, Australia*

and

Yoshiaki Iwamura

*Professor of Physiology, Toho University School of Medicine,
Tokyo, Japan*

CRC Press
Taylor & Francis Group
Boca Raton London New York

CRC Press is an imprint of the
Taylor & Francis Group, an **informa** business

First published 2001 by Harwood Academic Publishers

Published 2019 by CRC Press
Taylor & Francis Group
6000 Broken Sound Parkway NW, Suite 300
Boca Raton, FL 33487-2742

© 2001 by Taylor & Francis Group, LLC
CRC Press is an imprint of Taylor & Francis Group, an Informa business

First issued in paperback 2019

No claim to original U.S. Government works

ISBN 13: 978-0-367-45539-2 (pbk)
ISBN 13: 978-90-5702-384-2 (hbk)

Visit the Taylor & Francis Web site at
http://www.taylorandfrancis.com

and the CRC Press Web site at
http://www.crcpress.com

British Library Cataloguing in Publication Data

A catalogue record for this book is available from the British Library.

Cover design by Lee McLachlan.

Contents

Preface

This volume has been based around a symposium on *'Somatosensory Processing: From Single Neuron to Brain Imaging'* held as part of the last Congress of the International Union of Physiological Sciences (IUPS) in St Petersburg in 1997. In addition to the contributions from speakers in that symposium there are chapters from other leading international figures in the somatosensory field. The St Petersburg symposium was chaired by the editors of this volume and was organized under the auspices of the IUPS Commission for Somatosensory Physiology and Pain. This IUPS Commission was established by the IUPS Council at a meeting in Stockholm in 1973 under Ainsley Iggo as its foundation Chairman who served for approximately ten years before being succeeded first by Bill Willis, then Herman Handwerker, and the current chair, Mark Rowe. The Commission formally sponsored satellite symposia associated with the IUPS Congresses of 1977, 1980, 1983 and 1986, which resulted in the following publications:

1. *Active Touch*, ed G. Gordon, Pergamon Press: Oxford, 1978. (Meeting held in Beaune, France, 1977)
2. *Spinal Cord Sensation*, eds A.G. Brown and M. Rethely, Scottish Academic Press: Edinburgh, 1981. (Keszthely, Hungary, 1980)
3. *Development, Organization and Processing in Somatosensory Pathways*, eds Mark Rowe and William D. Willis Jr., Alan R. Liss: New York, 1985 (Hunter Valley, Australia, 1983)
4. *Fine Afferent Nerve Fibers and Pain*, eds R.F., H.-G. Schaible and C. Vahle-Hinz, VCH: Weinheim; Germany, 1987. (Lake Louise, Canada, 1986).
5. *Effects of Injury on Trigeminal and Spinal Somatosensory Systems*, eds Lillian M. Pubols & Barry J. Sessle; Alan R. Liss: New York 1987 (Timberline Lodge, Oregon 1986).

Other publications have received informal and active support over the years from the Commission for Somatosensory Physiology and Pain. The present volume maintains the tradition of publishing the scientific proceedings of somatosensory system symposia held either as part of the IUPS congresses or as satellite symposia of these congresses.

The material covered in the present volume presents data from major international research groups using a diversity of contemporary investigative approaches directed towards understanding the neural mechanisms responsible for sensory and perceptual experience in the areas of tactile, kinaesthetic and pain processing. These approaches range from singe neuron electrophysiological analyses through to magnetic imaging of cortical function. The first three chapters concentrate on peripheral neural mechanisms, in particular, for coding for three-dimensional tactile stimuli (Goodwin and Wheat); the peripheral factors responsible for pain associated with nerve injury (Tracey); and a review of a model peripheral nociceptive system arising from the cornea of the eye (Belmonte and Gallar). Remaining chapters concentrate on central neural mechanisms, first from Willis's group with evidence for a visceral pain pathway traversing the spinal dorsal columns, and second from Rowe who presents an analysis of transmission security at the dorsal column nuclei synaptic relay for inputs arising from single fibers of the different tactile and kinaesthetic sensory fiber classes. Cortical mechanisms for integrating somatosensory and visual information are addressed by Iwamura's group, and the mechanisms for sensorimotor integration within orofacial areas of the primate cortex are dealt with by Sessle and Murray and their colleagues. The capacity of the somatosensory pathways to adapt to partial deafferentation or nerve injury is addressed in a series of four chapters by Snow and Wilson, Schwark, Calford, and Florence and Kaas. In the remaining chapters, Favorov *et al.* examine the capacity for lateral interactions among cortical networks for the purposes of sensory detection; Nelson discusses the capacity of somatosensory cortical responsiveness to be modulated according to behavioural outcomes, and Kakigi describes his magnetic imaging of human somatosensory cortical function.

Mark Rowe
Chair, IUPS Commission for
Somatosensory Physiology and Pain

Contributors

E.D. Al-Chaer
Department of Anatomy and Neurosciences
Department of Internal Medicine
University of Texas Medical Branch
301 University Blvd.
Galveston, TX 77555-1069, USA

C. Belmonte
Instituto de Neurociencias
Universidad Miguel Hernandez
San Juan de Alicante, Spain

M.B. Calford
Psychobiology Laboratory
Division of Psychology
The Australian National University
Canberra, ACT 0200, Australia

O.V. Favorov
Department of Biomedical Engineering
University of North Carolina at Chapel Hill
Chapel Hill, NC 27599-7575, USA

S.L. Florence
Department of Psychology
Vanderbilt University
Nashville, TN 37240, USA

J. Gallar
Instituto de Neurociencias
Universidad Miguel Hernandez
San Juan de Alicante, Spain

A.W. Goodwin
Department of Anatomy and Cell Biology
University of Melbourne
Parkville, Victoria 3052, Australia

J.T. Hester
Department of Biomedical Engineering
University of North Carolina at Chapel Hill
Chapel Hill, NC 27599-7575, USA

A. Iriki
Department of Physiology
Toho University School of Medicine
5-21-16 Omori-Nishi
Otaku, Tokyo 143, Japan

Y. Iwamura
Department of Physiology
Toho University School of Medicine
5-21-16 Omori-Nishi
Otaku, Tokyo 143, Japan

J.H. Kaas
Department of Psychology
Vanderbilt University
Nashville, TN 37240, USA

R. Kakigi
Department of Integrative Physiology
National Institute for Physiological Sciences
Myodaiji, Okazaki 444, Japan

D.G. Kelly
Department of Statistics
University of North Carolina at Chapel Hill
Chapel Hill, NC 27599-7575, USA

L.-D. Lin
School of Dentistry
National Taiwan University
Taipei, Taiwan

G.M. Murray
Faculty of Dentistry
University of Sydney
Sydney, NSW 2006, Australia

R.J. Nelson
Department of Anatomy and Neurobiology
College of Medicine
University of Tennessee
Memphis, 875 Monroe Avenue,
Memphis, TN 38163, USA.

M.J. Quast
Department of Anatomy and Neurosciences
Department of Internal Medicine
University of Texas Medical Branch
301 University Blvd.
Galveston, TX 77555-1069, USA

M.J. Rowe
School of Physiology and Pharmacology
University of New South Wales
Sydney, NSW 2052, Australia

H.D. Schwark
Department of Biological Sciences
University of North Texas, USA

B.J. Sessle
Faculty of Dentistry
University of Toronto
Toronto M5G 1G6, Canada

P.J. Snow
Cerebral and Sensory Functions Unit
Department of Anatomical Sciences
University of Queensland
QLD 4072, Australia

M. Tanaka
Department of Physiology
Toho University School of Medicine
5-21-16 Omori-Nishi
Otaku, Tokyo 143, Japan

M. Taoka
Department of Physiology
Toho University School of Medicine
5-21-16 Omori-Nishi
Otaku, Tokyo 143, Japan

T. Toda
Department of Physiology
Toho University School of Medicine
5-21-16 Omori-Nishi
Otaku, Tokyo 143, Japan

M. Tommerdahl
Department of Biomedical Engineering
University of North Carolina at Chapel Hill
Chapel Hill, NC 27599-7575, USA

D.J. Tracey
School of Anatomy
University of New South Wales
Sydney, NSW 2052, Australia

K.N. Westlund
Department of Anatomy and Neurosciences
Department of Internal Medicine
University of Texas Medical Branch
301 University Blvd. Galveston
TX 77555-1069, USA

H.E. Wheat
Department of Anatomy and Cell Biology
University of Melbourne
Parkville
Victoria 3052, Australia

B.L. Whitsel
Departments of Biomedical Engineering
  and Physiology
University of North Carolina at Chapel Hill
Chapel Hill, NC 27599-7575, USA

W.D. Willis
Department of Anatomy and Neurosciences
Department of Internal Medicine
University of Texas Medical Branch
301 University Blvd.
Galveston, TX 77555-1069, USA

P. Wilson
Division of Human Biology
School of Biological Sciences
University of New England
Armidale, NSW 2351, Australia

D. Yao
Faculty of Dentistry
University of Toronto
Toronto M5G 1G6, Canada

CHAPTER 1
RESPONSES OF SLOWLY ADAPTING CUTANEOUS MECHANORECEPTIVE AFFERENT FIBRES TO THREE-DIMENSIONAL TACTILE STIMULI

A.W. Goodwin and H.E. Wheat

Department of Anatomy and Cell Biology, University of Melbourne, Parkville, Victoria, Australia

Humans use their hands to explore and manipulate a large variety of objects in their environment. In some cases active tactile exploration is used as a means to determine the form of an object, as occurs when an art lover explores the curves of a statue, a physician palpates a lump in a patient, or a spray painter assesses the smoothness of the finished surface. However, in most cases the acquisition of tactile information accompanies the purposive manipulation of an object, as occurs when using a computer mouse or when threading a needle. In all of these situations, tactile feedback is an integral part of the sensorimotor control loop and is essential for the appropriate precise movements of the fingers and thumb. Precision hand movements are not possible if tactile feedback is eliminated by blocking the digital nerves with local anaesthetic (Johansson and Westling, 1984). Similarly, in patients with neurological diseases that disrupt the sensory nerve signals, manual dexterity deteriorates markedly even though the motor system is not affected directly (Moberg, 1962; Rothwell *et al.*, 1982).

There are a number of parameters that must be relayed accurately to the central nervous system during a successful manipulation. One crucial parameter is the local shape of the object in contact with the skin. There have been a number of studies of digital cutaneous afferent responses to objects of different shape, either indented into the skin (Srinivasan and LaMotte, 1987; Goodwin *et al.*, 1995; Goodwin *et al.*, 1997; Dodson *et al.*, 1998; Khalsa *et al.*, 1998) or scanned across the skin (LaMotte and Srinivasan, 1987a; LaMotte and Srinivasan, 1987b; LaMotte *et al.*, 1994; LaMotte and Srinivasan, 1996). During the manipulation it is essential that information about two additional parameters, the position of the object on the skin and the contact force, is relayed to the central nervous system. The neural responses characterising these parameters have been analysed to some extent (Ray and Doetsch, 1990; Cohen and Vierck, Jr., 1993; Wheat *et al.*, 1995). Many earlier studies of cutaneous afferent response also have some bearing on our understanding of the neural representation of these parameters — see Darian-Smith (1984) for review.

Address for correspondence: A.W. Goodwin, Department of Anatomy and Cell Biology, University of Melbourne, Parkville, Victoria 3052, Australia. Tel: 61 9344 5814; Fax: 61 3 9347 5219; E-mail: awg@anatomy.unimelb.edu.au

OUR APPROACH

We have investigated the peripheral neural representation of the shape and position of an object on the finger, and the contact force between the digit and the object. In these studies, we used two classes of shapes that can be defined simply and accurately. The first class comprises spheres which are defined by a single parameter, the radius of the sphere or, equivalently, its curvature (the reciprocal of the radius). The second class of shapes, cylinders, are completely specified by two parameters, the radius (or curvature) and the orientation of the cylinder's axis.

We have chosen to study the signals arising in the glabrous skin of the fingers and thumb when objects contact the skin passively with no lateral movement. The fingerpad is immobilised and a stimulator is used to present the stimuli with accurately controlled parameters. This regimen of stimulus presentation guarantees that the only information relayed to the central nervous system is that transduced by the receptors in the glabrous skin. In most normal manipulations, there are two additional sources of input that must be taken into consideration. First, the movements of the digits may have a scanning component (tangential to the surface of the object) which will result in time-varying signals from the cutaneous afferents that can be integrated to provide more complex information about the object (like global shape and surface texture) and about the task dynamics. Second, cutaneous information can be combined with information from other sources, including receptors in the joints and muscles and descending motor commands (Gandevia *et al.*, 1992). Our current focus is a quantitative characterisation of the cutaneous afferent responses with the aim of extending these studies in the future to include the additional sensory information from the hand.

The first step in achieving our objective of determining how each of the stimulus parameters is represented in the peripheral nerve discharge requires reliable measurements of the capacities of humans. That is, it must be ascertained whether the cutaneous afferents do signal information about a particular stimulus parameter and if so, with what accuracy. The second step requires measurements of the response characteristics of the primary afferent fibres in the peripheral nerve. The third step consists of synthesising the responses of whole populations of primary afferent fibres. Finally, a comparison of the neural population responses with psychophysical performance allows specific neural coding mechanisms to be hypothesised and tested.

PSYCHOPHYSICAL MEASUREMENTS

Tactile capacities of healthy human subjects were measured. The subject's index finger was immobilised by fixing the dorsal surface of the finger in a bed of plasticine, and objects were applied passively to the fingerpad with a stimulator that provided rapid and accurate control of the stimulus parameters.

Two types of psychophysics experiments were conducted. In one, human capacities to scale specific aspects of a stimulus were quantified and in the other, their ability to discriminate small differences in one stimulus parameter was determined. Measuring the ability of a subject to scale the magnitude of the stimulus is the simpler of the two experiments. An example of scaling performance is shown in Figure 1.1A which depicts the relationship between the actual curvature of a sphere and the subject's perception of

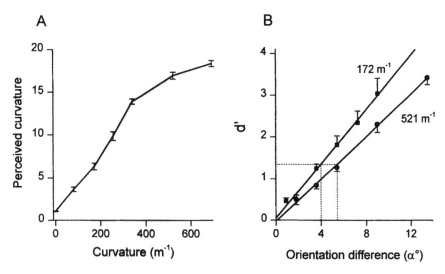

Figure 1.1 Human psychophysical performance for objects applied passively to the fingerpad. **A:** Scaling function for the curvature of a sphere. Mean ± SE for 6 subjects. **B:** Discrimination of the orientation of a cylinder. The orientation is the angle shown in Figure 2A. Mean ± SE for 6 subjects shown for cylinders of curvature 172 or 521 m^{-1}. The discrimination threshold, indicated by the dotted lines, corresponds to d'=1.35.

that curvature. The relationship is monotonic, indicating that the continuum of increasing curvature is mapped into a corresponding continuum of increasing sensation. Thus, the first requirement of any plausible neural representation of the curvature of the stimulus is that its value changes monotonically with increases in stimulus curvature. Scaling is often taken for granted, perhaps because observations equivalent to scaling experiments occur commonly in life. For example, an increased intensity of sunlight leads to a sensation of increased brightness. However, it is not obvious *a priori* how humans will perceive parameters like curvature. Also, before neural codes can be postulated it is important to determine whether the scaling function is monotonic or not, and whether or not saturation is present. We have shown that scaling functions for the curvature, position and contact force of a sphere contacting the finger are monotonic in humans (Goodwin *et al.*, 1991; Goodwin and Wheat, 1992).

Discrimination measurements allow us to determine the accuracy with which humans can distinguish stimuli that differ in one parameter. This is the minimum precision with which that parameter must be represented in the neural discharge. Accurate assessment of discrimination performance requires experimental methods that allow quantitative measurements free from any response biases on the subject's part. We use a two-alternative forced choice paradigm and base our analysis on signal detection theory (Johnson, 1980; Macmillan and Creelman, 1991). Figure 1.1B shows the index of discriminability d' for the orientation of a cylinder contacting the finger. The discrimination threshold, which in this case is equivalent to a difference in orientation that can be discriminated with a probability of 0.75, corresponds to a d' value of 1.35. For a cylinder of curvature 172 m^{-1}

(radius 5.81 mm) the discrimination threshold was 4.2° and for a cylinder of curvature 521 m^{-1} (radius 1.92 mm) it was 5.4° (Dodson *et al.*, 1998).

To measure the precision with which humans can determine the position of an object on the skin, difference thresholds were measured using spherical stimuli. The difference limen was 0.55 mm for a sphere of curvature 172 m^{-1} and 0.38 mm for a sphere of curvature 521 m^{-1} (Wheat *et al.*, 1995). We have also measured the accuracy with which the curvature of spherical objects can be distinguished. For spheres of curvature 144 and 287 m^{-1}, the difference limens were 14 and 32 m^{-1} respectively; a flat surface could be discriminated from a sphere of curvature 4.89 m^{-1} and from a concave surface of curvature –5.40 m^{-1} (Goodwin *et al.*, 1991).

SINGLE FIBRE RESPONSES

Because the stimuli are applied passively to the fingerpad in our experimental paradigm, all information is relayed to the central nervous system by the low-threshold cutaneous mechanoreceptive afferent fibres. In the human fingerpad there are four groups of such fibres classified as the slowly adapting type I and II fibres (SAI and SAII) and the fast adapting type I and II fibres (FAI and FAII) (Vallbo and Johansson, 1984). Monkey glabrous skin is innervated by SAIs, FAIs and FAIIs with properties similar to those in humans; but SAIIs have not been found in monkey glabrous skin. Our stimuli are applied to the skin with defined contact forces, and the movement of the stimulator is damped so that contact and subsequent indentation is smooth. Under these conditions, the only significant responses occur in the SAIs.

Responses of single SAIs in the median nerve were recorded in anaesthetised monkeys. This method enables long periods of stable recording with clearly isolated single fibre responses and thus it has been possible to quantify the responses in extensive parametric studies. All afferents used had receptive fields that were located on the central portion of the fingerpad. These single fibre studies were the first step in identifying the representation of stimulus parameters in the neural responses.

As an illustration, consider how the responses of SAIs change when the orientation of a cylinder contacting the finger changes. Figure 1.2A shows the conventions used for the *x* and *y* coordinates and for the orientation of the cylinder (α). The location of the receptive field centre of one SAI recorded from is shown by the square. For this afferent, the receptive field centre coincided with the centre of rotation of the cylinder. The responses of this afferent, shown by the square symbols in Figure 1.2B, were essentially independent of orientation; thus, by itself, this afferent conveys no information about the orientation of the cylinder. The responses of a second SAI, with a receptive field centre located at the triangular symbol (Fig. 1.2A), are shown by the triangular symbols in Figure 1.2B. This afferent's receptive field centre was located at an orientation of –27°. Thus the response of the unit increased monotonically as the orientation of the cylinder decreased to –27°, and further decreases in orientation resulted in a decrease in response (not shown on the figure). This afferent, by itself, did provide information about the orientation of the cylinder but it was ambiguous. For example, the response at an orientation of –20° was the same as the response at –34°.

Single fibre responses are also ambiguous in the sense that they do not determine which stimulus parameter or parameters have changed. In the case of a cylindrical stimulus, the

A   B

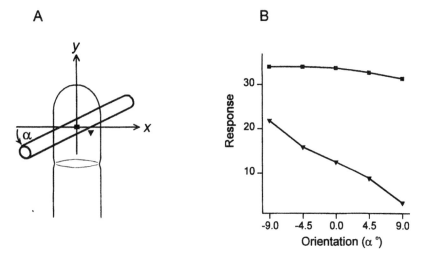

Figure 1.2   Responses of single SAIs to changes in the orientation of a contacting cylinder. **A:** Coordinate conventions for the *x* and *y* axes. The orientation of the cylinder (α) is positive in the counterclockwise direction. The square and the triangle show the location of the receptive field centres of two SAIs. **B:** Responses of those two SAIs as a function of the orientation of the cylinder.

response of an SAI will also change if the position of the cylinder changes, if its curvature changes, or if the contact force changes.

With spherical stimuli there are similar ambiguities in single fibre responses including an inability to distinguish between changes in curvature, position and contact force.

POPULATION RESPONSES

Each of the stimulus parameters must be represented unambiguously in the SAI population response because this was the only source of information available to our human subjects and they were able to make accurate judgements about these parameters.

It is not possible to record simultaneously from large numbers of primary afferent fibres. Thus the SAI population response has to be synthesised from the responses of a number of individually recorded and well characterised SAIs. The principle is illustrated for cylindrical stimuli. It was found that when a cylinder with an orientation of 0° was located at various positions in the receptive field, the response *r* was given by $r = kae^{-by2}$ where the constants *a* and *b* depended on the curvature of the cylinder. All SAIs had the same spatial response characteristics and the only difference between fibres was their sensitivity; this is reflected by the constant *k*. Thus the function ae^{-by2} may be interpreted as the response of an ideal population of SAIs in which all afferents have the same sensitivity. Rotating the cylinder results in a corresponding rotation of the response profile.

In Figure 1.3A the response is shown for an ideal population of SAIs responding to a cylinder with a curvature of 172 m⁻¹ at an orientation of 0°. When the orientation of the cylinder changes, the population response rotates by a corresponding amount as

A. Curvature 172 m⁻¹; α = 0°   B. Curvature 172 m⁻¹; α = 15°

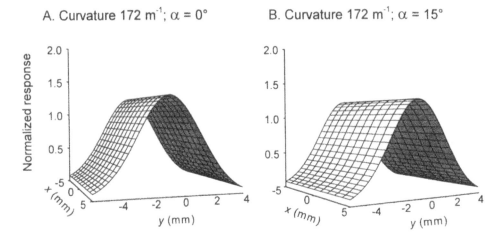

C. Curvature 340 m⁻¹; α = 0°

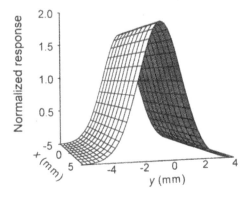

Figure 1.3 Response of an ideal population of SAIs to a cylinder contacting the fingerpad. The effect of changing the orientation of a cylinder (curvature 172 m⁻¹) by 15° can be seen by comparing **A** and **B.** Comparison of **A** and **C** shows the effect of changing the curvature of the cylinder from 172 to 340 m⁻¹ while maintaining the orientation constant.

illustrated in Figure 1.3B for a rotation of 15°. In contrast, if the orientation of the cylinder is unchanged but its curvature increases to 340 m⁻¹ then, as shown in Figure 1.3C, the profile shape changes, increasing in height and decreasing in width (because of changes in the constants a and b). It is simple to distinguish rotation of the profile from a change in the profile shape in Figure 1.3, and it is clear how the population response represents these two parameters independently even though there is ambiguity in the single fibre responses. If the contact force were to change, then the profiles would be scaled correspondingly, and it is obvious that this parameter is also represented independently as was found for spherical stimuli (Goodwin *et al.*, 1995).

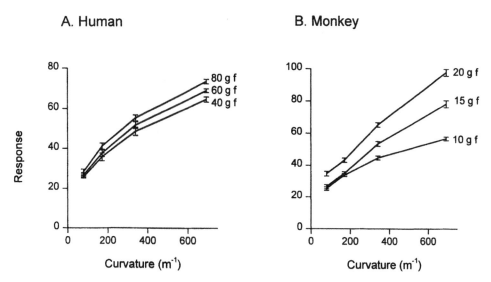

Figure 1.4   Comparison of responses of a human SAI **A** and a monkey SAI **B** to spheres with different curvatures applied to the centre of the receptive field. The range of contact forces used (40 to 80 g f in the human and 10 to 20 g f in the monkey) are approximately equivalent because of the different mechanics of the finger in the two species.

For spherical stimuli, the ideal population response is a function of the form $ae^{-bx2-cy2}$ where the constants a, b and c vary with the curvature of the sphere (see Fig. 1.6). For spheres, it is clear how the curvature and position of the stimulus are represented independently. Comparison of Figures 1.3 and 6 also shows that it is evident from the profile shape whether the stimulus is a cylinder or a sphere.

SPECIES ASSUMPTIONS

The psychophysics experiments were performed in human subjects because it is the mechanisms of human hand function that we wish to explain. By assuming that single fibre responses in humans are similar to single fibre responses in monkeys, we were able to synthesise human SAI population responses for comparison with the psychophysical performance. The validity of this comparison rests on the degree to which single unit properties are similar in the two species. There have been a number of experimental designs for which single fibre responses have been recorded both in humans and in monkeys. In most of those experiments the stimulus was a punctate probe; there was a close correspond-ence in the properties of SAIs, FAIs and FAIIs in humans and in monkeys — for review see Darian-Smith (1984). In other experiments where Braille dots or embossed dot arrays were scanned across the fingerpad, the responses in the two species were also similar (Phillips *et al.*, 1992; Connor *et al.*, 1990; Phillips *et al.*, 1990; Johnson and Lamb, 1981).

In a recent study it was shown that the responses of SAIs to spherical stimuli are also comparable in humans and monkeys (Goodwin *et al.*, 1997). Responses of single SAIs

Somatosensory Processing

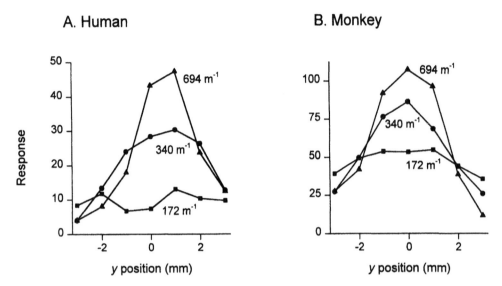

Figure 1.5 Comparison of receptive field profiles for a human SAI **A** and a monkey SAI **B**. A sphere of curvature 172, 340 or 694 m^{-1} was located on the skin at successive positions along the *y* axis which passed through the receptive field centre and was parallel to the axis of the finger.

to spheres of different curvature applied at three different contact forces are shown in Figure 1.4 for the two species. Receptive field response profiles are shown for three different spheres in Figure 1.5. In both figures, the response functions for humans and monkeys are similar and thus the assumption that human and monkey SAIs respond similarly to our stimuli is reasonable. The major difference between humans and monkeys appears to be in the distribution of SAIIs. They are present in glabrous skin of the human hand, albeit at low density, but are absent in monkey glabrous skin (Johansson and Vallbo, 1979). In human glabrous skin, SAIIs do respond to our stimuli but their role in this context is not clear (Goodwin *et al.*, 1997).

POTENTIAL NEURAL CODES

Inspection of the synthesised neural population responses suggests a number of potential codes for the various stimulus parameters. For example, in Figure 1.6 ideal population responses are shown for a sphere of curvature 256 m^{-1} located at either one of two positions on the fingerpad with a separation of 1.5 mm. It is apparent that the locus of activity shifts as the position of the stimulus changes. A potential neural code for the position of the stimulus is simply the 'central point' of that neural activity. This can be calculated formally as the centroid of the 3-dimensional profile, and a change in the position of the stimulus would then be signalled by a change in the centroid.

The first test of the viability of this or any neural code is a comparison with the human scaling performance (Blake *et al.*, 1997). The function relating the perceived position of

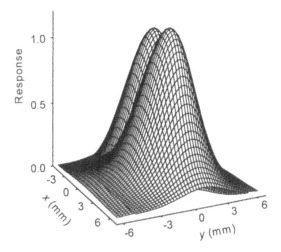

Figure 1.6 Ideal SAI population responses for a sphere of curvature 256 m⁻¹ applied to the fingerpad. The sphere has been applied at one of two positions on the skin which are separated by a distance of 1.5 mm along the *y* axis (parallel to the axis of the finger).

a stimulus on the skin to its actual position (Fig. 1.7A) is monotonic as expected. For the centroid of the response to be a potential code for position, it must satisfy the reciprocal condition that two stimuli which evoke different sensations must give rise to different values of the neural code and, conversely, two stimuli that give rise to different values of the neural code must evoke different sensations. As seen in Figures 1.7B and C, the position of the response centroid is a monotonic function of the position of the stimulus and thus the human sensation and the neural code have a monotonic relationship and satisfy the above condition.

A neural code must satisfy a number of additional conditions in order to be a valid candidate. As we have pointed out, an important aspect of tactile feedback is the presence of independent information about the different stimulus parameters. For example, humans are able to distinguish changes in the orientation of a grasped object from changes in the location of the object, and distinct neural signals for these parameters are obviously crucial for the hand movements required to manipulate such objects with precision. Thus a viable neural code must mirror the values of one stimulus parameter independently of the values of other stimulus parameters which may be varying concomitantly.

REALISTIC POPULATION RESPONSES

The population responses illustrated above are based on ideal populations of SAIs. In reality, however, there are a number of factors that have a significant effect on the population response profiles. For instance, although the shape of the receptive field profile is common to all SAIs on the fingerpad, different afferents have different sensitivities; the range of sensitivities follows a normal distribution with a coefficient of variation of 0.388. Profiles like those shown in Figures 1.3 and 6 will therefore be 'distorted' by the fluctuation

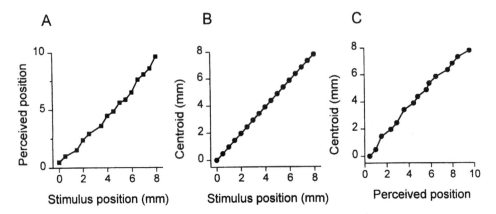

Figure 1.7 A candidate neural code for the position of a stimulus on the skin. **A:** Function for one subject scaling the position of a sphere of curvature 340 m^{-1} (unpublished data, Lucas and Wheat). **B:** The position of the centroid of the population response of an ideal population of SAIs. **C:** Relationship between perceived sensation and the neural code.

in sensitivity among the afferents. This, in turn, will have an effect on the neural codes extracted from the population responses. In addition, the smooth ideal profiles shown imply an infinite innervation density whereas the estimated innervation density of SAIs in the human fingertip is only 0.7 mm^{-2} (Johansson and Vallbo, 1979). Finite innervation density has the effect of sampling the profiles at a relatively small number of points. This will also impact on the neural codes extracted from the population responses.

The consequences of variation in sensitivity and of a finite sampling density are greatest when considering resolution. Assuming that a change in the stimulus produces a change in neural responses, this will lead to a 'signal' consisting of a corresponding change in the neural code. But, neural noise will cause the responses of neurons to vary randomly even when the stimulus parameters remain constant, and therefore any associated neural code will be 'noisy'. The ability of a candidate neural code to resolve two different stimuli will depend on the signal to noise ratio which depends on the innervation density and on the distribution of sensitivities. For a code to be viable, the resolution must be at least as great as the capacity of human subjects to discriminate the two stimuli. We have modelled the resolution of realistic populations accounting for the limitations imposed by the known properties of SAIs.

CONCLUSION

Our experiments, and those of others, have shown that when objects contact the fingers passively, humans are able to make accurate judgements about properties of the stimulus like its shape, its position on the finger, the contact force and its surface texture. Matching neurophysiological experiments in monkeys have shown that this information is relayed by the cutaneous mechanoreceptive afferents, with the slowly adapting type I afferents playing a major role (e.g. Phillips *et al.*, 1992; Wheat *et al.*, 1995; Khalsa *et al.*, 1998).

These same signals are present when humans manipulate objects in their everyday lives and, together with additional proprioceptive information from joint, muscle and skin receptors, form the sensory feedback that is known to be crucial for precise hand movements.

The information conveyed by single cutaneous afferent fibres about each of the stimulus parameters is ambiguous. The same response may result from two different values of a single parameter that can be clearly discriminated by humans, and identical responses may result from several different combinations of the multiple stimulus parameters. These ambiguities are resolved in the responses of the entire population of active fibres. Comparison of the neural population responses with human psychophysical performance allows candidate neural codes to be postulated (Blake *et al.*, 1997).

REFERENCES

Blake, D.T., Hsiao, S.S. and Johnson, K.O. (1997). Neural coding mechanisms in tactile pattern recognition: The relative contributions of slowly and rapidly adapting mechanoreceptors to perceived roughness. *Journal of Neuroscience*, **17**, 7480–7489.

Cohen, R.H. and Vierck, C.J., Jr. (1993). Population estimates for responses of cutaneous mechanoreceptors to a vertically indenting probe on the glabrous skin of monkeys. *Experimental Brain Research*, **94**, 105–119.

Connor, C.E., Hsiao, S.S., Phillips, J.R. and Johnson, K.O. (1990). Tactile roughness: Neural codes that account for psychophysical magnitude estimates. *Journal of Neuroscience*, **10**, 3823–3836.

Darian-Smith, I. (1984). The sense of touch: performance and peripheral neural processes. In *Handbook of Physiology – The Nervous System III*, edited by J.M. Brookhart, V.B. Mountcastle, I. Darian-Smith and S.R. Geiger, pp. 739–788. Bethesda, USA: American Physiological Society.

Dodson, M.J., Goodwin, A.W., Browning, A.S. and Gehring, H.M. (1998). Peripheral neural mechanisms determining the orientation of cylinders grasped by the digits. *Journal of Neuroscience*, **18**, 521–530.

Gandevia, S.C., McCloskey, D.I. and Burke, D. (1992). Kinaesthetic signals and muscle contraction. *Trends in Neuroscience*, **15**, 62–65.

Goodwin, A.W., Browning, A.S. and Wheat, H.E. (1995). Representation of curved surfaces in responses of mechanoreceptive afferent fibers innervating the monkey's fingerpad. *Journal of Neuroscience*, **15**, 798–810.

Goodwin, A.W., John, K.T. and Marceglia, A.H. (1991). Tactile discrimination of curvature by humans using only cutaneous information from the fingerpads. *Experimental Brain Research*, **86**, 663–672.

Goodwin, A.W., Macefield, V.G. and Bisley, J.W. (1997). Encoding of object curvature by tactile afferents from human fingers. *Journal of Neurophysiology*, **78**, 2881–2888.

Goodwin, A.W. and Wheat, H.E. (1992). Magnitude estimation of force when objects with different shapes are applied passively to the fingerpad. *Somatosensory and Motor Research*, **9**, 339–344.

Johansson, R.S. and Vallbo, A.B. (1979). Tactile sensibility in the human hand: relative and absolute densities of four types of mechanoreceptive units in glabrous skin. *Journal of Physiology*, **286**, 283–300.

Johansson, R.S. and Westling, G. (1984). Roles of glabrous skin receptors and sensorimotor memory in automatic control of precision grip when lifting rougher or more slippery objects. *Experimental Brain Research*, **56**, 550–564.

Johnson, K.O. (1980). Sensory discrimination: decision process. *Journal of Neurophysiology*, **43**, 1771–1792.

Johnson, K.O. and Lamb, G.D. (1981). Neural mechanisms of spatial tactile discrimination: neural patterns evoked by braille-like dot patterns in the monkey. *Journal of Physiology*, **310**, 117–144.

Khalsa, P.S., Friedman, R.M., Srinivasan, M.A. and LaMotte, R.H. (1998). Encoding of shape and orientation of objects indented into the monkey fingerpad by populations of slowly and rapidly adapting mechanoreceptors. *Journal of Neurophysiology*, **79**, 3238–3251.

LaMotte, R.H. and Srinivasan, M.A. (1987a). Tactile discrimination of shape: Responses of rapidly adapting mechanoreceptive afferents to a step stroked across the monkey fingerpad. *Journal of Neuroscience*, **7**, 1672–1681.

LaMotte, R.H. and Srinivasan, M.A. (1987b). Tactile discrimination of shape: responses of slowly adapting mechanoreceptive afferents to a step stroked across the monkey fingerpad. *Journal of Neuroscience*, 7, 1655–1671.

LaMotte, R.H. and Srinivasan, M.A. (1996). Neural encoding of shape: Responses of cutaneous mechanoreceptors to a wavy surface stroked across the monkey fingerpad. *Journal of Neurophysiology*, 76, 3787–3797.

LaMotte, R.H., Srinivasan, M.A., Lu, C. and Klusch-Petersen, A. (1994). Cutaneous neural codes for shape. *Canadian Journal of Physiology and Pharmacology*, 72, 498–505.

Macmillan, N.A. and Creelman, C.D. (1991). *Detection theory: a user's guide*. Cambridge,U.K.: Cambridge University Press.

Moberg, E. (1962). Criticism and study of methods for examining sensibility in the hand. *Neurology*, 12, 8–19.

Phillips, J.R., Johansson, R.S. and Johnson, K.O. (1990). Representation of braille characters in human nerve fibres. *Experimental Brain Research*, 81, 589–592.

Phillips, J.R., Johansson, R.S. and Johnson, K.O. (1992). Responses of human mechanoreceptive afferents to embossed dot arrays scanned across fingerpad skin. *Journal of Neuroscience*, 12, 827–839.

Ray, R.H. and Doetsch, G.S. (1990). Coding of stimulus location and intensity in populations of mechanosensitive nerve fibers of the raccoon: II. Across– fiber response patterns. *Brain Research Bulletin*, 25, 533–550.

Rothwell, J.C., Traub, M.M., Day, B.L., Obeso, P.K., Thomas, P.K. and Marsden, C.D. (1982). Manual motor performance in a deafferented man. *Brain*, 105, 515–542.

Srinivasan, M.A. and LaMotte, R.H. (1987). Tactile discrimination of shape: responses of slowly and rapidly adapting mechanoreceptive afferents to a step indented into the monkey fingerpad. *Journal of Neuroscience*, 7, 1682–1697.

Vallbo, A.B. and Johansson, R.S. (1984). Properties of cutaneous mechanoreceptors in the human hand related to touch sensation. *Human Neurobiology*, 3, 3–14.

Wheat, H.E., Goodwin, A.W. and Browning, A.S. (1995). Tactile resolution: peripheral neural mechanisms underlying the human capacity to determine positions of objects contacting the fingerpad. *Journal of Neuroscience*, 15, 5582–5595.

CHAPTER 2
PAIN DUE TO NERVE INJURY: THE ROLE OF NERVE GROWTH FACTOR

David J. Tracey

School of Anatomy, University of New South Wales, Sydney, Australia

INTRODUCTION

Nerve damage due to trauma or disease often leads to chronic pain, for reasons that are still poorly understood. Nerve damage often results in increased sensitivity to painful stimuli (hyperalgesia) and the perception of innocuous stimuli as painful (allodynia); it may also result in spontaneous pain. In man, these symptoms are referred to as neuropathic pain, or causalgia if the damage is traumatic in origin. Disorders of pain sensation due to nerve damage are common, debilitating and difficult to treat (Kingery, 1997). Animal models of peripheral neuropathy have recently been developed in which the mechanisms underlying hyperalgesia due to nerve injury can be analysed (Bennett and Xie, 1988; Seltzer *et al.*, 1990; Kim and Chung, 1992). These animal models are well established (Bennett, 1993; Seltzer, 1995) and have already provided useful information about the central mechanisms of hyperalgesia (Woolf and Doubell, 1994).

Less is known about peripheral mechanisms underlying neuropathic hyperalgesia, although lesioned afferent fibres in the damaged nerve appear to become more sensitive to mechanical, chemical and probably thermal stimuli (Simone, 1992; Devor, 1994). These increases in sensitivity are due in part to upregulation of Na^+ and Ca^{2+} channels (Devor, 1991; Dib-Hajj *et al.*, 1999). Primary afferents may also be sensitised by increased levels of algesic mediators such as bradykinin, serotonin or prostaglandins (Mizumura and Kumazawa, 1996; Kress and Reeh, 1996; Rang *et al.*, 1991; Wood and Docherty, 1997). These increased levels are characteristic of inflamed tissue. While nerve injury does not lead to a typical inflammatory response, it is clear that immune cells known to release inflammatory mediators are recruited to damaged nerves, and are likely to make a significant contribution to neuropathic hyperalgesia (Tracey and Walker, 1995).

Recently it has been suggested that nerve growth factor plays a key role in inflammatory hyperalgesia (Woolf and Doubell, 1994; Woolf, 1996; Lewin, 1995; McMahon, 1996), and the aim of this chapter is to review evidence on the role of nerve growth factor in regulating sensitivity to painful stimuli, with particular reference to data from our own laboratory suggesting that nerve growth factor contributes to the hyperalgesia which results from nerve injury.

Address for correspondence: Dr David J. Tracey, School of Anatomy, University of New South Wales, Sydney NSW 2052, Australia. Tel: (+61-2) 9385-2471; Fax: (+61-2) 9313-6252; E-mail: d.tracey@unsw.edu.au

Nerve growth factor (NGF) is best known for its role in protecting certain classes of neurons from cell death during development (Davies, 1996; Lindsay, 1996). In recent years it has become clear that NGF has other roles as well in the nervous system (Lewin and Barde, 1996); examples are promotion of sprouting of peripheral sensory neurons (Diamond *et al.*, 1987) and regulation of connectivity and plasticity in the cortex (Shieh and Ghosh, 1997). The actions of NGF are not confined to the nervous system, since it mediates interactions between the nervous system and immune and endocrine functions (Levi-Montalcini *et al.*, 1996; Scully and Otten, 1995).

An early indication that NGF might play a role in regulating sensitivity to painful stimuli was the finding by Levine's laboratory that the amino-terminal octapeptide of nerve growth factor elicited hyperalgesia in the rat when applied to injured tissue (Taiwo *et al.*, 1991). These experiments were motivated by apparent similarities between bradykinin, which directly activates nociceptors, and this fragment of NGF. More recently it was found that transgenic mice which overexpress NGF in the skin are hyperalgesic (Davis *et al.*, 1993) and that peripheral injection of NGF causes hyperalgesia (Andreev *et al.*, 1995; Lewin *et al.*, 1993; Theodosiou *et al.*, 1999). Furthermore, experimental inflammation induced by intraplantar injection of complete Freund's adjuvant causes hyperalgesia which is accompanied by elevated levels of NGF (Woolf *et al.*, 1994), and administration of agents which sequester NGF prevents the hyperalgesia which normally accompanies inflammation (McMahon *et al.*, 1995; Lewin *et al.*, 1994; Woolf *et al.*, 1994).

The mechanism for the hyperalgesic action of NGF is not completely understood, but there is good evidence for a peripheral mechanism, in which NGF may act directly on trkA or p75 receptors located on the nociceptor itself (Bennett *et al.*, 1998a; Bevan and Winter, 1995; Lewin *et al.*, 1993), or indirectly via other cells. NGF may regulate the sensitivity of nociceptors indirectly by an action on immune cells or sympathetic neurons, which would then release an algesic mediator acting directly on the nociceptor. One possibility is that NGF acts on mast cells (Lewin *et al.*, 1994; Rueff and Mendell, 1996; Woolf *et al.*, 1996) to elicit the release (in the rodent at least) of serotonin (Lewin *et al.*, 1994), which is believed to act directly on nociceptors (Rueff and Dray, 1993; Taiwo and Levine, 1992). A second possibility is an action of NGF on neutrophils. NGF is chemotactic for neutrophils (Boyle *et al.*, 1985), and there is recent evidence that the hyperalgesia elicited by NGF is dependent on circulating neutrophils (Bennett *et al.*, 1998b). While the algesic mediator released by neutrophils was not specified by these authors, earlier work suggests that a likely candidate is 8R, 15S diHETE, an eicosanoid product of the 15-lipoxygenase pathway which is released by neutrophils (Levine *et al.*, 1985), activates nociceptors (White *et al.*, 1990) and causes hyperalgesia (Levine *et al.*, 1986). Postganglionic sympathetic neurons may also mediate sensitisation of nociceptors by NGF (Andreev *et al.*, 1995), although the role of sympathetic neurons is less important than that of mast cells (Woolf *et al.*, 1996). It is also likely that retrograde transport of NGF by sensory axons to their cell bodies in the dorsal root ganglia leads to changes in gene expression, leading to elevation in the levels of certain neuropeptides (and amplification of nociceptive sensory signals) as well as upregulation of growth-related molecules, resulting in sprouting of sensory neurons to hyperinnervate injured tissue (Woolf, 1996).

NEW FINDINGS

In our laboratory, we set out to determine whether NGF and other neurotrophins may contribute to neuropathic hyperalgesia in an animal model of peripheral nerve injury (Theodosiou *et al.*, 1999). Our experiments were carried out on inbred male Wistar rats in which about one third of the diameter of the sciatic nerve was tightly ligated in the proximal thigh (Seltzer *et al.*, 1990). This model of neuropathic pain has proven to be reliable in our hands (Tracey *et al.*, 1995a; Tracey *et al.*, 1995b). Hyperalgesia was estimated by measuring the latency to withdrawal of the hindpaw from a noxious thermal stimulus (Hargreaves *et al.*, 1988) or the threshold to hindpaw withdrawal from a mechanical stimulus, using an analgesymeter based on a standard test for mechanical nociception (Randall and Selitto, 1957). Tests were carried out prior to surgery to establish a baseline and accustom the rats to the testing equipment; this reduces the variability of the measurements (Taiwo *et al.*, 1989). The extent of hyperalgesia was estimated by comparing values obtained from the operated hindpaw (usually the left) with those obtained from the contralateral paw. We believe that use of the contralateral paw as an internal control gives more consistent and reliable data than estimates based on comparison with a baseline value established prior to surgery; such estimates may change with time or behavioural parameters such as attention.

Before surgery, there was no significant difference in sensitivity of the left and right hindpaws to thermal and mechanical stimuli. Hyperalgesia usually developed within two days after surgery, since mechanical thresholds and thermal latencies for withdrawal were significantly less for the operated hindpaw than for the contralateral hindpaw (Fig. 2.1A, C).

NEUROTROPHINS

Subcutaneous injection of nerve growth factor (500 ng) into the hindpaws of normal rats resulted in hyperalgesia in response to both mechanical and thermal stimuli (Fig. 2.2). Subcutaneous injection of brain-derived neurotrophic factor (BDNF, 500 ng) or neurotrophin-3 (NT-3, 500 ng) resulted in thermal hyperalgesia, but not mechanical hyperalgesia (not shown). In rats with hyperalgesia resulting from partial ligation of the sciatic nerve, significant relief of mechanical and thermal hyperalgesia was produced by intraperitoneal injection of antiserum against NGF (5 µL/g on each of day 11 and day 12 after nerve injury; Fig. 2.3). Injection of antiserum against BDNF using the same doses of antiserum on days 16 and 17 after nerve injury also relieved both mechanical and thermal hyperalgesia (Fig. 2.4). Antiserum against NT-3 had no effect on hyperalgesia.

MAST CELLS

Treatment of hyperalgesic rats with the mast cell degranulating compound 48/80 on four successive days (Coderre *et al.*, 1989; Lewin *et al.*, 1994) produced significant relief of hyperalgesia about 48 hours after the first injection of 48/80, with maximal relief after the last injection, about 80 hours after the initial injection (Fig. 2.5). Degranulation was confirmed by comparing the histological profiles of mast cells in the sciatic nerves of treated and control rats (Tal and Liberman, 1997).

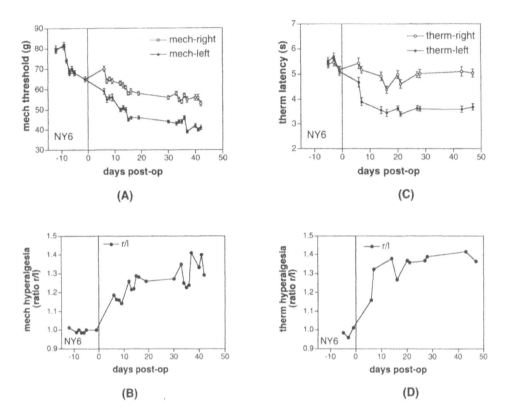

Figure 2.1 Hyperalgesia following partial ligation of sciatic nerve. **A:** Withdrawal thresholds for left and right hindpaws in response to mechanical stimuli after partial ligation of the sciatic nerve. **B:** Timecourse of mechanical hyperalgesia following partial ligation of the sciatic nerve. Data points represent the threshold for the right paw divided by that for the left paw (r/l). **C:** Withdrawal latencies for left and right hindpaws in response to thermal stimuli after partial ligation of the sciatic nerve. **D:** Timecourse of thermal hyperalgesia following partial ligation of the sciatic nerve. Data points represent right paw latency divided by left paw latency (r/l). Data points are means ± S.E.M. Vertical line indicates day of nerve injury; n=12.

SEROTONIN RECEPTOR ANTAGONISTS

We confirmed that a single subcutaneous injection of serotonin (1 μg) into the hindpaw of normal (unoperated) rats elicited mechanical hyperalgesia as previously reported (Taiwo and Levine 1992). Subcutaneous injection of NAN-190 HBr, ketanserin and ICS-205,930 (antagonists to the $5HT_{1A}$, $5HT_{2A}$ and $5HT_3$ receptor subtypes respectively) alleviated mechanical hyperalgesia in a dose-dependent manner (Fig. 2.6). In each case hyperalgesia was relieved if the drug was injected into the hyperalgesic hindpaw, but not when injected into the contralateral paw, suggesting that the site of action was localised to the hyperalgesic hindlimb and the effect was not systemic.

NGF-mechanical

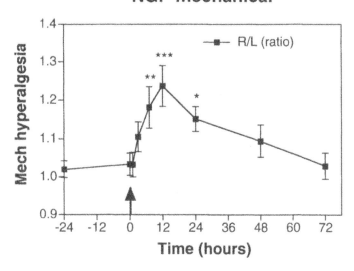

NGF-thermal

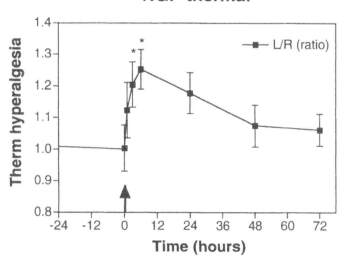

Figure 2.2 Hyperalgesic effect of nerve growth factor injected into the hindpaw. Top: Hyperalgesic index (mechanical, R/L) following subcutaneous injection of 500 ng/10 μL NGF into the left hindpaw of a normal (unoperated) rat. Note peak hyperalgesia at 12 hours. n=12. **Bottom:** Hyperalgesic index (thermal, L/R) following subcutaneous injection of 500 ng/10 μL NGF into the right hindpaw of a normal (unoperated) rat. Note peak hyperalgesia at 6 hours. n=12. In each case 10 μL vehicle (saline) was injected into the contralateral paw. Data points are means ± S.E.M. Asterisks indicate a significant difference between that data point and pre-operative values as determined by posthoc comparisons (Fisher's PLSD). Arrows indicate time of injection. n=12, ·p<0.05, ··p<0.01, ···p<0.001.

anti NGF-mechanical

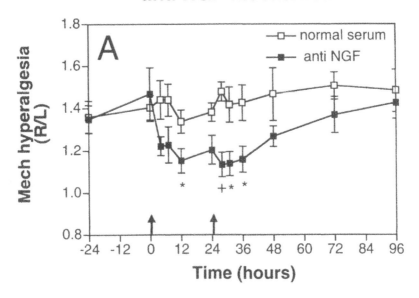

anti NGF-thermal

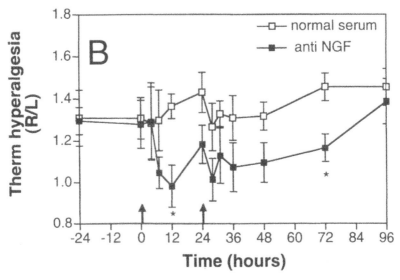

Figure 2.3   Relief of hyperalgesia by systemic anti-NGF. Anti-NGF (5 µL/g, i.p.) relieved the mechanical **(top)** and thermal **(bottom)** hyperalgesia induced by partial ligation of the sciatic nerve (n=6), whereas injection of normal sheep serum had no effect (n=6). Arrows indicate times of injection. Data points are means ± S.E.M. Asterisks indicate a significant difference between the two groups of rats at the same time points (Fisher's PLSD). ·p<0.05, +p<0.001.

anti BDNF-mechanical

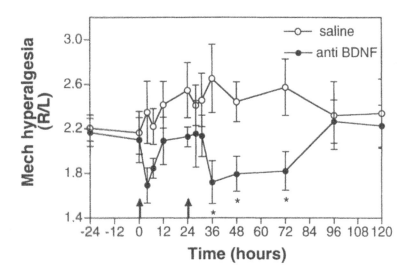

anti BDNF-thermal

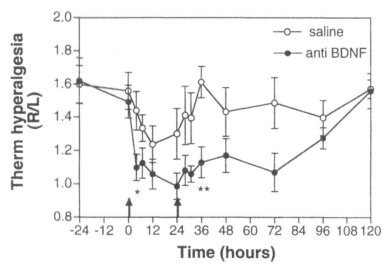

Figure 2.4 Relief of hyperalgesia by systemic anti-BDNF. Anti-BDNF (5 µL/g, i.p.) relieved the mechanical (A) and thermal (B) hyperalgesia induced by partial ligation of the sciatic nerve (n=6), whereas injection of saline had no effect (n=6). Arrows indicate times of injection. Data points are means ± S.E.M. Asterisks indicate a significant difference between the two groups of rats at the same time points (Fisher's PLSD). ·p<0.05, ··p<0.01.

Compound 48/80

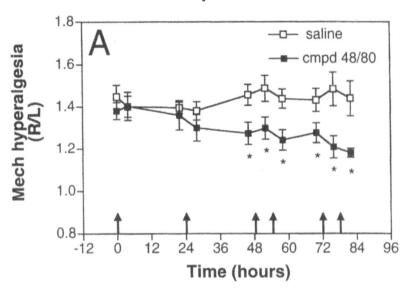

Compound 48/80

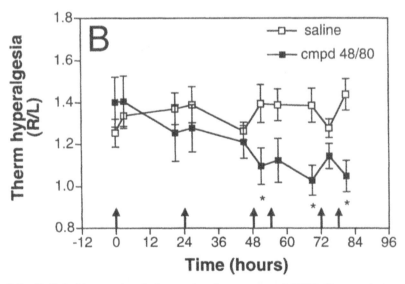

Figure 2.5   Relief of hyperalgesia by systemic compound 48/80. Degranulation of mast cells with compound 48/80 relieved the mechanical **(top)** and thermal **(bottom)** hyperalgesia induced by partial ligation of the sciatic nerve (n=10) whereas treatment with vehicle (saline) had no effect. Arrows indicate times of injection. Data points are means ± S.E.M. Asterisks indicate a significant difference between the two groups of rats at the same time points (Fisher's PLSD). ·p<0.05.

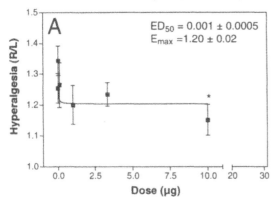

NAN-190 HBr (5HT$_{1A}$ blocker)

A

$ED_{50} = 0.001 \pm 0.0005$
$E_{max} = 1.20 \pm 0.02$

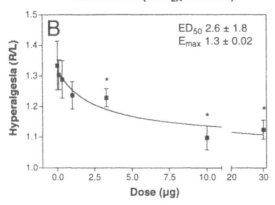

Ketanserin (5HT$_{2A}$ blocker)

B

ED_{50} 2.6 ± 1.8
E_{max} 1.3 ± 0.02

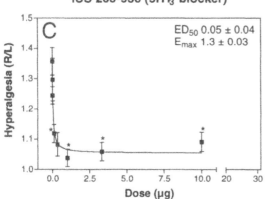

ICS 205-930 (5HT$_3$ blocker)

C

ED_{50} 0.05 ± 0.04
E_{max} 1.3 ± 0.03

Figure 2.6. Dose-dependent relief of mechanical hyperalgesia by local administration of 5HT receptor antagonists. Top: NAN-190 HBr, dose-response plot. Effect of different doses of the 5HT$_{1A}$ receptor anta-gonist NAN-190 HBr are shown on hyperalgesia induced by nerve injury, measured 40 min after injection. NAN-190 was injected into the left (hyperalgesic) hindpaw. Maximal relief was with 10 µg NAN-190. Asterisk indicates a significant difference between the effect of that dose and treatment with saline (n=11). **Centre:** Ketanserin, dose-response plot. Effect of different doses of the 5HT$_{2A}$ receptor antagonist ketanserin are shown on hyperalgesia induced by nerve injury, measured 40 min after injection. Ketanserin was injected into the left (hyperalgesic) hindpaw. Maximal relief was with 10 µg ketanserin. Asterisks indicate significant differences between the effects of marked doses and treatment with saline (n=11). **Bottom:** ICS 205–930 methiodide, dose-response plot. Effect of different doses of the 5HT$_3$ receptor antagonist ICS 205–930 are shown on hyperalgesia induced by nerve injury, measured 40 min after injection. ICS was injected into the left (hyperalgesic) hindpaw. Maximal relief was with 1 µg ICS. Asterisks indicate significant differences between the effects of marked doses and treatment with saline (n=11).

Data points are means±S.E.M. Posthoc comparisons carried out using Fisher's PLSD. ·p<0.05.

DISCUSSION

Our data confirm that NGF elicits hyperalgesia in normal rats, and show that antisera against NGF alleviate both mechanical and thermal hyperalgesia in rats with injured sciatic nerves. We also found that degranulation of mast cells alleviates hyperalgesia in nerve-injured rats, and that specific serotonin receptor blockers also alleviate this neuropathic hyperalgesia. These data are consistent with the hypothesis that NGF makes a significant contribution to peripheral mechanisms of neuropathic hyperalgesia, and are in agreement with other work implicating NGF in the hyperalgesia induced by chronic constriction of the sciatic nerve (Herzberg *et al.*, 1997; Ro *et al.*, 1999). However, NGF may also *relieve* hyperalgesia when applied directly to the nerve immediately after ligation (Ren *et al.*, 1995; Ro *et al.*, 1999). This apparent discrepancy is probably due to NGF acting in the periphery to sensitize nociceptors, and also as a retrogradely transported neurotrophic signal to reduce central hyperexcitability (Ro *et al.*, 1999). In the periphery, nerve injury would lead to elevated levels of NGF, presumably in the injured nerve or in the skin that it innervates. NGF could then act directly on nociceptors, or indirectly by acting on non-neural cells such as mast cells to elicit the release of algesic mediators. Nociceptors bear trkA receptors which have a high-affinity for NGF (Averill *et al.*, 1995; Verge *et al.*, 1989) as well as p75 receptors, which have a low affinity for neurotrophins (Wright and Snider, 1995). NGF could sensitise nociceptors directly by an action on trkA receptors (McMahon, 1996; Bennett *et al.*, 1998a). The intracellular mechanism is not clear, but may involve upregulation of B1 or B2 bradykinin receptors (Rueff *et al.*, 1996; Kasai *et al.*, 1998), increased sensitivity to protons or low pH (Bevan and Winter, 1995), augmentation of sodium channels (Gould *et al.*, 2000) or upregulation of the ion flux carried by the tetrodotoxin-insensitive sodium channels of nociceptive sensory neurones (Akopian *et al.*, 1996; Cohen and Barchi, 1993). NGF is also likely to produce hyperalgesia indirectly, by eliciting the release of algesic mediators from cells in the environment of nociceptors. NGF activates the inflammatory transcription factor NF-kappa B in Schwann cells (Carter *et al.*, 1996), exerts trophic influences on various inflammatory cells (Otten and Gadient, 1995; Scully and Otten, 1995), and elicits the release of inflammatory mediators from basophils (Bischoff and Dahinden, 1992) and mast cells (Horigome *et al.*, 1993).

Our evidence suggests that some common mechanisms underlie neuropathic and inflammatory hyperalgesia. In inflammation, NGF levels are elevated in the skin (Woolf *et al.*, 1994); in nerve injury, it has been shown that macrophages invading the proximal stump and distal segment of the transected nerve induce NGF expression by Schwann cells (Heumann *et al.*, 1987; Matsuoka *et al.*, 1991). Mast cells have already been implicated in the hyperalgesia induced by NGF (Lewin *et al.*, 1994; Rueff and Mendell, 1996; Woolf *et al.*, 1996), NGF has been shown to elicit the release of serotonin from rodent mast cells (Horigome *et al.*, 1993), and the serotonin receptor blockers methiothepin (now considered as selective for 5-HT_{1E} and 5-HT_{1F} receptors) and ICS 205-930 (selective for 5-HT_3 receptors) block or delay the onset of NGF-induced hyperalgesia (Lewin *et al.*, 1994). When a peripheral nerve is lesioned, there is a five-fold increase in the number of endoneurial mast cells throughout the part of the nerve distal to the lesion (Enerbäck *et al.*, 1965) and serotonin is known to sensitise nociceptors (Rueff and Dray, 1993) and elicit hyperalgesia (Taiwo and Levine, 1992), lending support to the idea that release of serotonin by mast cells may contribute to neuropathic hyperalgesia.

The mechanism outlined above is by no means the only peripheral mechanism operating to produce hyperalgesia following nerve injury. Other immune cells and algesic mediators are also likely to contribute as well (Tracey and Walker, 1995; Woolf *et al.*, 1996). Recent evidence supports a role for neutrophils (Bennett *et al.*, 1998b), although it is surprising that these authors found that depletion of circulating neutrophils completely abolished NGF-induced hyperalgesia in view of the evidence implicating mast cells cited above. They also noted that the mast cell amine antagonists mepyramine and methysergide had no effect on NGF-induced hyperalgesia, but mepyramine targets histamine receptors, while methysergide is a partial agonist for certain 5-HT_1 receptor subtypes.

TREATMENT OF PERIPHERAL NEUROPATHY WITH NEUROTROPHINS

It may seem paradoxical that excess NGF is implicated in the pain and hyperalgesia which often result from nerve injury, while a deficit of neurotrophins such as NGF is suggested as a cause of peripheral neuropathies which often lead to abnormal pain sensitivity (Anand, 1996; Tomlinson *et al.*, 1996; McMahon and Priestley, 1995). Presumably this apparent paradox is due to the multiple roles played by NGF, which is necessary for normal nerve function on the one hand, but can also lead to activation of nociceptors.

ACKNOWLEDGMENTS

I would like to thank Mr Michael Theodosiou, Dr Robert Rush, Dr X-F. Zhou, Dr Daping Hu and Dr. J.S. Walker for their contributions to this work, and Professor Kazue Mizumura for constructive comments on the manuscript.

REFERENCES

Akopian, A.N., Abson, N.C. and Wood, J. N. (1996). Molecular genetic approaches to nociceptor development and function. *Trends in Neurosciences*, **19**, 240–246.

Anand, P. (1996). Neurotrophins and peripheral neuropathy. *Philosophical Transactions of the Royal Society of London Series B: Biological Sciences*, **351**, 449–454.

Andreev, N.Y., Dimitrieva, N., Koltzenburg, M. and McMahon, S.B. (1995). Peripheral administration of nerve growth factor in the adult rat produces a thermal hyperalgesia that requires the presence of sympathetic post–ganglionic neurones. *Pain*, **63**, 109–115.

Averill, S., McMahon, S.B., Clary, D.O., Reichardt, L.F. and Priestley, J.V. (1995). Immunocytochemical localization of trkA receptors in chemically identified subgroups of adult rat sensory neurons. *European Journal of Neuroscience*, **7**, 1484–1494.

Bennett, D.L., Koltzenburg, M., Priestley, J.V., Shelton, D.L. and McMahon, S.B. (1998a). Endogenous nerve growth factor regulates the sensitivity of nociceptors in the adult rat. *European Journal of Neuroscience*, **10**, 1282–1291.

Bennett, G., al-Rashed, S., Hoult, J.R. and Brain, S.D. (1998b). Nerve growth factor induced hyperalgesia in the rat hind paw is dependent on circulating neutrophils. *Pain*, **77**, 315–322.

Bennett, G.J. (1993). An animal model of neuropathic pain: a review. *Muscle and Nerve*, **16**, 1040–1048.

Bennett, G.J. and Xie, Y.K. (1988). A peripheral mononeuropathy in rat that produces disorders of pain sensation like those seen in man. *Pain*, **33**, 87–107.

Bevan, S. and Winter, J. (1995). Nerve growth factor (NGF) differentially regulates the chemosensitivity of adult rat cultured sensory neurons. *Journal of Neuroscience*, **15**, 4918–4926.

Bischoff, S.C. and Dahinden, C.A. (1992). Effect of nerve growth factor on the release of inflammatory mediators by mature human basophils. *Blood*, **79**, 2662–2669.

Boyle, M.D., Lawman, M.J., Gee, A.P. and Young, M. (1985). Nerve growth factor: a chemotactic factor for polymorphonuclear leukocytes *in vivo. Journal of Immunology,* **134,** 564–568.

Carter, B.D., Kaltschmidt, C., Kaltschmidt, B., Offenhauser, N., Bohm-Matthaei, R., Baeuerle, P.A. and Barde, Y.A. (1996). Selective activation of NF-kappa B by nerve growth factor through the neurotrophin receptor p75. *Science,* **272,** 542–545.

Coderre, T.J., Basbaum, A.I. and Levine, J.D. (1989). Neural control of vascular permeability: interactions between primary afferents, mast cells, and sympathetic efferents. *Journal of Neurophysiology,* **62,** 48–58.

Cohen, S.A. and Barchi, R.L. (1993). Voltage-dependent sodium channels. *International Review of Cytology,* **137C,** 55–103.

Davies, A.M. (1996). The neurotrophic hypothesis: where does it stand? *Philosophical Transactions of the Royal Society, London B Biological Science,* **351,** 389–394.

Davis, B.M., Lewin, G.R., Mendell, L.M., Jones, M.E. and Albers, K.M. (1993). Altered expression of nerve growth factor in the skin of transgenic mice leads to changes in response to mechanical stimuli. *Neuroscience,* **56,** 789–792.

Devor, M. (1991). Neuropathic pain and injured nerve: peripheral mechanisms. *British Medical Bulletin,* **47,** 619–630.

Devor, M. (1994). The pathophysiology of damaged peripheral nerves. In *Textbook of Pain,* 3rd edn, edited by P.D. Wall. and R. Melzack, pp. 79–100. Edinburgh: Churchill Livingstone.

Diamond, J., Coughlin, M., Macintyre, L., Holmes, M. and Visheau, B. (1987). Evidence that endogenous beta nerve growth factor is responsible for the collateral sprouting, but not the regeneration, of nociceptive axons in adult rats. *Proceedings of the National Academy of Science USA,* **84,** 6596–6600.

Dib-Hajj, S.D., Fjell, J., Cummins, T.R., Zheng, Z., Fried, K., La Motte, R., Black, J.A., Waxman, S.G. (1999). Plasticity of sodium channel expression in DRG neurons in the chronic constriction injury model of neuropathic pain. *Pain,* **83,** 591–600.

Enerbäck, L., Olsson, Y. and Sourander, P. (1965). Mast cells in normal and sectioned peripheral nerve. *Zeitschrift für Zellforschung,* **66,** 596–608.

Gould, H.J. 3rd., Gould, T.N., England, J.D., Paul, D., Liu, Z.P., Levinson, S.R. (2000). A possible role for nerve growth factor in the augmentation of sodium channels in models of chronic pain. *Brain Research,* **854,** 19–29.

Hargreaves, K., Dubner, R., Brown, F., Flores, C. and Joris, J. (1988). A new and sensitive method for measuring thermal nociception in cutaneous hyperalgesia. *Pain,* **32,** 77–88.

Herzberg, U., Eliav, E., Dorsey, J.M., Gracely, R.H. and Kopin, I.J. (1997). NGF involvement in pain induced by chronic constriction injury of the rat sciatic nerve. *Neuroreport,* **8,** 1613–1618.

Heumann, R., Korsching, S., Bandtlow, C. and Thoenen, H. (1987). Changes of nerve growth factor synthesis in nonneuronal cells in response to sciatic nerve transection. *Journal of Cell Biology,* **104,** 1623–1631.

Horigome, K., Pryor, J.C., Bullock, E.D. and Johnson, E.M., Jr. (1993). Mediator release from mast cells by nerve growth factor. Neurotrophin specificity and receptor mediation. *Journal of Biological Chemistry,* **268,** 14881–14887.

Kasai, M., Kumazawa, T. and Mizumura, K. (1998). Nerve growth factor increases sensitivity to bradykinin, mediated through B2 receptors, in capsaicin-sensitive small neurons cultured from rat dorsal root ganglia. *Neuroscience Research,* **32,** 231–239.

Kim, S.H. and Chung, J.M. (1992). An experimental model for peripheral neuropathy produced by segmental spinal nerve ligation in the rat. *Pain,* **50,** 355–363.

Kingery, W.S. (1997). A critical review of controlled clinical trials for peripheral neuropathic pain and complex regional pain syndromes. *Pain,* **73,** 123–139.

Kress, M. and Reeh, P.W. (1996). Chemical excitation and sensitization in nociceptors. In *Neurobiology of Nociceptors,* edited by C. Belmonte, C. and F. Cervero, pp. 258–297. Oxford: Oxford University Press.

Levi–Montalcini, R., Skaper, S.D., Dal Toso, R., Petrelli, L. and Leon, A. (1996). Nerve growth factor: from neurotrophin to neurokine. *Trends in Neurosciences,* **19,** 514–520.

Levine, J.D., Gooding, J., Donatoni, P., Borden, L. and Goetzl, E.J. (1985). The role of the polymorphonuclear leukocyte in hyperalgesia. *Journal of Neuroscience,* **5,** 3025–3029.

Levine, J.D., Lam, D., Taiwo, Y., Donatoni, P. and Goetzl, E.J. (1986). Hyperalgesic properties of 15–lipoxygenase products of arachidonic acid. *Proceeding of the National Academy of Science USA,* **83,** 5331–5334.

Lewin, G.R. (1995). Neurotrophic factors and pain. *Seminars in the Neurosciences,* **7,** 227–232.

Lewin, G.R. and Barde, Y.A. (1996). Physiology of the neurotrophins. *Annual Review of Neuroscience*, **19**, 289–317.

Lewin, G.R., Ritter, A.M. and Mendell, L.M. (1993). Nerve growth factor-induced hyperalgesia in the neonatal and adult rat. *Journal of Neuroscience*, **13**, 2136–2148.

Lewin, G.R., Rueff, A. and Mendell, L.M. (1994). Peripheral and central mechanisms of NGF-induced hyperalgesia. *European Journal of Neuroscience*, **6**, 1903–1912.

Lindsay, R.M. (1996). Role of neurotrophins and trk receptors in the development and maintenance of sensory neurons: an overview. *Philosophical Transactions of the Royal Society, London B Biological Science*, **351**, 365–373.

Matsuoka, I., Meyer, M. and Thoenen, H. (1991). Cell-type-specific regulation of nerve growth factor (NGF) synthesis in non–neuronal cells: comparison of Schwann cells with other cell types. *Journal of Neuroscience*, **11**, 3165–3177.

McMahon, S.B. (1996). NGF as a mediator of inflammatory pain. *Philosophical Transactions of the Royal Society, London B Biological Science*, **351**, 431–440.

McMahon, S.B., Bennett, D.L.H., Priestley, J.V. and Shelton, D.L. (1995). The biological effects of endogenous nerve growth factor on adult sensory neurons revealed by a trkA-IgG fusion molecule. *Nature Medicine*, **1**, 774–780.

McMahon, S.B. and Priestley, J.V. (1995). Peripheral neuropathies and neurotrophic factors: animal models and clinical perspectives. *Current Opinion in Neurobiology*, **5**, 616–624.

Mizumura, K. and Kumazawa, T. (1996). Modification of nociceptor responses by inflammatory mediators and second messengers implicated in their action — a study in canine testicular polymodal receptors. *Progress in Brain Research*, **113**, 115–141.

Otten, U. and Gadient, R.A. (1995). Neurotrophins and cytokines-intermediaries between the immune and nervous systems. *International Journal of Developmental Neuroscience*, **13**, 147–151.

Randall, L.O. and Selitto, J.J. (1957). A method for measurement of analgesic activity on inflamed tissue. *Archives Internationales de Pharmacodynamie et de Therapie*, **111**, 409–419.

Rang, H.P., Bevan, S. and Dray, A. (1991). Chemical activation of nociceptive peripheral neurones. *British Medical Bulletin*, **47**, 534–548.

Ren, K., Thomas, D.A. and Dubner, R. (1995). Nerve growth factor alleviates a painful peripheral neuropathy in rats. *Brain Research*, **699**, 286–292.

Ro, L-S., Chen, S-T., Tang, L-M. and Jacobs, J.M. (1999). Effect of NGF and anti-NGF on neuropathic pain in rats following chronic constriction injury of the sciatic nerve. *Pain*, **79**, 265–274.

Rueff, A., Dawson, A.J. and Mendell, L.M. (1996). Characteristics of nerve growth factor induced hyperalgesia in adult rats: dependence on enhanced bradykinin-1 receptor activity but not neurokinin-1 receptor activation. *Pain*, **66**, 359–372.

Rueff, A. and Dray, A. (1993). Pharmacological characterization of the effects of 5-hydroxytryptamine and different prostaglandins on peripheral sensory neurons *in vitro*. *Agents and Actions*, **38**, C13–C5.

Rueff, A. and Mendell, L.M. (1996). Nerve growth factor and NT–5 induce increased thermal sensitivity of cutaneous nociceptors *in vitro*. *Journal of Neurophysiology*, **76**, 3593–3596.

Scully, J.L. and Otten, U. (1995). NGF: not just for neurons. *Cell Biology International*, **19**, 459–469.

Seltzer, Z. (1995). The relevance of animal neuropathy models for chronic pain in humans. *Seminars in the Neurosciences*, **7**, 211–219.

Seltzer, Z., Dubner, R. and Shir, Y. (1990). A novel behavioral model of neuropathic pain disorders produced in rats by partial sciatic nerve injury. *Pain* , **43**, 205–218.

Shieh, P.B. and Ghosh, A. (1997). Neurotrophins: new roles for a seasoned cast. *Current Biology*, **7**, R627–630.

Simone, D.A. (1992). Neural mechanisms of hyperalgesia. *Current Opinion in Neurobiology*, **2**, 479–483.

Taiwo, Y.O., Coderre, T.J. and Levine, J.D. (1989). The contribution of training to sensitivity in the nociceptive paw-withdrawal test. *Brain Research*, **487**, 148–151.

Taiwo, Y.O. and Levine, J.D. (1992). Serotonin is a directly-acting hyperalgesic agent in the rat. *Neuroscience*, **48**, 485–490.

Taiwo, Y.O., Levine, J.D., Burch, R.M., Woo, J.E. and Mobley, W.C. (1991). Hyperalgesia induced in the rat by the amino-terminal octapeptide of nerve growth factor. *Proceedings of the National Academy of Sciences of the United States of America*, **88**, 5144–5148.

Tal, M. and Liberman, R. (1997). Local injection of nerve growth factor (NGF) triggers degranulation of mast cells in rat paw. *Neuroscience Letters*, **221**, 129–132.

Theodosiou, M., Rush, R.A., Zhou, X-F., Hu, D., Walker, J.S. and Tracey, D.J. (1999). Hyperalgesia due to nerve damage_role of nerve growth factor. *Pain*, **81**, 245–255.

Tomlinson, D.R., Fernyhough, P. and Diemel, L.T. (1996). Neurotrophins and peripheral neuropathy. *Philosophical Transactions of the Royal Society, London B Biological Science*, **351**, 455–462.

Tracey, D.J., Cunningham, J.E. and Romm, M.A. (1995a). Peripheral hyperalgesia in experimental neuropathy: mediation by α_2 adrenoreceptors on postganglionic sympathetic terminals. *Pain*, **60**, 317–327.

Tracey, D.J., Romm, M.A. and Yao, N.N.L. (1995b). Peripheral hyperalgesia in experimental neuropathy: Exacerbation by neuropeptide Y. *Brain Research*, **669**, 245–254.

Tracey, D.J. and Walker, J.S. (1995). Pain due to nerve damage: are inflammatory mediators involved? *Inflammation Research*, **44**, 407–411.

Verge, V.M., Richardson, P.M., Benoit, R. and Riopelle, R.J. (1989). Histochemical characterization of sensory neurons with high-affinity receptors for nerve growth factor. *Journal of Neurocytology*, **18**, 583–591.

White, D.M., Basbaum, A.I., Goetzl, E.J. and Levine, J.D. (1990). The 15-lipoxygenase product, 8R,15S-diHETE, stereospecifically sensitizes C-fiber mechanoheat nociceptors in hairy skin of rat. *Journal of Neurophysiology*, **63**, 966–970.

Wood, J.N. and Docherty, R. (1997). Chemical activators of sensory neurons. *Annual Review of Physiology*, **59**, 457–482.

Woolf, C.J. (1996). Phenotypic modification of primary sensory neurons: the role of nerve growth factor in the production of persistent pain. *Philosophical Transactions of the Royal Society, London B Biological Science*, **351**, 441–448.

Woolf, C.J. and Doubell, T.P. (1994). The pathophysiology of chronic pain-increased sensitivity to low threshold A beta-fibre inputs. *Current Opinion in Neurobiology*, **4**, 525–534.

Woolf, C.J., Ma, Q.P., Allchorne, A. and Poole, S. (1996). Peripheral cell types contributing to the hyperalgesic action of nerve growth factor in inflammation. *Journal of Neuroscience*, **16**, 2716–2723.

Woolf, C.J., Safieh-Garabedian, B., Ma, Q.P., Crilly, P. and Winter, J. (1994). Nerve growth factor contributes to the generation of inflammatory sensory hypersensitivity. *Neuroscience*, **62**, 327–331.

Wright, D.E. and Snider, W.D. (1995). Neurotrophin receptor mRNA expression defines distinct populations of neurons in rat dorsal root ganglia. *Journal of Comparative Neurology*, **351**, 329–338.

CHAPTER 3

THE PRIMARY NOCICEPTIVE NEURON: A NERVE CELL WITH MANY FUNCTIONS

Carlos Belmonte and Juana Gallar

Instituto de Neurociencias, Universidad Miguel Hernandez,
CSIC San Juan de Alicante, Spain

Since the pioneering studies of Adrian and Zotterman (1926), the characteristics of the impulse response of primary somatosensory neurons to peripheral stimuli, encoding external forces into a message intelligible for the CNS, have been extensively studied. However, the cellular and molecular mechanisms involved in the initial steps of this process (transduction, generation of propagated impulses) are still incompletely known.

This is particularly true for primary nociceptive neurons, the sensory ganglion neurons involved in the detection of injurious stimuli in the skin and deep tissues. Primary nociceptive neurons are not homogeneous in their ability to transduce different forms of energy. A variety of nociceptive neuron subtypes have been distinguished based upon the capability of their peripheral endings to respond to one or several modalities of injurious stimuli (mechanical, chemical, thermal). The bases of these differences are not fully understood. Moreover, transduction of noxious stimuli is not the sole role played by these cells. Nociceptive neurons maintain complex interactions with their target tissues, apparently directed to the mutual maintenance of functional and structural integrity.

Figure 1 depicts schematically the interactions between primary nociceptive neurons and their peripheral target cells. These include the detection of energy changes that may cause tissue damage and the encoding of this information into a discharge of nerve impulses that contains the main spatio-temporal characteristics of the stimulus. Peripheral nociceptive neurons also contribute to the tissue reaction to injury through the release of substances (neuropeptides) which participate in local inflammation. Finally, nociceptive neurons seem to contribute to the integrity of peripheral tissues and to the onset and development of healing processes following injury. For their part, cells of peripheral target tissues play a role in modulating the sensitivity of nociceptive terminals through the release of substances that mediate sensitization. They also contribute with their chemical signals to the early development and determination of the modality of primary nociceptive neurons and to the maintenance of their normal structure and function in the adult state. Finally, signals originating in peripheral tissue cells participate in neuronal reactions to axonal injury and regeneration.

An important reason for our present ignorance of the cellular mechanisms involved in the functions of nociceptive neurons depicted above is the absence of appropriate experimental models. Most somatic and visceral tissues offer a marked structural complexity and are innervated by low- and high-threshold primary sensory neurons that in spite of their functional differences are morphologically homogeneous. Moreover, nociceptor sensory

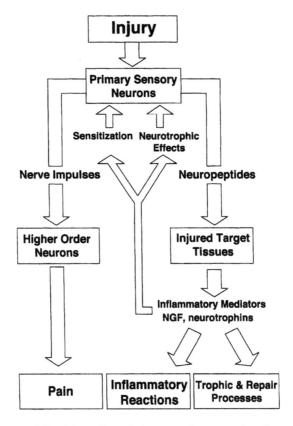

Figure 3.1 Scheme of the interactions between primary nociceptive neurons and their peripheral target cells.

nerve terminals in mammals have a very small size ($< 0.2\ \mu m$) and are found in a relatively low density in the peripheral tissues. These circumstances have precluded the application to identified endings or somas of nociceptive neurons of the conventional biophysical and biochemical techniques, employed successfully to analyze the function of other, more accessible peripheral sensory cells like olfactory, cochlear or photoreceptor cells.

The purpose of this chapter is to describe the functional characteristics of the primary sensory neurons innervating the cornea of the eye, a simple tissue that is exclusively supplied by nociceptive neurons. The experimental use of this model is providing answers to some of the unsolved questions regarding the functional roles of primary nociceptive neurons and their interactions with target cells.

THE CORNEA, A MODEL FOR THE STUDY OF NOCICEPTIVE INNERVATION

The cornea is an avascular tissue, composed of a superficial, polystratified epithelium containing 3–5 cell layers, a stroma of connective tissue and a single-layered endothelium,

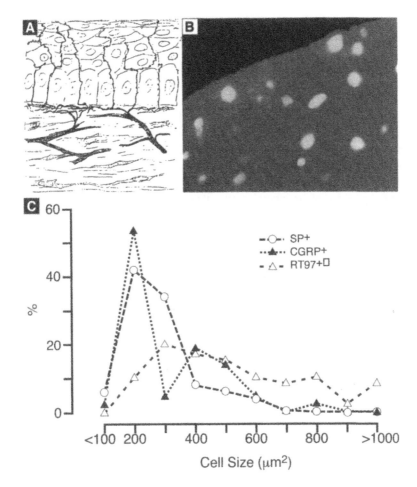

Figure 3.2 Trigeminal neurons innervating the cornea. **A:** Innervation of the corneal epithelium by terminal buttons and beaded nerve filaments ascending from the subepithelial plexus (from Ramón y Cajal, 1899). **B:** Trigeminal neurons retrogradely labeled after application of fluorogold to the mouse cornea. **C:** Size distribution of corneal trigeminal neurons. Proportion of SP-, CGRP-, and RT97-positive corneal neurons, represented as percentage of the total number of neurons in each subpopulation. (B and C, from De Felipe *et al.*, 1999).

facing the anterior chamber of the eye. Sensory innervation is provided by trigeminal ganglion neurons, whose peripheral axons end as free nerve endings in the outermost layers of the epithelium, only a few microns from the surface (Ramón y Cajal, 1899; Zander and Weddell, 1951) (Fig. 3.2A). The density of corneal sensory endings is high, varying from 1500 to 6000 endings /mm^2 depending on the studied species (Rózsa and Beuerman, 1982; De Castro *et al.*, 1997; Brock *et al.*, 1998; De Felipe *et al.*, 1999). This innervation density has been estimated in the rabbit to be 20 times higher than in the tooth pulp and about

300 times higher than in the skin (Rózsa and Beuerman, 1982). Cornea nerve terminals lie between corneal epithelium cells and are relatively accessible to experimental manipulations.

Due to the permanent defoliation of corneal epithelium cells (life span: 48 h), sensory endings undergo continuous renewal in the outermost epithelial layers of the normal cornea (Müller *et al.*, 1996). Very active short-term passive rearrangements, produced by the outward migration of differentiating epithelial cells as well as long-term, nerve-directed reorganization of corneal axons have been reported (Harris and Purves, 1989). When injured, the cornea is reinnervated first by collateral sprouts arising from neighboring intact axons and later by regenerating growing cones from the injured axons (Beuerman and Kupke, 1982; Beuerman and Rózsa, 1984). Thus, the cornea is also highly suitable to study regeneration processes of peripheral nerve fibers.

Corneal neurons in the trigeminal ganglion can be marked selectively with fluorogold applied to their peripheral nerve terminals either by iontophoresis or through a selective epithelial wound that heals in about 48 h (De Felipe and Belmonte, 1999; De Felipe *et al.*, 1999) (Fig. 3.2B). There are in the mouse approximately 250 corneal trigeminal neurons; virtually all of small or medium size (77% C and 23% A-delta, based on the reactivity to the antibody RT-97). About 20% of these neurons, mostly small, are SP+, and nearly 60% CGRP+, with an extensive overlap (Fig. 3.2C). Both types of neuropeptides have also been found in peripheral corneal nerve terminals in the same proportion (Tervo *et al.*, 1981; Stone *et al.*, 1986).

A useful feature of the corneal innervation is that this tissue is innervated only by a limited number of functional types of sensory neurons. Roughly the same proportion of A-delta and C corneal neurons obtained with immunocytological methods, was found when conduction velocity of the axon was measured in cat, rabbit, guinea pig or mouse (20–30% A-delta, 70–80% C, depending on the species) (Belmonte *et al.*, 1997; Brock *et al.*, 1998; Lopez de Armentia *et al.*, 1999).

FUNCTIONAL TYPES OF CORNEAL SENSORY NEURONS

Based on extensive electrophysiological studies analyzing the responsiveness of corneal sensory axons to controlled mechanical, thermal and chemical stimulation of the cornea, corneal neurons were classified functionally as mechanosensory, polymodal and cold sensitive neurons (Belmonte and Giraldez, 1981; Tanelian and Beuerman, 1984; Belmonte *et al.*, 1991; Gallar *et al.*, 1993; McIver and Tanelian, 1993a,b; for review see Belmonte and Gallar, 1996; Belmonte *et al.*, 1997). Irritation and pain are apparently the sole sensations that were evoked in humans by application of the different qualities of stimuli to the cornea (Kenshalo, 1960; Beuerman and Tanelian, 1979; Chen *et al.*, 1995; Acosta *et al.*, 1999). Therefore, corneal trigeminal ganglion neurons can be considered mainly and perhaps exclusively nociceptive. In the guinea pig, a fraction of about 30% of corneal sensory axons did not respond to any form of natural stimulation of the intact cornea (Brock *et al.*, 1998) and may correspond to 'silent' nociceptive neurons described in other territories (Schaible and Schmidt, 1985; Schmidt *et al.*, 1995).

Figure 3.3A shows an example of the impulse discharge in a mechano-sensory axon evoked by mechanical stimulation of the cornea. Figure 3.3B depicts the response of a polymodal fiber to mechanical stimulation, to heat and to acid while Figure 3.4A shows

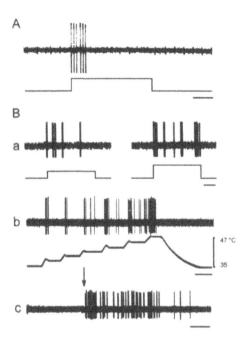

Figure 3.3 Impulse responses of corneal mechanosensory and polymodal nerve fibers (upper traces) evoked by mechanical, thermal and chemical stimulation of the cornea (lower traces). **A**. Response of a pure mechanosensory unit to a suprathreshold square-wave indentation. **B**. Response of a polymodal nerve fiber to: **(a)** two square-wave corneal indentations of 80 and 150 μm. **(b)** stepwise heating from 35° up to 47°C. **(c)** topical application of 10 mM acetic acid (arrow). **Time scales**: A and Ba = 1 s; Bb = 15 s; Bc = 0.5 s.

the activation of this same type of polymodal fiber by capsaicin and by a mixture of endogenous inflammatory substances ('inflammatory soup'), as an example of the extended chemical sensitivity of polymodal nociceptive neurons of the cornea. Corneal polymodal nociceptive terminals also sensitize strongly in response to heat or to previous injury (Fig. 3.4B), and effect mediated at least in part by local release of prostaglandins (Belmonte *et al.*, 1994) (Fig. 3.4C). No substantial functional differences appear to exist between A-delta and C polymodal nociceptive neurons, except in the size of their peripheral receptive field, thus suggesting that transduction mechanisms are common for thin myelinated and unmyelinated polymodal nociceptors (Belmonte *et al.*, 1991; Gallar *et al.*, 1993).

A separate functional type of corneal sensory neurons is the cold-sensitive neuron (Beuerman and Tanelian, 1979; Gallar *et al.*, 1993). Their corneal endings are exquisitely responsive to an air jet or to cold saline applied to the corneal surface and behave in many respects like cold sensory fibers of other territories, including the ocular conjunctiva, although their response to sustained cold stimuli is more phasic and appears to have a weak sensitivity to chemicals, as is shown in Figure 3.5.

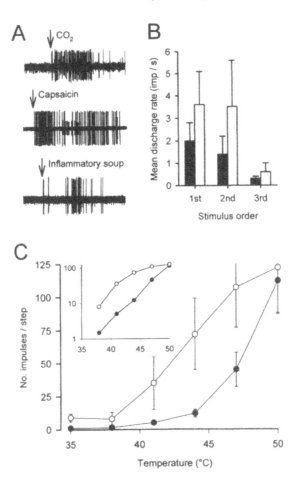

Figure 3.4 Sensitization of corneal polymodal nociceptive neurons. **A:** Impulse response of three separate polymodal units evoked by application (arrows) of 98.5% CO_2, 0.33 mM capsaicin and a mixture of inflammatory substances. **B:** Mean discharge rate of the impulse response of polymodal neurons to 10 mM acetic acid before (black bars) and after (open bars) topical application of 10^{-5} M PGE_2 (from Belmonte *et al.*, 1994). **C:** Stimulus-response relationship of corneal polymodal units elicited by two repetitions of consecutive stepwise heating cycles (from 35°C to 50°C, 15 s steps) applied with a 3 min interval (first, black circles; second, open circles). **3** Inset: log-log representation of the same data (from Belmonte and Giraldez, 1981). Data are mean ± SEM, n = 6 (B) and 8 (C).

CHEMICAL SENSITIVITY OF CORNEAL NOCICEPTIVE NEURONS

The accessibility of corneal nerve endings to exogenous stimulation is a distinct advantage of this preparation and allows one to study the effects of natural stimuli or drugs on nociceptor activity without the interference of vascular effects or unwanted tissue damage caused by noxious stimulation. Acid has been proposed as a natural excitatory mediator

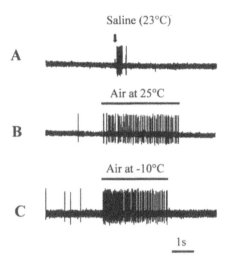

Figure 3.5 Firing response of a single corneal cold-sensitive neuron in response to a drop of isotonic saline solution at 23°C (arrow) and two jets of air (bars) at 25° and −10°C.

after injury and inflammation (Steen and Reeh, 1993). When a mixture of air and CO_2 is applied to the cornea, an activation of corneal polymodal nociceptors by locally formed carbonic acid is obtained (Chen *et al.*, 1995). As this controlled acidic stimulation with CO_2 does not cause apparent damage to nociceptive nerve endings it is useful for evaluating the effect of various drugs on nociceptor activity. Figure 3.6 exemplifies the effects of NSAIDs on the response to acid of corneal polymodal nociceptors (Chen *et al.*, 1997).

Transduction of the various forms of energy (mechanical, thermal, chemical) by periph-eral nerve endings of nociceptive neurons occurs presumably through separate membrane mechanisms. Experimental support for this hypothesis was obtained first in the cornea (Fig. 3.7) where topical application of a high dose of capsaicin inactivated the response of a single polymodal neuron to acid and to heat without affecting its response to mechani-cal stimulation (Belmonte *et al.*, 1991). This result suggested that detection of mechanical forces is made through a separate transduction mechanism but also that acid, capsaicin and heat share at least in part their transduction processes (Belmonte, 1996; Chen *et al.*, 1997). Likewise, diltiazem, a calcium antagonist that at high concentrations (> 0.5 mM) blocks the cyclic nucleotide gated channels of photoreceptors and olfactory receptor cells (Haynes, 1992), when applied to the cat's cornea (1 mM) reduced the response of corneal polymodal terminals to acid and to endogenous chemicals (Fig. 3.6), without affecting their mechanosensitivity (Pozo *et al.*, 1992). This observation pointed toward the possibility that ligand-gated channels (presumably non-selective cation channels) were involved in chemotransduction in nociceptors (Belmonte *et al.*, 1994; Belmonte, 1996).

The proposition that chemical and heat stimuli in nociceptors act through a common membrane mechanism has received direct experimental confirmation through the elegant studies of Julius and co-workers (Caterina *et al.*, 1997; Tominaga *et al.*, 1998). These authors characterized a capsaicin-gated, non-selective cation channel (the vanilloid receptor

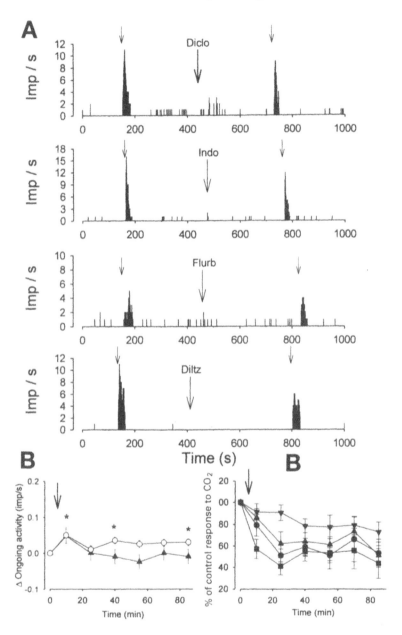

Figure 3.6 Effects of non-steroidal antiinflammatory drugs (NSAIDs) and diltiazem on the response of corneal polymodal nociceptors. A-Impulse frequency histograms of the neural response to CO_2 (small arrows) before and after application of drugs at the time indicated by the large arrow. B-(a) Ongoing activity elicited by repeated CO_2 stimulation and effect of application of the vehicle (▼) and of 0.1% sodium diclofenac (●). (b) Time course of the impulse response of corneal polymodal units evoked by CO_2 before and at different times after 0.03% flurbiprofen (Flurb. ▼) 0.1% sodium diclofenac (diclo. ▲) 0.1% indomethacin (indo.●) and 0.045% diltiazem hydrochloride (dilt. ■). Data are expressed as percentage of the control response (before treatment) and are means±SEM (from Chen *et al.*, 1997)

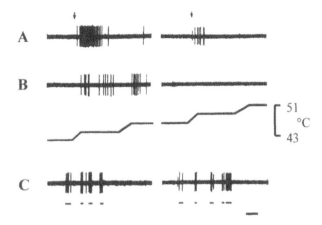

Figure 3.7 Impulse recordings illustrating the effect of 0.33 mM capsaicin on the response of corneal polymodal neurons to: **A:** 10 mM acetic acid (arrow). **B:** Stepwise heating up to 51°C (lower trace). **C:** Mechanical stimulation (small horizontal bars). **Left part:** before capsaicin treatment; **right part:** 5 min after capsaicin.

VR1) in small diameter neurons of sensory ganglia that could be expressed in *Xenopus* oocytes or transfected mammalian cells, showing that the channel is activated by capsaicin and acid but also by noxious heat. These novel approaches open new avenues to explore how irritant chemicals, protons and heat interact in nociceptive cells to finally generate nerve impulse discharges.

The processes involved in mechanical transduction in nociceptive terminals are still ignored. In preliminary experiments performed in our laboratory, mechanical responses of corneal polymodal nociceptors were not modified by topical application of drugs claimed to block mechanosensory channels in other sensory cells (amiloride, gentamicine, gadolinium). Although this may be associated with poor accessibility of the drugs to the nerve endings, it is also possible that the membrane mechanisms involved in mechanotransduction in polymodal nociceptors are in part different from those described in low threshold mechanosensory cells (baroreceptors, hair cells).

ACTIVITY OF SINGLE NOCICEPTIVE NERVE TERMINALS

A more precise idea of the ionic mechanisms involved in the generation of nerve impulses by peripheral axon terminals of primary nociceptive neurons may be obtained from the electrical changes evoked in single nerve terminals by noxious stimuli. The detection of such changes has been achieved recently in the cornea, where the activity of single nociceptive terminals has been recorded, using a micropipette placed on the corneal surface of the excised eye of the guinea pig superfused *in vitro* (Brock *et al.*, 1998). When sufficient suction is applied, a seal is produced, allowing the detection of the electrical activity in sometimes two or three and in general a single corneal nociceptive terminal. In this preparation (Fig. 3.8), nerve endings were stimulated mechanically by pushing gently the electrode against the cornea with the micromanipulator, or chemically by adding substances

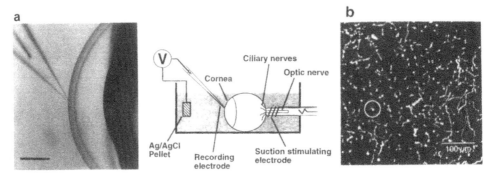

Figure 3.8 Extracellular focal recording from corneal epithelial nerve terminals. **A:** Photomicrograph showing the location of the recording electrode and schematic diagram of the recording set-up. **B:** Confocal micrograph of corneal nerve terminals (from Brock *et al.*, 1998).

like capsaicin or acid or drugs to the bathing solution or to the perfusion fluid inside the pipette. Thermal stimulation was also performed, changing the temperature of the bath.

With this methodology, different categories of nerve terminals were identified, corresponding to the same functional types of nociceptive neurons found when extracellular nerve fiber recordings were used (i.e. mechanosensory, polymodal and cold sensitive terminals). Antidromic electrical stimulation of the parent axon evoked the same impulse in the nerve terminal that was generated by natural stimulation (Fig. 3.9). However, when TTX was added to the bath, this evoked potential was blocked while spontaneous or natural stimulus-evoked impulses remained substantially unaltered (Fig. 3.10). This indicated that impulses were conducted in the nerve by TTX-sensitive channels but in the nerve endings they were supported by TTX resistant channels.

The production of nerve impulses through TTX-resistant Na+ channels was confirmed by perfusion with lignocaine, which eliminated evoked and spontaneous impulses when applied outside the recording pipette and slowed the rate of change of the downstroke component of the nerve terminal impulse (NTI), without significantly affecting the initial upstroke when applied inside the pipette. This latter finding is consistent with the local blockade of a voltage-activated increase in Na+ conductance and suggests that impulses were actively propagated into the terminals. Activity still occurred in the presence of cadmium, a calcium channel blocker, thus confirming that nerve terminal impulses were not calcium potentials.

In general, the shape of the spontaneously recorded NTIs from polymodal and/or mechanosensitive receptors was similar to those of the antidromic NTIs recorded from the same terminal. However, in a few instances in polymodal and mechanosensory endings, the initial rate of rise of the spontaneous nerve terminal impulse was slower than that of the antidromic impulse and the negative-going component was reduced in amplitude (Fig. 3.9). The simplest explanation for these shape changes is that the spontaneous nerve terminal impulses were superimposed on a relatively slow transient outward current which is driven by the generator potential triggering spontaneous firing of the nerve terminal.

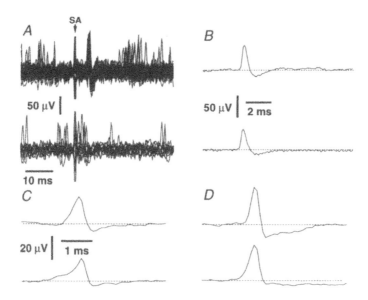

Figure 3.9 Spontaneous and evoked activity in corneal single nerve terminals. **A:** Overlaid traces in which nerve terminal impulses (NTIs) evoked by electrical stimulation (SA, stimulus artifact) blocked spontaneous firing (upper trace). Spontaneous NTIs fired just before or after SA caused the failure of the electrically evoked NTIs (lower trace). **B–D:** Averaged electrically evoked (upper) and spontaneous (lower) NTIs recorded **(B)** in the same attachment as in A, from a mechano-nociceptor **(C)**, and from a polymodal nociceptive nerve terminal **(D)** (from Brock *et al.*, 1998).

In summary, recording of extracellular currents near pure mechano-, polymodal and cold nociceptive corneal nerve terminals provided evidence that propagated nerve impulses supported by TTX-resistant Na+ channels are produced at these endings, possibly closely similar to those recorded in the soma of small primary sensory neurons (Yoshida and Matsuda, 1979; Akopian *et al.*, 1996; Villiere and McLachlan, 1996). The capacity of nociceptive terminals to generate propagated impulses that will reach unstimulated terminal branches of the same parent axon, triggering the local secretion of neuropeptides by nociceptive nerve endings, provides an explanation for the extension of neurogenic inflammation to nearby areas in damaged tissues (Lewis, 1927, 1937).

MEMBRANE PROPERTIES OF CORNEAL NOCICEPTIVE NEURONS

Peripheral terminals of nociceptive neurons appear to be equipped with a variety of membrane mechanisms for the transduction of the various forms of stimulus energy. However, the presence in the soma of the same transduction mechanisms has not been firmly established. In cultured neurons, membrane currents and intracellular calcium concentration changes evoked by protons, heat, capsaicin or endogenous chemicals have been detected (Bevan and Yeats, 1991; Petersen and LaMotte, 1993; Liu and Simon, 1994;

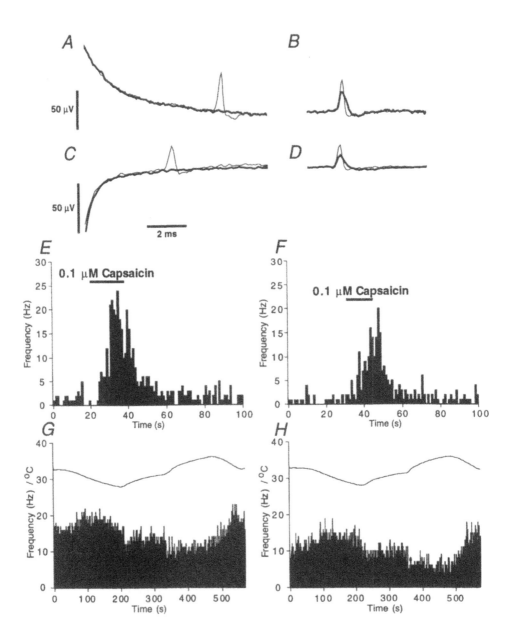

Figure 3.10 Effects of TTX on corneal single nerve terminal impulses. **A–D:** Averaged electrically evoked **(A, C)** and spontaneously occurring **(B, D)** NTIs recorded before (thin line) and in the presence of TTX (1 μM for 30 min, thick line) from a mechano-nociceptor **(A, B)** and a polymodal nociceptor **(C, D)**. **E–H:** Effects of capsicin (0.1 μm) on the frequency of occurrence of NTIs recorded, before **(E, G)** and during **(F, H)** application of TTX, from a polymodal nociceptor **(E, F)** and from a cold-sensitive nerve terminal responding to temperature changes **(G, H)** to illustrate that normal responsiveness of nerve terminals persisted after TTX treatment (from Brock *et al.*, 1998).

Garcia-Hirschfeld *et al.*, 1995; Cesare and McNaughton, 1996) suggesting that the various ionic channels and modulatory mechanisms found in nerve terminals may also be present in the cell soma.

The influence of the heterogeneity of nerve terminals on the passive and active membrane properties of the soma of primary sensory neurons is not established. Previous studies in myelinated chemo- and baroreceptor neurons of the petrosal ganglion had shown that chemosensory neurons had a broad spike with a clear hump in the descending limb and a long post-spike after-hyperpolarization, while mechanosensory, baroreceptor neurons had a narrow spike, a shorter or no inflexion in the descending phase of the spike and a short-lasting hyperpolarization (Belmonte and Gallego, 1983). Mendell and co-workers (Rose *et al.*, 1986; Koerber *et al.*, 1988; Ritter and Mendell, 1992), observed that low-threshold, mechanosensory neurons of the dorsal root ganglion innervating the skin had fast spikes while high-threshold mechanosensory neurons exhibited broader spikes with a hump in the falling phase of the spike. These differences were apparent in thin myelinated neurons but not in C neurons, in spite of their functional heterogeneity.

The existence of only three main types of sensory neurons innervating the cornea offers a good opportunity to analyze the correspondence between membrane properties and functional modality in a well-defined population of primary nociceptive neurons. Lopez de Armentia *et al.* (1997, 1999) developed a preparation of the excised trigeminal ganglion of the mouse attached by its nerves to the eye and superfused *in vitro*. Corneal neurons had been labelled one week in advance with fluorogold applied to the sensory terminals. After this time, corneal neurons were visualized with a fluorescence microscope on the surface of the ganglion and then impaled with microelectrodes. They were also identified by their response to electrical stimulation of the corneal surface and in some instances, to natural stimuli (mechanical or chemical stimulation of the cornea with acid). The differences in passive membrane properties (membrane potential, input resistance, membrane time constant and capacitance) found among corneal neurons were mainly associated with their conduction velocity grouping (A delta or C). This was also the case for the characteristics of the action potential, which in C neurons was broader and always presented an inflexion in the repolarization phase of the action potential (Fig. 3.11). However, among A delta corneal neurons with comparable conduction velocity, subtle differences were found in the rate of depolarization and duration of the action potential. A subpopulation of corneal A-delta neurons had longer action potentials and a slower depolarizing rate, similar to C corneal neurons (Fig. 3.11). They were identified as corneal polymodal neurons. A smaller group of A-delta corneal neurons exhibited a faster depolarization rate and a narrower action potential, and were considered mechanonociceptive neurons (Lopez de Armentia *et al.*, 1997, 1999). It is conceivable that the differences between these subclasses of A-delta neurons are associated with a greater presence of TTX-resistant channels detected in small diameter primary sensory neurons, presumed to be polymodal (Villiere and McLachlan, 1996).

TROPHIC INTERACTIONS BETWEEN CORNEAL NOCICEPTIVE NEURONS AND TARGET CELLS

Primary sensory neurons depend on the neurotrophin family of growth factors for their survival (for review see Snider, 1994). The different neurotrophins support different sets

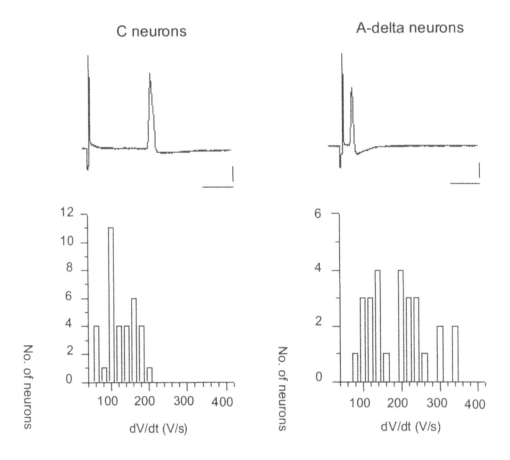

Figure 3.11 Action potentials evoked by electrical stimulation of the cornea (upper part) and frequency distribution of maximum rate of depolarization (dV/dt) (lower part) for C- (left) and A-delta (right) corneal trigeminal neurons recorded *in vitro*. Scale bars: left = 20 mV and 2 ms; right = 20 mV and 1 ms   (from López de Armentia *et al.*, 1999).

of primary sensory neurons on the basis of the sensory modality they subserve (Wright and Snider, 1995). Small dark neurons in sensory ganglia presumed to be nociceptive, express TrkA, the high affinity specific tyrosine kinase receptor for nerve growth factor (NGF). Moreover, mice in which specific genes for NGF or TrkA were inactivated, do not react to deep pinpricks or the hot plate test (Crowley *et al.*, 1994; Smeyne *et al.*, 1994). Thus, it has been suggested that small peptidergic neurons subserving pain are dependent on NGF/*trkA* signaling and are selectively lost in mice lacking *trkA* receptors or NGF. Most primary sensory neurons innervating the cornea and other ocular tissues seem to be NGF-dependent (Unger *et al.*, 1988). Nevertheless, in *trkA* knockout mice a sparse number of corneal sensory fibers is still present; they also show a response albeit greatly reduced, to topical application of acid, heat or mechanical stimuli to the eye (Fig. 3.12), thus

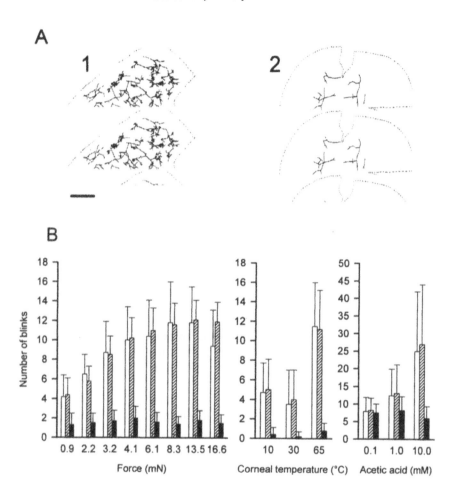

Figure 3.12 Density of nerve branches and sensitivity to noxious stimulation of the cornea in trkA knockout mice. **A:** Camera lucida reconstruction of gold chloride stained, whole mount corneas from wild type **(1)** and knockout **(2)** mice, showing the stromal and sub-epithelial nerve branches. **B:** Number of blinks evoked by mechanical (von Frey hairs of increasing force), thermal (a drop of saline at different temperatures), and chemical (a drop of acetic acid at different concentrations) stimulation of the cornea. Responses of knockout animals (–/–, black bars) were statistically different from wild type (+/+, white bars) and heterozygous (+/–, hatched bars) (from de Castro *et al.*, 1998).

suggesting that a small proportion of trigeminal corneal ganglion cells, possibly polymodal nociceptive neurons, are NGF independent (de Castro *et al.*, 1998).

Corneal sensory neurons offer an extraordinary example of the dynamic relationships between peripheral sensory nerve endings and the target tissues they innervate, both in the intact tissue and following injury. The arrangement of corneal nerve terminals is influenced by the continuous shedding of corneal epithelial cells. When intact nerve afferents innervating the corneal surface of the mouse were examined *in vivo* over a period of 24 h, they showed an extensive rearrangement of nerve terminals although an overall similarity to the initial configuration was retained. After one week, the architecture of terminal arborisations bore no resemblance to the initial branching pattern. In contrast, stromal nerves maintained a constant position (Harris and Purves, 1989). Thus, there is a continuous displacement and renewal of nerve terminals induced by the migration and desquamation of epithelial cells.

Rearrangement and outgrowth of corneal sensory axons is more profound when nerves are damaged by corneal injury. Small epithelial wounds of the cornea are covered in about 48 h by basal cells that migrate from the wound margin over the denuded area. In this time, a few terminals arriving from the subepithelial plexus outside the wound penetrate the wounded area. Newly formed terminals originating at long distances occur irregularly around the wound margin in a radial fashion. These fibers result from sprouting of fibers of the subepithelial plexus and deeper stromal nerves located outside the wound (Beuerman and Kupke, 1982; Beuerman and Rózsa, 1984; Leeuw and Chan, 1989). In stromal corneal wounds, the wound margins are completely surrounded within 24 h by sprouts originating in neighboring intact axons that course perpendicularly to the wound border. A second wave of regeneration takes place at around 7 days, now originating in transected axons of the subepithelial plexus. This process is accompanied by secondary degeneration of the collateral sprouts of intact axons (Rózsa *et al.*, 1983; Beuerman and Rózsa, 1984). A reduced nerve density in relation to the intact cornea was still observed months afterwards.

The cellular processes implicated in the normal growth and rearrangement of intact nerve endings appear to be different from those involved in axonal regeneration following nerve injury. The proto-oncogene c-*Jun* is induced in a variety of neurons by axotomy and also when these develop collateral sprouting, suggesting that c-*Jun* expression occurs not only in response to axonal damage but also when the processes underlying axonal outgrowth are activated during nerve sprouting and regeneration (Jenkins and Hunt, 1991; De Felipe and Hunt, 1994). No c-*Jun* expression was observed in identified intact sensory neurons of the mice innervating the normal cornea, thus suggesting that the active rearrangement of epithelial nerve terminals taking place continuously in the intact tissue does not induce the proto-oncogene (De Felipe and Belmonte, 1999). In contrast, lesion of stromal corneal axons by a corneal wound or by eye enucleation induced expression of c-*Jun* in about half of the population of corneal neurons, predominantly thin myelinated neurons (Fig. 3.13). Moreover, when only the epithelial terminals were destroyed, expression of c-*Jun* was not induced. On the other hand, about the same proportion of intact corneal neurons sending sprouts to the injured area expressed c-*Jun* (Fig. 3.13) (De Felipe and Belmonte, 1999). These observations suggest that expression of the proto-oncogene in response to injury or during regeneration is restricted to a portion of the primary sensory neurons innervating the cornea, perhaps those associated with Schwann cells capable of forming myelin, thus providing evidence of heterogeneity among primary nociceptive neurons in their responses to peripheral axotomy and regeneration signals.

Target cells seem to be also under the influence of chemical signals produced by peripheral nerve terminals of nociceptive neurons. In the cornea, as in other territories, sensory nerve excitation by injury causes a local inflammatory response (neurogenic

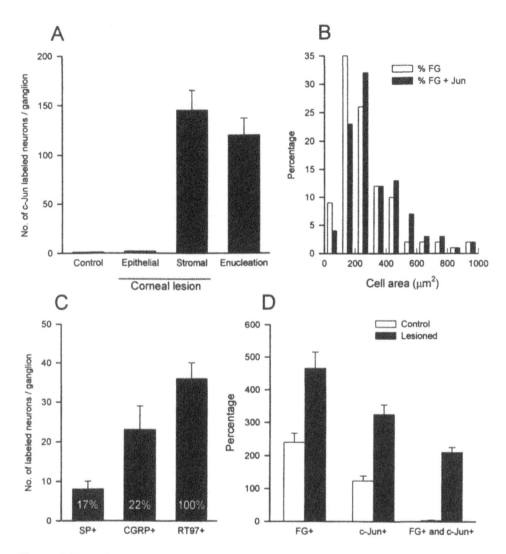

Figure 3.13 c-Jun expression in corneal trigeminal neurons after injury. **A:** Number of corneal neurons expressing c-Jun protein after intraepithelial nerve terminal lesion or after parent axon injury induced by corneal stromal wounding or by transection of the ciliary nerves with eye enucleation. In control (unlesioned) corneas, few neurons express c-Jun protein. **B:** Size distribution of the corneal neuron population in controls and after corneal stromal lesions. Size was measured in terms of cross-sectional area (the cell soma in µm²) **C:** Immunohistochemical characterization of corneal (FG+) neurons after injury. Numbers inside the bars represent the percentage within the population of SP-, CGRP- and RT97-positive corneal neurons that also expressed c-Jun after stromal injury. **D:** c-Jun expression in sprouting neurons. The number of FG-positive and of FG and c-Jun-positive trigeminal neurons were significantly increased after stromal wounding. Data are mean±SEM (from De Felipe and Belmonte, 1999).

inflammation) mediated by neuropeptides released by sensory nerve terminals (Camras and Bito, 1980; Krootila *et al.*, 1988; Gallar *et al.*, 1995; Gonzalez *et al.*, 1995).

In addition to these acute effects, sensory axons seem to have a more subtle, trophic action on corneal epithelium cells. It is well established from long standing clinical observations that damage of the corneal innervation either for therapeutic purposes (Gasserian ganglionectomy) or by accident leads to the appearance of severe lesions in the corneal epithelium (keratitis neuroparalytica) (Pannabecker, 1944). Destruction of corneal sensory neurons in more controlled experimental conditions was also followed by several morphological and biochemical disturbances in the corneal epithelium and recurrent corneal erosions (Sigelman and Friedenwald, 1954; Alper, 1976; Beuerman and Schimmelpfennig, 1980; Araki *et al.*, 1994).

The existence of trophic influences of sensory neurons on corneal epithelial cells has been confirmed experimentally in co-cultures of trigeminal neurons with corneal epithelial cells. In the presence of trigeminal sensory neurons, epithelial cells increased their mitotic rate and number (Garcia-Hirschfeld *et al.*, 1994; Fig. 3.14A) and expressed type VII collagen, a component of anchoring fibrils that adhere epithelial cells of the cornea to their basement membrane (Baker *et al.*, 1993).

Indirect evidence suggests that these neurotrophic influences may be mediated by neuropeptides contained in corneal sensory terminals. Capsaicin administered neonatally, destroyed a large amount of small peptidergic neurons in the trigeminal ganglion, including those innervating the cornea; in these animals, the loss of an important part of SP and CGRP containing nerves was accompanied by corneal signs of neuroparalitic keratitis; the severity of corneal lesions went in parallel with the density of peptide-containing nerves (Buck *et al.*, 1983; Fujita *et al.*, 1984; Ogilvy and Borges, 1990; Marfurt *et al.*, 1993). Moreover, the healing rate of corneal epithelium wounds was delayed after retrobulbar injection of capsaicin (Fig. 3.14B) (Gallar *et al.*, 1990), a maneuver that blocks axoplasmic transport of neuropeptides in corneal nerves (Bynke, 1983), while it was promoted by exogenous application of substance P (Reid *et al.*, 1990). Finally, SP enhanced the mitotic rate of corneal epithelial cells in culture (as did their co-culture with trigeminal neurons), while CGRP reduced it (Garcia-Hirschfeld *et al.*, 1994). Moreover, in mice lacking the receptor for substance P (De Felipe *et al.*, 1998) neurogenic inflammation was diminished and corneal wound healing rate was slightly retarded (Aracil *et al.*, 1999). Therefore, there is experimental support for the possibility that substance P and CGRP contained in sensory afferents modulate the functional activity of epithelial cells in the cornea, perhaps through antagonistic effects, thus contributing to the integrity of the normal cornea.

CONCLUDING REMARKS

Primary sensory neurons signaling pain appear to be a heterogeneous population of neurons in their morphology and functional properties. They have been classified in several groups according to their firing activity in response to one or to various forms of energy acting on their peripheral terminals (mechanical, thermal, chemical). Peripheral nerve endings appear to be equipped with a variety of molecular membrane mechanisms for the transduction of the different modalities of stimuli that when activated, lead to subthreshold depolarization, becoming sensitized or to suprathreshold depolarization giving rise to propagated action potentials. Possibly, many of the differences among the various func-

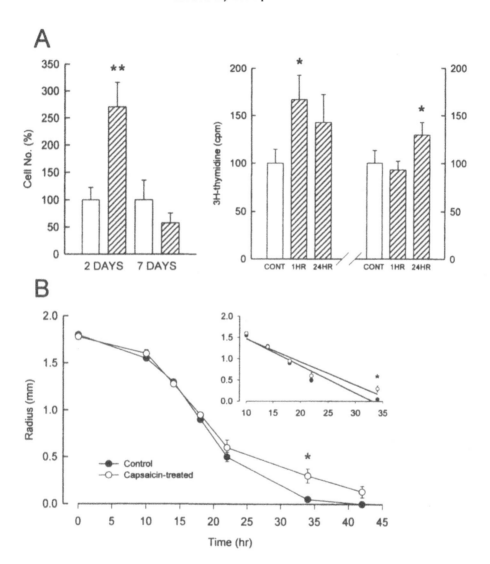

Figure 3.14 Trophic influences of trigeminal ganglion neurons on corneal epithelium cells. **A:** (left) Number of corneal epithelium cells cultured alone (open columns) or co-cultured with trigeminal neurons for 2 or 7 days (hatched columns). Right: ³H-thymidine incorporation and corneal epithelium cell number 1 hr and 24 hr after treatment with 1 μM substance P. Data represent percentage of values of control experiments (form Garcia-Hirschfeld *et al.*, 1994). **B:** Effect of neuropeptide depletion on corneal wound healing rate. Time course of radius reduction of corneal epithelial wound performed with n-heptanol in control (black circles) and capsaicin-treated (33 mM retrobulbar + 3.3 mM topical) rabbits. Inset: Regression lines of wound radius values between 10 and 34 hours after wounding, being the slopes of the epithelial migration rate (from Gallar *et al.*, 1990).

tional classes of primary nociceptive neurons reflect the prevalence of some of these transduction mechanisms over the others, making the cell more sensitive to a particular form of energy. In that respect, it appears more appropriate to classify nociceptive neurons according to the molecular processes for the transduction and amplification of the stimulus present in their nerve endings rather than by the propagated responses finally evoked by one or several types of stimuli.

Primary nociceptive neurons have a dependence upon growth factors and cytokines produced by cells of target tissues during development that presumably remains also in the adult state. This property plays an important role in the regeneration of injured axons and in the outgrowth of neural processes from intact neurons after local tissue damage and does not seem to be homogeneous for all categories of nociceptive neurons. On the other hand, at least some primary nociceptive neurons release substances including neuropeptides, that contribute to acute local inflammatory reactions but possibly also to the modulation of long term repair processes in wounded tissues to restore their integrity.

ACKNOWLEDGMENTS

This article is a revised version of the Adrian and Zotterman Lecture, given by Carlos Belmonte at the XXXIII International Congress of Physiological Sciences, IUPS, St. Petersburg, Russia in June, 1997. Supported by Grants from the Comisión Nacional de Ciencia y Tecnología, SAF 99-0066-C02.

REFERENCES

Acosta, M.C., Belmonte, C. and Gallar, J. (1999). Sensory experiences in humans and single unit activity in cats evoked by selective mechanical, chemical and thermal stimulation of the cornea. *Journal of Physiology,* (submitted).

Adrian, E.D. and Zotterman, Y. (1926). The impulses produced by sensory nerve endings. Part II. The response of a single organ. *Journal of Physiology*, **61**, 151–171.

Akopian, A.N., Sivilotti, L. and Wood, J.N. (1996). A tetrodotoxin-resistant voltage-gated sodium channel expressed in sensory neurons. *Nature*, **379**, 257–262.

Alper, M.G. (1976). The anesthesic eye: an investigation of change in the anterior ocular segment of the monkey caused by interrupting the trigeminal nerve at various levels along its course. *Transactions of the American Ophthalmological Society*, **72**, 323–365.

Aracil, A., De Felipe, C., Belmonte, C. and Gallar, J. (1999). Corneal wound healing and neurogenic inflammation in mice lacking the receptor for substance P (NK1). *European Journal of Neuroscience*, (submitted).

Araki, K., Ohashi,Y., Kinoshita, S., Hayashi, K., Kuwayama, Y. and Tano, Y. (1994). Epithelial wound healing in the denervated cornea. *Current Eye Resesearch*, **13**, 203–211.

Baker, K.S., Anderson, S.C., Romanowski, E.G., Thoft, R.A. and SundarRaj, N. (1993). Trigeminal ganglion neurons affect corneal epithelial phenotype. Influence on type VII collagen expression in vitro. *Investigative Ophthalmology and Visual Science*, **34**, 137–144.

Belmonte, C. and Gallar, J. (1996). Corneal nociceptors. In *Neurobiology of nociceptors*. edited by C. Belmonte and F. Cervero, pp. 146–183. New York: Oxford University Press.

Belmonte, C. and Gallego, R. (1983). Membrane properties of cat sensory neurones with chemoreceptor and baroreceptor endings. *Journal of Physiology, (London)*, **342**, 603–614.

Belmonte, C. (1996). Signal transduction in nociceptors: general principles. In *Neurobiology of nociceptors*. edited by C. Belmonte and F. Cervero, pp. 243–257. New York: Oxford University Press.

Belmonte, C. and Giraldez, F. (1981). Responses of cat corneal sensory receptors to mechanical and thermal stimulation. *Journal of Physiology, (London)*, **321**, 355–368.

Belmonte, C., Gallar, J., Pozo, M.A. and Rebollo, I. (1991). Excitation by irritant chemical substances of sensory afferent units in the cat's cornea. *Journal of Physiology (London)*, **437**, 709–725.

Belmonte, C., Gallar, J., Lopez-Briones, L.G. and Pozo, M.A. (1994). Polymodality in nociceptive neurons: Experimental models of chemotransduction. In *Cellular mechanisms of sensory processing*, edited by L.Urban, NATO Series, vol. H 79, pp. 87–117. Berlin Heidelberg: Springer-Verlag.

Belmonte, C., Garcia-Hirschfeld, J. and Gallar, J. (1997). Neurobiology of ocular pain. *Progress in Retinal and Eye Research*, **16**, 117–156.

Beuerman, R.W. and Kupke, K. (1982). Neural regeneration following experimental wounds of the cornea in the rabbit. In *The structure of the eye*, edited by J.G. Hollyfield, pp. 319–330. Amsterdam: Elsevier.

Beuerman, R.W. and Rózsa, A.J. (1984). Collateral sprouts are replaced by regenerating neurites in the wounded corneal epithelium. *Neuroscience Letters*, **44**, 99–104.

Beuerman, R.W. and Schimmelpfennig, B. (1980). Sensory denervation of the rabbit cornea affects epithelial properties. *Experimental Neurology*, **69**, 196–201.

Beuerman, R.W. and Tanelian, D.L. (1979). Corneal pain evoked by thermal stimulation. *Pain*, **7**, 1–14.

Bevan, S. and Yeats, J. (1991). Protons activate a cation conductance in a sub-population of rat dorsal root ganglion neurones. *Journal of Physiology, (London)*, **433**, 145–161.

Brock, J.A., McLachlan, E.M. and Belmonte, C. (1998). Tetrodotoxin-resistant impulses in single nociceptor nerve terminals in the guinea-pig cornea. *Journal of Physiology, (London)*, **512**, 211–217.

Buck, S.H., Walsh, J.H., Davis, T.P., Brown, M.R., Yamamura, H.I. and Burks, T.F. (1983). Characterization of the peptide and sensory neurotoxic effects of capsaicin in the guinea pig. *Journal of Neuroscience*, **3**, 2064–2074.

Bynke, G. (1983). Capsaicin pretreatment prevents disruption of the blood-aqueous barrier in the rabbit eye. *Investigative Ophthalmology and Visual Science*, **24**, 744–748.

Camras, C.B. and Bito, L.Z (1980). The pathophysiological effects of nitrogen mustard on the rabbit eye. II. The inhibition of the initial hypertensive phase by capsaicin and the apparent role of substance P. *Investigative. Ophthalmology and Visual Science*, **19**, 423–428.

Caterina, M.J., Schumacher, M.A., Tominaga, M., Rosen, T.A., Levine, J.D. and Julius, D. (1997). The capsaicin receptor: a heat-activated ion channel in the pain pathway. *Nature*, **389**, 816–24.

Cesare, P. and McNaughton, P. (1996). A novel heat-activated current in nociceptive neurons and its sensitization by bradykinin. *Proceedings of the National Academy of Science, (USA)*, **93**, 15435–15439.

Chen, X., Gallar, J., Pozo, M.A., Baeza, M. and Belmonte, C. (1995). CO_2 stimulation of the cornea: A comparison between human sensation and nerve activity in polymodal nociceptive afferents of the cat. *European Journal of Neuroscience*, **7**, 1154–1163.

Chen, X., Belmonte, C. and Rang, H.P. (1996). Capsaicin and CO_2 act by distinct mechanisms on sensory nerve terminals in the cat's cornea. *Pain*, **70**, 23–29.

Chen, X., Gallar, J. and Belmonte, C. (1997). Reduction by anti-inflammatory drugs of the response of corneal sensory nerve fibers to chemical irritation. *Investigative Ophthalmology and Visual Science*, **38**, 1944–1953.

Crowley, C., Spencer, S.D., Nishimura, M.C., Chen, K.S., Pitts-Meek, S., Armanini, M.P., Ling, L.H., MacMahon, S.B., Shelton, D.L., Levinson, A.D. and Phillips, H.S. (1994). Mice lacking nerve growth factor display perinatal loss of sensory and sympathetic neurons yet develop basal forebrain cholinergic neurons. *Cell*, **76**, 1001–1011.

de Castro, F., Silos-Santiago, I., López de Armentia, M., Barbacid, M. and Belmonte, C. (1998). Corneal innervation and sensitivity to noxious stimuli in trkA knockout mice. *European Journal of Neuroscience*, **10**, 146–152.

De Felipe, C. and Hunt, S.P. (1994). The differential control of c-jun expression in regenerating sensory neurons and their associated glial cells. *Journal of Neuroscience*, **14**, 2911–2923.

De Felipe, C. and Belmonte, C. (1999). c-Jun expression after axotomy of corneal trigeminal ganglion neurons is dependant on the site of injury. *European Journal of Neuroscience*, **11** (in press).

De Felipe, C., Herrero, J.F., O'Brien, J.A., Palmer, J.A., Doyle, C.A., Smith, A.J.H., Laird, J.M.A., Belmonte, C., Cervero, F. and Hunt, S.P. (1998). Altered nociception, analgesia and aggression in mice lacking the receptor for substance P. *Nature*, **392**, 394–397.

De Felipe, C., Gonzalez, G.G., Gallar, J. and Belmonte, C. (1999). Number and immunocytochemical characteristics of corneal trigeminal ganglion neurons: effect of corneal wounding. *European Journal of Pain* (in press).

Fujita, S., Shimizu, T., Izumi, K., Fukuda, T., Sameshima, M. and Ohba, N. (1984). Capsaicin-induced neuroparalytic-like corneal changes in the mouse. *Experimental Eye Research*, **38**, 165–175.

Gallar, J., Pozo, M.A., Rebollo, I. and Belmonte, C. (1990). Effects of capsaicin on corneal wound healing. *Investigative Ophthalmology and Visual Science*, **31**, 1968–1974.

Gallar, J., Pozo, M.A., Tuckett, R.P. and Belmonte, C. (1993). Response of sensory units with unmyelinated fibres to mechanical, thermal and chemical stimulation of the cat's cornea. *Journal of Physiology, (London)*, **468**, 609–622.

Gallar, J., Garcia de la Rubia, P., Gonzalez, G.G. and Belmonte, C. (1995). Irritation of the anterior segment of the eye by ultraviolet radiation: influence of nerve blockade and calcium antagonists. *Current Eye Research*, **14**, 827–835.

Garcia-Hirschfeld, J., Lopez-Briones, L.G. and Belmonte, C. (1994). Neurotrophic influences on corneal epithelial cells. *Experimental Eye Research*, **59**, 597–605.

Garcia-Hirschfeld, J., López-Briones, L.G., Belmonte, C. and Valdeolmillos, M. (1995). Intracellular free calcium responses to protons and capsaicin in cultured trigeminal neurons. *Neuroscience*, **67**, 234–235.

Gold, M.S., Reichling, D.B., Shuster, M.J. and Levine, J.D. (1996). Hyperalgesicagents increase a tetrodotoxin-resistant Na$^+$ current in nociceptors. *Proceedings of the National Academy of Science, USA*, **93**, 1108–1112.

Gonzalez, G.G., Gallar, J. and Belmonte, C. (1995). Influence of diltiazem on the ocular irritative response to nitrogen mustard. *Experimental Eye Research*, **61**, 205–212.

Harris, L.W. and Purves, D. (1989). Rapid remodelling of sensory endings in the corneas of living mice. *Journal of Neuroscience*, **9**, 2210–2214.

Haynes, L.W. (1992). Block of the cyclic GMP-gated channel of vertebrate rod and cone photoreceptors by l-cis-diltiazem. *Journal of General Physiology*, **100**, 783–801.

Jenkins, R. and Hunt, S.P. (1991). Long-term increase in the levels of c-jun mRNA and jun protein-like immunoreactivity in motor and sensory neurons following axon damage. *Neuroscience Letters*, **129**, 107–110.

Kenshalo, D.R. (1960). Comparison of thermal sensitivity of the forehead, lip, conjunctiva and cornea. *Journal of Applied Physiology*, **15**, 987–991.

Koerber, H.R., Druzinsky, R.E. and Mendell, L.M. (1988). Properties of somata of spinal dorsal root ganglion cells differ according to peripheral receptor innervated. *Journal of Neurophysiology*, **60**, 1584–1596.

Krootila, K., Uusitalo, H. and Palkama, A. (1988). Effects of neurogenic irritation and calcitonin-gene related peptide (CGRP) on ocular blood flow in the rabbit. *Current Eye Research*, **7**, 695–703.

Leeuw, M. de, and Chan, K.Y. (1989). Corneal nerve regeneration. Correlation between morphology and restoration of sensitivity. *Investigative Ophthalmology and Visual Science*, **30**, 1980–1990.

Lewis, T. (1927). *The blood vessels of the human skin and their responses*. London: Shaw and Sons.

Lewis, T. (1937). The nocicensor system of nerves and its reactions. *British Medical Journal*, **1**, 431–435.

Liu, L. and Simon, S.A. (1994). A rapid capsaicin–activated current in rat trigeminal ganglion neurons. *Proceedings of the National Academy of Science (USA)*, **91**, 738–741.

López de Armentia, M., Gallego, R. and Belmonte, C. (1997). Electrophysiological properties of trigeminal ganglion neurons innervating the cornea. *Society of Neuroscience Abstracts*, **23**, 1526.

López de Armentia, M. and Belmonte, C. (1999). Membrane properties of identified trigeminal ganglion corneal nociceptive neurons. *Journal of Neurophysiology* (submitted).

MacIver, M.B. and Tanelian, D.L. (1993a). Free nerve ending terminal morphology is fiber type specific for A-delta and C fibers innervating rabbit corneal epithelium. *Journal of Neurophysiology*, **69**, 1779–1783.

MacIver, M.B. and Tanelian, D.L. (1993b). Structural and functional specialization of A-delta and C fiber free nerve endings innervating rabbit corneal epithelium. *Journal of Neuroscience*, **13**, 4511–4524.

Marfurt, C.F., Ellis, L.C. and Jones, M.A. (1993). Sensory and sympathetic nerve sprouting in the rat cornea following neonatal administration of capsaicin. *Somatosensory and Motor Research*, **10**, 377–398.

Müller, L.J., Pels, L. and Vrensen, G.F.J.M. (1996). Ultrastructural organization of human corneal nerves. *Investigative Ophthalmology and Visual Science*, **37**, 476–488.

Ogilvy, C.S. and Borges, L.F. (1990). Changes in corneal innervation during postnatal development in normal rats and in rats treated at birth with capsaicin. *Investigative Ophthalmology and Visual Science*, **31**, 1810–1815.

Pannabecker, C.L. (1944). Keratitis neuroparalytica. *Archives of Ophthalmology*, **32**, 456–463.

Petersen, M. and LaMotte, R.H. (1993). Effect of protons on the inward current evoked by capsaicin in isolated dorsal root ganglion cells. *Pain*, **54**, 37–42.

Pozo, M.A., Gallego, R., Gallar, J. and Belmonte, C. (1992). Blockade by calcium antagonists of chemical excitation and sensitization of polymodal nociceptors in the cat's cornea. *Journal of Physiology*, **450**, 179–189.

Ramón y cajal, S. (1899). *Textura del sistema nervioso del hombre y de los vertebrados.* Alicante: Vidal Leuca.

Reid, T.W., Murphy, C.J., Twahashi, C., Malfroy, B. and Mannis, M.J. (1990). The stimulation of DNA synthesis in epithelial cells by substance P and CGRP. *Investigative Ophthalmology and Visual Science,* **31** (suppl) 2.

Ritter, A.M. and Mendell, L.M. (1992). Somal membrane properties of physiologically identified sensory neurons in the rat: effects of nerve growth factor. *Journal of Neurophysiology,* **68,** 2033–2041.

Rose, R.D., Koerber, H.R., Sedivec, M.J. and Mendell, L.M. (1986). Somal action potential duration differs in identified primary afferents. *Neuroscience Letters,* **63,** 259–264.

Rózsa, A.J. and Beuerman, R.W. (1982). Density and Organization of Free Nerve Endings in the Corneal Epithelium of the Rabbit. *Pain,* **14,** 105–120.

Rózsa, A.J., Guss, R.B. and Beuerman, R.W. (1983). Neural remodelling following experimental surgery of the rabbit cornea. *Investigative Ophthalmology and Visual Science,* **24,** 1033–1051.

Schaible H-G. and Schmidt, R. (1985). Effect of an experimental arthritis on the sensory properties of fine articular afferent units. *Journal of Neurophysiology,* **54,** 1109–1122.

Schmidt, R., Schmels, M., Forster, C., Ringkamp, M., Torebjörk, H.E. and Handwerker, H.O. (1995). Novel classes of responsive and unresponsive C nociceptors in human skin. *Journal of Neuroscience,* **15,** 333–341.

Sigelman, S. and Friedenwald, J.S. (1954). Mitotic and wound–healing activities of the corneal epithelium. *Archives of Ophthalmology,* **52,** 46–57.

Smeyne, R.J., Klein, R., Schnapp, A., Long, L.K., Bryant, S., Lewin, A., Lira, S.A. and Barbacid, M. (1994). Severe sensory and sympathetic neuropathies in mice carrying a disrupted Trk/NGF receptor gene. *Nature,* **368,** 246–249.

Snider, W.D. (1994). Functions of neurotrophins during nervous system development: what the knockouts are teaching us. *Cell,* **77,** 627–638.

Steen, K.H. and Reeh, P.W. (1993). Sustained graded pain and hyperalgesia from harmless experimental tissue acidosis in human skin. *Neuroscience Letters,* **154,** 113–116.

Stone, R.A., Kuwayama, Y., Terenghi, G. and Polak, J.M. (1986). Calcitonin gene-related peptide: Occurrence in corneal sensory nerves. *Experimental Eye Research,* **43,** 279–283.

Tanelian, D.L. and Beuerman, R.W. (1984). Responses of rabbit corneal nociceptors to mechanical and thermal stimulation. *Experimental Neurology,* **84,** 165–178.

Tervo, K., Tervo, T., Eränkö, L. and Eränkö, O. (1981). Substance P immunoreactive nerves in the rodent cornea. *Neuroscience Letters,* **25,** 95–97.

Tominaga, M., Caterina, M.J., Malmberg, A.B., Rosen, T.A., Gilbert, H., Skinner, K., Raumann, B.E., Basbaum, A.I. and Julius, D. (1998). The cloned capsaicin receptor integrates multiple pain-producing stimuli. *Neuron,* **21,** 531–43.

Unger, W.G., Terenghi, G., Zhang, S-Q. and Polak, J.M. (1988). Alteration in the histochemical presence of tyrosine hydroxylase and CGRP-immunoreactivities in the eye following chronic sympathetic or sensory denervation. *Current Eye Research,* **7,** 761–769.

Villiere, V. and McLachlan, E.M. (1996). Electrophysiological properties of neurons in intact dorsal root ganglia classified by conduction velocity and action potential duration. *Journal of Neuroscience,* **76,** 1924–1941.

Wright, D.E. and Snider, W.D. (1995). Neurotrophin receptor mRNA expression defines distinct populations of neurons in rat dorsal root ganglia. *Journal of Comparative Neurology,* **351,** 329–338.

Yoshida, S. and Matsuda, Y. (1979). Studies on the sensory neurons of the mouse with intracellular-recording and horseradish peroxidase-injection techniques. *Journal of Neurophysiology,* **42,** 1134–1145.

Zander, E. and Weddell, G. (1951). Observations on the innervation of the cornea. *Journal of Anatomy, (London),* **85,** 68–99.

CHAPTER 4

EVIDENCE FOR THE PRESENCE OF A VISCERAL PAIN PATHWAY IN THE DORSAL COLUMN OF THE SPINAL CORD

William D. Willis, Elie D. Al-Chaer, Michael J. Quast and Karin N. Westlund

Department of Anatomy and Neurosciences and Department of Internal Medicine, University of Texas Medical Branch

INTRODUCTION

Long ascending axons in the dorsal column of the mammalian spinal cord are generally thought to arise either from dorsal root ganglion cells or from neurons whose cell bodies are located in the gray matter of the dorsal horn (postsynaptic dorsal column neurons). These axons transmit mechanoreceptive information from skin and deep somatic tissue to the dorsal column nuclei (reviewed in Willis and Coggeshall, 1991; in humans, the term 'posterior' is more appropriate than 'dorsal,' but 'dorsal' will be used in this review for convenience in describing both clinical and animal studies). Clinically, lesions of the dorsal column result in deficits in two-point discrimination, the ability to judge the intensity of pressure stimuli, recognition of objects by palpation, graphesthesia, vibratory sensation and proprioception (Head and Thompson, 1906; Noordenbos and Wall, 1976; Wall and Noordenbos, 1977; Nathan *et al.*, 1986). Interruption of the dorsal column in patients does not interfere with cutaneous pain and temperature sensations. In fact, cutaneous pain and temperature sensations may become enhanced (Nathan *et al.*, 1986; cf. Brown-Séquard, 1960). Stimulation of the dorsal column can relieve pain (Shealy *et al.*, 1967; see review by Barolat, 1998).

Experimental work is generally consistent with clinical reports (Willis and Coggeshall, 1991). However, at least some neurons in the dorsal column nuclei can be activated by noxious stimuli (Angaut-Petit, 1975), and many of these have been shown to project to the contralateral thalamus (Ferrington *et al.*, 1988; Cliffer *et al.*, 1992). Because many postsynaptic dorsal column neurons respond to noxious cutaneous stimuli (Uddenberg, 1968; Angaut-Petit, 1975; Brown *et al.*, 1983; Lu *et al.*, 1983; however, cf. Giesler and Cliffer, 1985), nociceptive responses in the gracile nucleus have been attributed to activation of nociceptive postsynaptic dorsal column neurons (Angaut-Petit, 1975; Brown

Address for Correspondence: Dr Wm. D. Willis, Chairman, Department of Anatomy and Neurosciences, Cecil H. & Ida M. Green Professor and Director of the Marine Biomedical Institute, 301 University Boulevarde, Galveston, TX 77555-1069, USA. Tel: (409) 772-2103; Fax: (409) 772-4687; E-mail: wdwillis@utmb.edu

et al., 1983). However, the dorsal columns contain not only myelinated axons but also unmyelinated, peptide-containing axons, many of which originate from dorsal root ganglia (Patterson *et al.*, 1989; 1990; Tamatani *et al.*, 1989; Conti *et al.*, 1990; Fabri and Conti, 1990; cf., Giuffrida and Rustioni, 1992). The function of these primary afferent fibers is unknown, but it is possible that some are nociceptive. Thus, nociceptive responses in the dorsal column nuclei could be mediated by nociceptive postsynaptic dorsal column neurons and also by the ascending collaterals of primary afferent nociceptors (Cliffer *et al.*, 1992).

The role of nociceptive responses of dorsal column neurons in human sensation is unclear. One suggestion is that these neurons contribute to the recurrence of pain after an initially successful cordotomy (Lu *et al.*, 1983; Willis, 1985; cf., White and Sweet, 1969). Another is that nociceptive neurons in the dorsal column nuclei help activate descending antinociceptive pathways (Melzack and Wall, 1965; Lu *et al.*, 1983; Rees and Roberts, 1993).

There have been advocates for a visceral afferent function of the dorsal column system. For example, White (1943) proposed that the dorsal column mediates the sensations of colon and bladder distention, since these sensations are undisturbed by bilateral ventro-lateral cordotomies but are lost in the subacute combined degeneration in vitamin B_{12} deficiency, in which the dorsal column degenerates. Sarnoff *et al.* (1948) made a similar suggestion concerning duodenal distention. A number of investigators have described responses of neurons in the dorsal column-medial lemniscus system following electrical stimulation of visceral afferents (Amassian, 1951; Aidar *et al.*, 1952; Rigamonti and Hancock, 1974, 1978). Amassian (1951) suggested that splanchnic nerve A-β fibers might function to signal visceral distention. However, he thought that the cortical action of splanchnic A-δ fibers might relate to alerting or some function other than the sensation of distention. Slowly adapting responses of dorsal column axons to bladder distention have been described (Yamamoto and Sugihara, 1956). Rigamonti and Hancock (1974) thought that the response of gracile neurons to A-β fibers in the greater splanchnic nerve might relate to vibratory sensation mediated by Pacinian corpuscles in the abdomen. Kuo and De Groat (1985) were able to trace the projections of primary afferent fibers in the greater splanchnic nerve directly to the gracile nucleus in cats. They mention the possibility that the afferents might be involved in visceral distention, as well as in monitoring of vascular pulsations (cf. Gammon and Bronk, 1935). Recently, it has been shown that many neurons in the gracile nucleus can be excited following innocuous or noxious distention of gastrointestinal viscera or female reproductive organs (Berkley and Hubscher, 1995).

None of these studies has suggested that the dorsal column contains a specific visceral pain pathway. Berkley and Hubscher (1995) discouraged the view that the dorsal column pathway functions as a visceral pain pathway but instead proposed a cooperative role of the dorsal column pathway and the spinothalamic tract in mediating this sensation.

CLINICAL EVIDENCE FOR A DORSAL COLUMN VISCERAL PAIN PATHWAY

Visceral pain can be relieved, at least temporarily, by cordotomy performed to interrupt the spinothalamic tract (White and Sweet, 1969; Gybels and Sweet, 1989). However, visceral pain is often bilateral, and so bilateral cordotomies may be required, with the attendant danger of serious complications (Gybels and Sweet, 1989). The procedure of commissural myelotomy was designed to sever the crossing axons of the spinothalamic tracts over several segments, thereby providing bilateral relief of visceral pain (Armour, 1927; Putnam, 1934; Leriche, 1936). However, this procedure can also lead to severe complications (Gybels and Sweet, 1989). Curiously, some of the early trials of commissural myelotomy involved incisions that extended only 2–3 mm into the dorsal column (Mansuy *et al.*, 1944), a depth in the human spinal cord that would not reach the anterior white commissure, where spinothalamic axons decussate. Furthermore, commissural myelotomy often produced pain relief in areas of the body beyond those that showed a sensory deficit on clinical testing (King, 1977; Cook *et al.*, 1984; see Gybels and Sweet, 1989).

Hitchcock (1970, 1972a, b) introduced a less destructive procedure, limited midline myelotomy, which resulted in fewer complications than did bilateral cordotomy or commissural myelotomy. His lesion was placed in the midline at the C1 level and this procedure, like commissural myelotomy, was often followed by pain relief over much of the body (see also Papo and Luongo, 1976; Schvarcz, 1976, 1978). Schvarcz (1984) reported that this procedure produced satisfactory pain relief, defined as no pain or infrequent pain relieved by non-narcotic analgesics, in 76% of a series of 79 patients with midline or bilateral intractable pelvic cancer pain. Most of the patients died within 6 months, but some survived as long as 30 months. For the treatment of pelvic cancer pain, Gildenberg and Hirshberg (1984) moved the site of their limited midline myelotomy procedure to the T10 segmental level, since visceral afferents from pelvic structures enter the spinal cord over dorsal roots below this level. The relief of visceral pain afforded by limited midline myelotomy led Gybels and Sweet (1989) to write that this 'compels a major revision in our thinking anent the pathways for pain in the spinal cord of man.'

Six of the cases reported by Gildenberg and Hirshberg (1984), as well as two more recent cases, were summarized in Hirshberg *et al.* (1996). The spinal cord containing the lesion in the most recent case was available for postmortem examination. The patient was a 44-year-old man who had a carcinoma of the colon, for which the colon was resected, followed by radiotherapy and chemotherapy. Spread of the cancer caused extreme pain that was inadequately controlled by intravenous morphine delivered though a patient-controlled analgesia pump. The visceral pain was completely relieved following a limited midline myelotomy at T10 that interrupted axons in the medial part of the fasciculus gracilis bilaterally (Fig. 4.1). Morphine was discontinued over the course of 3 days. Postoperatively, a cutaneous abscess occurred below the umbilicus. This cutaneous site, on the other hand, was painful until drained. The surgery produced no neurological deficit. The patient required no more pain medication and died 3 months later.

Another case was reported by Nauta *et al.* (1997). The patient was a 39-year-old woman who had been successfully treated for cancer of the uterine cervix. However, the irradiation caused inflammation of the bowel and other organs, with the consequences of severe colorectal pain, despite morphine, and progressive weight loss. A punctate midline myelotomy was made at T8. She subsequently was able to reduce her pain medication and to return home. She regained some of her lost weight. After other complications of her condition, she had a colostomy and she developed a sacral decubitus ulcer. There was postoperative pain at the colostomy site and pain in the upper abdomen due to peritonitis. She also had pain from the decubitus ulcer. However, at the time of a follow-up examination of this patient 30 months after the surgery, she reported never to have had a recurrence of the severe lower abdominal pain that led to the punctate midline myelotomy.

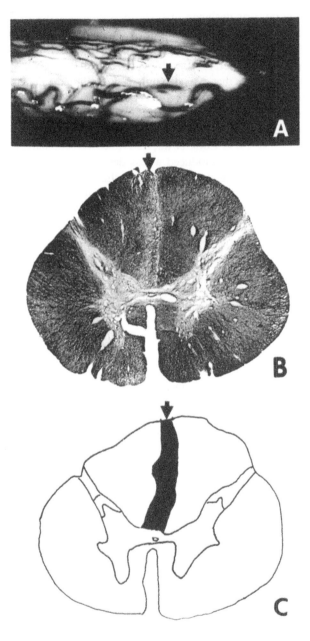

Figure 4.1 Limited midline myelotomy lesion of the T10 spinal cord of a patient treated for colon cancer pain. In **A** is shown an intraoperative photograph of the spinal cord of the patient. The arrow indicates the site of the myelotomy. **B** shows a cross-section of the postmortem spinal cord of this patient just rostral to the lesion. The section was stained for myelin. The arrow points to a bilateral area of demyelination in the medial parts of the gracile fascicli. **C** is a drawing of the same section; the demyelinated area is shown in black. (From Hirshberg *et al.*, 1996).

EVIDENCE FROM EXPERIMENTAL STUDIES IN RATS

The widespread pain relief seen following commissural or limited midline myelotomy at thoracic levels in these patients cannot be explained by interruption of the crossing axons of the spinothalamic tract. The explanation for the pain relief that formed the rationale for these surgical procedures was interruption of a hypothetical polysynaptic extralemniscal pain pathway that ascends in the central gray matter of the spinal cord or in the adjacent white matter (Schvarcz, 1976, 1978, 1984). Evidence for such a pathway near the center of the spinal cord is available from experiments on cats by Karplus and Kreidl (1914) and on rats by Basbaum (1973). These investigators found that these animals continue to show supraspinally mediated nocifensive responses despite double hemisections at different levels of the spinal cord that would interrupt all long ascending pathways. However, in monkeys such double hemisections did eliminate supraspinally mediated nocifensive responses, indicating that such responses in primates depend on the long sensory tracts in the spinal cord (Karplus and Kreidl, 1914). Thus, the mechanism for the relief of pelvic cancer visceral pain by lesions of the dorsal column in humans is unclear. For this reason, studies were initiated in rats to investigate other mechanisms that might explain the relief of visceral pain by a dorsal column lesion.

ANATOMICAL STUDIES SUPPORTING A DORSAL COLUMN PROJECTION SYSTEM FOR VISCERAL PAIN

Anatomical studies in rats were initiated to establish which neurons in the spinal cord were the origins of the fibers traveling in the dorsal column that might be mediating the transmission of the visceral pain. The retrograde tracer, WGA-HRP, was injected into the dorsal column at the upper cervical spinal cord (Hirshberg *et al.*, 1996). The retrograde tracer was absorbed into the axons and transported to cell bodies located in the central region throughout the length of the spinal cord and in particular at the level innervating the pelvic viscera (Fig. 4.2). The central region of the spinal cord is known to be involved in the processing of visceral information (Honda, 1985; Ness and Gebhart, 1987). The fiber projections of the cells located in central region of the L6-S1 spinal cord were then traced using an anterograde tracer (Hirshberg *et al.*, 1996; Wang *et al.*, 1996, 1999). Their fiber projections were followed throughout the length of the spinal cord to the gracile nucleus (Fig. 4.14). Thus, a direct anatomical projection was established arising from spinal visceral processing regions, traveling in the dorsal column and terminating in the dorsal column nuclei. This observation established the availability of a neuronal pathway in the dorsal column that could relay visceral information from neurons in the central region of the sacral spinal cord where visceral information is pro-cessed. Another component of the postsynaptic dorsal column pathway previously described has its cells of origin primarily in lamina III and IV (Bennett *et al.*, 1983) and is believed not to carry information about pain in rats (Giesler and Cliffer, 1984). However, this view was based on the lack of responses of these neurons to cutaneous noxious heat stimuli. The neurons did respond to strong mechanical stimulation of the skin.

To establish a role of the dorsal column and, in particular, of postsynaptic dorsal column neurons, in transmitting visceral nociceptive information from the spinal cord to the brain, a series of electrophysiological studies were designed.

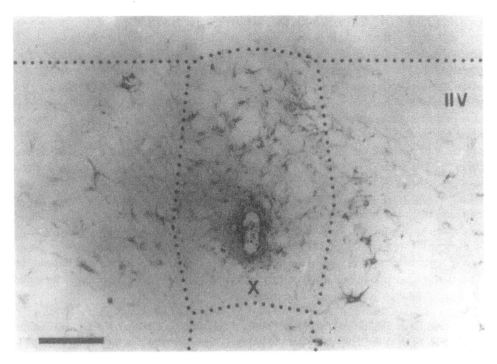

Figure 4.2 Retrogradely labeled neurons in the central region of the gray matter in the S2 spinal cord segment. WGA-HRP was injected bilaterally into the dorsal column at a cervical level 7 days before the animal was sacrificed. Many labeled neurons are seen in lamina X and the adjacent parts of laminae V and VII. The calibration bar represents 130 μm. (From Hirshberg *et al.*, 1996).

EFFECTS OF LESIONS OF THE DORSAL COLUMN AND THE SPINOTHALAMIC TRACT ON THE RESPONSES OF THALAMIC NEURONS TO VISCERAL STIMULI

The strategy that was chosen for the initial experiments was to record electrophysiological responses to noxious visceral stimuli from neurons in the ventral posterolateral (VPL) nucleus of the thalamus and the gracile nucleus (Al-Chaer *et al.*, 1996a, b; Hirshberg *et al.*, 1996). In previous work, neurons that respond to visceral stimuli were found in the VPL nucleus of rats and monkeys (Chandler *et al.*, 1992; Berkley *et al.*, 1993; Brüggemann *et al.*, 1994) and in the gracile nucleus of rats (Berkley and Hubscher, 1995). For our experiments we chose noxious visceral stimuli that we thought might mimic the pain experienced by patients with pelvic cancer. However, a constraint on the experiments planned was that the pain had to develop acutely so that the responses of the neurons being recorded could be compared before and after a lesion of the dorsal column or of the spinothalamic tract. We decided to employ two types of noxious visceral stimuli, mechanical and chemical. The mechanical stimulus was colorectal distention. Experimental work and studies on human subjects have shown that distention of the colon by pressures less

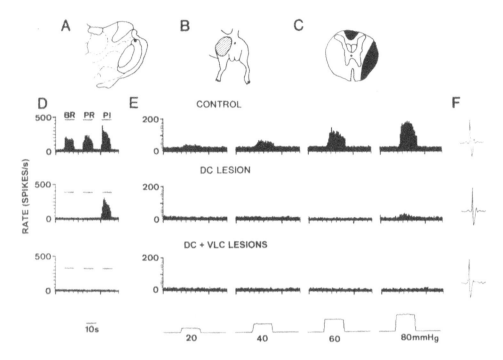

Figure 4.3 Alterations in the responses of a neuron in the ventral posterolateral (VPL) nucleus of the thalamus of the rat by successive lesions of the dorsal column (DC) and of the ventrolateral column (VLC). The drawings in **A–C** show, respectively, the recording site, the cutaneous receptive field and the maximum extents of the lesions. **D** shows responses to mechanical stimuli applied to the skin (BR, brush; PR, pressure; PI, pinch). **E** shows responses to colorectal distention (to pressures of 20, 40, 60 and 80 mmHg, monitored as shown by the traces at the bottom). The upper row of records are the controls, the middle row were taken after a DC lesion, and the lower row after an additional VLC lesion. The action potentials in **F** show that the recording conditions remained the same throughout the experiment. (From Al-Chaer *et al.*, 1996a).

than 40 mmHg are innocuous, whereas distention of the colon by pressures over 40 mmHg are noxious in rats and clearly painful in humans (Ness and Gebhart, 1988; Ness *et al.*, 1990). The noxious chemical stimulus employed was the injection of mustard oil into the lumen of the colon to produce acute inflammation. Innocuous and noxious cutaneous stimuli were also used.

Figure 4.3 illustrates the effects of a dorsal column lesion at T10 on the responses of a neuron in the VPL nucleus. The upper rows of peristimulus time histograms in Figures 4.3D and E show the initial responses to graded intensities of mechanical stimuli applied to the skin in the receptive field (Fig. 4.3B) and the responses to graded intensities of colorectal distention. Following the dorsal column (DC) lesion (extent shown in Fig. 4.3C), the responses of the cell to weak mechanical stimulation of the skin and to all but the most intense colorectal distention were eliminated (Fig. 4.3D and E, middle rows of histograms).

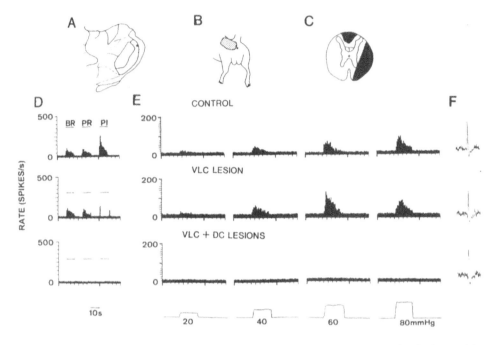

Figure 4.4   Alterations in the responses of a rat VPL neuron by successive lesions of the VLC and DC. The format is as in Figure 4.3 (From Al-Chaer *et al.*, 1996a).

A lesion was then made of the ventrolateral column (VLC) of the spinal cord that should have interrupted most of the spinothalamic tract (Fig. 4.3C). The response of the cell to noxious stimulation of the skin and to an 80 mmHg colorectal distention were now eliminated (Fig. 4.3D and E, lower rows of histograms). The action potential of the neuron was unchanged throughout these procedures (Fig. 4.3F). Similar observations were made on a total of 10 VPL cells.

A series of experiments was also done in which the sequence of lesions was reversed. Figure 4 illustrates one of these experiments. A lesion of the VLC of the spinal cord (Fig. 4.4C) eliminated the response of the VPL neuron to noxious stimulation of the skin, but not the responses to weak mechanical stimulation of the skin or to graded colorectal distention (Fig. 4.4D and E, upper two rows of histograms). An additional lesion of the dorsal column (Fig. 4.4C) eliminated the responses that remained after the VLC lesion. Similar results were obtained while recording from 10 different VPL neurons.

The responses of 20 VPL neurons to cutaneous stimulation and to colorectal distention were normalized and combined (Fig. 4.5). It can be seen that a lesion of the DC eliminated the responses to brushing the skin and greatly reduced the responses to a pressure stimulus, but had only a small effect on the responses to noxious pinch. On the other hand, the VLC lesion did not change the responses to brushing and had only a small effect on the pressure responses. However, it nearly eliminated the responses to noxious pinch. The DC lesion reduced the responses to colorectal distention to about 20% of their original values, but the VLC lesion reduced these responses to only about 60–80%.

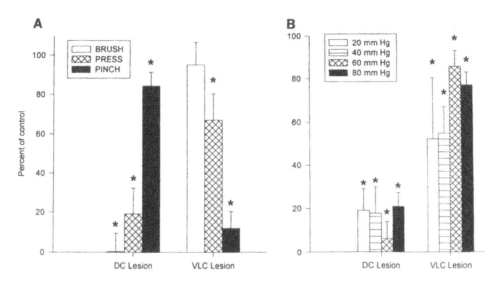

Figure 4.5 Changes in the responses of a set of 20 VPL neurons to cutaneous and colorectal stimulation following DC and VLC lesions. In **A** are shown the mean responses (±S.E.) to BRUSH, PRESS and PINCH stimuli following a DC or a VLC lesion (irrespective of the order of the lesions). In **B** are shown the responses to colorectal distentions of 20, 40, 60 and 80 mmHg following DC and VLC lesions. Changes that are significantly different from control values are indicated by asterisks. (From Al-Chaer *et al.*, 1996a).

These observations suggest that in rats the dorsal column plays the major role in mediating the responses of VPL neurons to noxious distention of the colon. The dorsal column also mediates responses to weak mechanical stimulation of the skin. On the other hand, the spinothalamic tract in the ventrolateral white matter is only partly responsible for the responses of VPL neurons to noxious colorectal distention but is the main pathway mediating the responses to noxious mechanical cutaneous stimuli.

Similar conclusions were made about the importance of the dorsal column in mediating visceral nociceptive activity when noxious chemical stimulation of the colon was employed. The background activity of VPL neurons became elevated after acute inflammation of the colon with mustard oil. The background activity was profoundly reduced by a DC lesion but was much less affected by a VLC lesion (Al-Chaer *et al.*, 1996a). When a VLC lesion was made first, the elevated background activity produced by mustard oil inflammation of the colon was temporarily reduced, but then continued to increase (Fig. 4.6). A later DC lesion reduced the background activity to the control level.

RESPONSES OF VPL NEURONS TO COLORECTAL DISTENTION DEPEND ON A RELAY IN THE GRACILE NUCLEUS

From these experiments in which it was shown that the responses of neurons in the VPL nucleus to noxious colorectal distention depended largely on signals transmitted through the dorsal column, it seemed likely that this information would have been relayed through

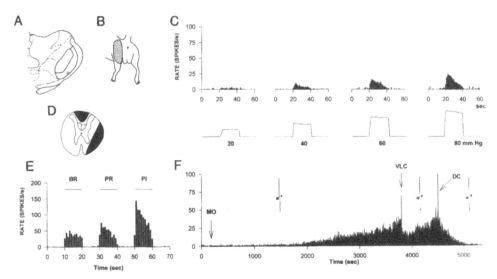

Figure 4.6 Effect of VLC and DC lesions on the activity of a VPL neuron after acute inflammation of the colon by mustard oil. **A**, **B** and **D** show the location of the neuron, its cutaneous receptive field and the lesions in the VLC and DC. The histograms in **C** show the responses of the cell to graded colorectal distentions and in **E** the responses to brush (BR), press (PR) and pinch (PI) stimuli applied to the skin. In **F**, the histogram shows the background activity of the cell before and after the injection of mustard oil (MO) into the colon. After the activity became enhanced, a lesion of the VLC was made, followed by a DC lesion. The inserts show action potentials recorded at the times indicated by the x-axis. (From Al-Chaer *et al.*, 1996a).

the gracile nucleus. To check that this was indeed the case, a study was done in which the responses of neurons in the VPL nucleus of rats to noxious colorectal distention were recorded before and after small lesions were placed in the gracile nucleus (Al-Chaer *et al.*, 1997). The lesions were made either by passing an electric current through a metal electrode inserted into the gracile nucleus or by microinjection of kainic acid into the nucleus. Even though the lesions were incomplete, destroying only a part of the gracile nucleus, the responses of VPL neurons to distentions of 80 mmHg were reduced on average by 66% following electrolytic lesions and by 51% after chemical lesions. The same lesions reduced the responses to brushing the skin by 84% and 81%, but did not affect the responses to pinching the skin. Presumably, more extensive lesions of the gracile nucleus would have reduced the responses of VPL neurons to noxious colorectal distention somewhat more. However, the maximum reduction expected, based on the results following dorsal column lesions (Fig. 4.5) would have been by 80%.

ROLE OF POSTSYNAPTIC DORSAL COLUMN NEURONS IN TRANSMITTING VISCERAL NOCICEPTIVE INFORMATION

If visceral nociceptive signals from the colon were transmitted to the VPL nucleus by way of the gracile nucleus, it would be important to know if these signals are conducted by directly projecting collaterals of dorsal root ganglion cells or by axons belonging to the postsynaptic dorsal column pathway. Experiments were therefore devised to test these possibilities (Al-Chaer *et al.*, 1996b). The objective of the experiments was to block synaptic transmission from colonic afferents to relevant postsynaptic dorsal column neurons by local administration of morphine or the non-N-methyl-D-asparate glutamate receptor antagonist, CNQX. These agents should not affect action potential conduction in the ascending collaterals of primary afferents within the dorsal column. It is known that morphine administered intrathecally can block the responses of dorsal horn neurons to visceral stimuli (Omote *et al.*, 1994), whereas morphine has little effect on action potential conduction (Hu and Rubly, 1983; Yuge *et al.*, 1985). CNQX has been shown by our group to reduce dramatically the responses of spinothalamic tract neurons to both innocuous and noxious stimuli (Dougherty *et al.*, 1992). Our plan was to administer either morphine or CNQX through a microdialysis fiber implanted in the spinal cord gray matter of the sacral spinal cord while recording from gracile neurons that responded to noxious colorectal distention. This approach would guarantee that the site of drug action was restricted to the spinal cord very close to the location of the microdialysis fiber. However, this plan depended on access by the drugs to the relevant spinal cord neurons. Drugs administered by microdialysis would not spread longitudinally more than a segment or two, and so they could not be expected to block synaptic transmission to postsynaptic dorsal column neurons distributed over lower thoracic to sacral segments. Since visceral afferents from the colon enter the spinal cord through both the pelvic and the hypogastric nerves, we chose to interrupt the hypogastric nerves at the start of the experiments. This limited colon input to segments L6-S1. The experimental arrangement is shown in Figure 4.7.

Records from one of these experiments are shown in Figure 4.8. The location of the unit in the gracile nucleus is indicated in Figure 4.8A. The responses of the neuron to graded colorectal distention are shown in Figure 4.8D before drug administration, during morphine administration, after systemic administration of naloxone, and during CNQX administration. The visceral responses were eliminated by either morphine or CNQX, and the effects of morphine were reversed by the opiate receptor antagonist, naloxone. This result is consistent with the idea that the responses of gracile neurons to colorectal distention depend on a synaptic relay in the spinal cord, presumably involving postsynaptic dorsal column neurons. In the same experiment, the responses of the gracile neuron to mechanical stimulation of the receptive field on the skin (Fig. 4.8B) were relatively unaffected (Fig. 4.8C). Presumably these responses depended on transmission of action potentials by the ascending collaterals of primary afferent axons and not on synaptically relayed activity. Similar results were obtained from 10 gracile neurons. As expected, when a dorsal column lesion was made in some of the experiments, all of the responses were eliminated. Thus, the responses of the gracile neurons depended on information carried by axons in the dorsal column and not on information reaching the nucleus by a circuitous route (cf. Berkley and Hubscher, 1995).

If postsynaptic dorsal column neurons mediate the responses of gracile neurons to colorectal distention, then it should be possible to block the responses of postsynaptic dorsal column neurons using the same drugs administered by microdialysis. Postsynaptic dorsal column neurons in the L6-S1 segments were identified by antidromic activation from the cervical fasciculus gracilis near the gracile nucleus. An example is shown in Figure 4.9. The location of the cell is indicated in Figure 4.9A. It is important to note that the postsynaptic dorsal column neurons that responded to colorectal distention were

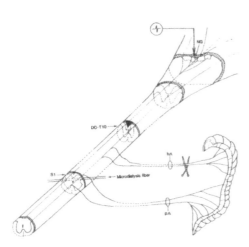

Figure 4.7 Experimental arrangement for determining if postsynaptic dorsal column neurons or directly projecting primary afferent axons are responsible for transmitting visceral nociceptive signals to the gracile nucleus. The hypogastric nerves were sectioned bilaterally at the start of the experiment. A microdialysis fiber was inserted into the gray matter of the S1 segment so that morphine or CNQX could be administered to block nociceptive synaptic transmission (but not action potential conduction in primary afferent collaterals). Recordings were made from neurons of the gracile nucleus that responded to colorectal distention. In some experiments, a DC lesion was made to verify that the input was carried by axons in the DC. (From Al–Chaer *et al.*, 1996b).

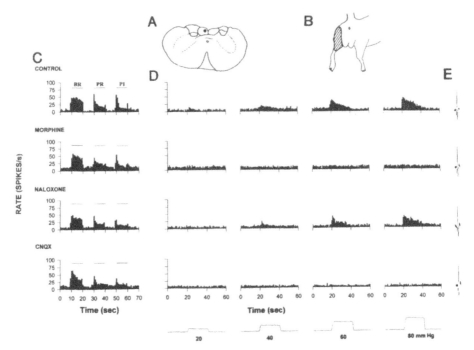

Figure 4.8 Alterations in the responses of a gracile neuron following microdialysis administration of morphine or CNQX into the rat sacral spinal cord. The drawings in **A** and **B** show the location of the neuron and its cutaneous receptive field, respectively. **C** shows the responses to mechanical stimulation of the skin and **D** the responses to colorectal distention. The upper row of records in C and D are the controls. The second row shows the effects of morphine. The third row was made after systemic administration of naloxone. The lower row shows the effect of CNQX. (From Al-Chaer *et al.*, 1996b).

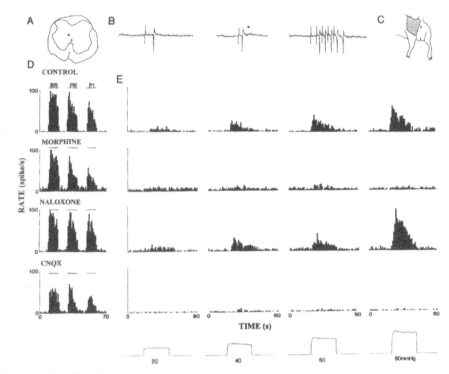

Figure 4.9 Alterations in the responses of a postsynaptic dorsal column neuron following microdialysis administration of morphine or CNQX into the rat sacral spinal cord. The location of the cell is shown in **A**. **B** illustrates the collision test used to identify the neuron. The antidromic stimulus was delivered just caudal to the gracile nucleus. The cutaneous receptive field is shown in **C**. **D** and **E** are the responses to mechanical stimuli applied to the skin and to colorectal distention. The upper row of records are the controls. The second row shows the effect of morphine. The third row was taken after systemic administration of naloxone. The fourth row was after CNQX. (From Al-Chaer *et al.*, 1996b).

concentrated around the central gray and not in laminae III and IV of the dorsal horn, the locations of previously investigated postsynaptic dorsal column neurons that respond just to cutaneous stimuli (Giesler *et al.*, 1984; Willis and Coggeshall, 1991). The responses of this postsynaptic dorsal column neuron to cutaneous stimuli are shown in Figure 4.9D and to colorectal distention in Figure 4.9E. Morphine reduced the responses to cutaneous pinch and to colorectal distention, and this effect was reversed by naloxone. All of the responses were reduced by CNQX. Similar effects of morphine and CNQX were seen on the responses of 10 postsynaptic dorsal column neurons.

NOCICEPTIVE INFORMATION FROM ABDOMINAL VISCERA IS ALSO TRANSMITTED IN THE DORSAL COLUMN

More recent experiments have addressed the possible role of the dorsal column in transmitting visceral nociceptive information from abdominal viscera and the effects of dorsal column lesions on nociceptive behaviors.

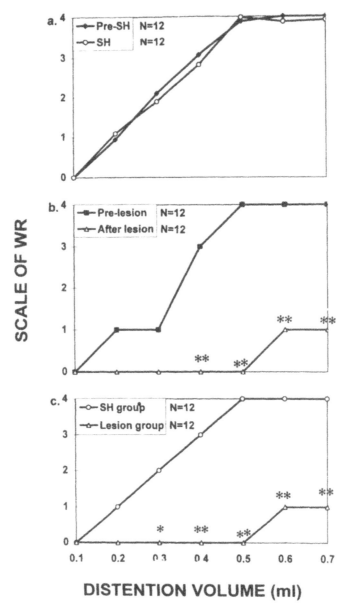

Figure 4.10 Behavioral alterations produced by distention of the duodenum following a lesion of the DC at C2. A balloon catheter was placed in the duodenum through gastrotomy, and the catheter was led through the skin so that volumes of up to 0.7 ml fluid could be injected to distend the duodenum. Writhing responses of different intensities were observed in proportion to the amount of distention; these were graded from 1–4. The graphs plot the intensity of the writhing responses against duodenal distention. In **a** are the responses in control animals (pre-lesion) and in animals after sham surgery (SH); **b** and **c** show that the DC lesion reduced the writhing responses (as compared to the controls and animals with sham surgery). (From Feng *et al.*, 1998).

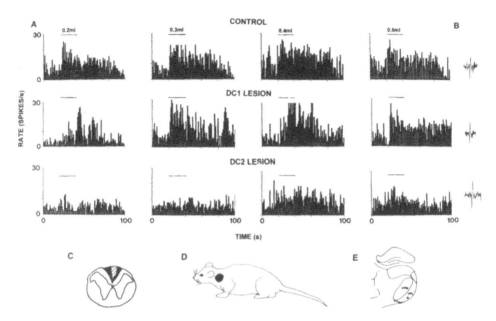

Figure 4.11 Reduction in the responses of a VPL neuron to colorectal distention following a DC lesion. The records in the upper row in **A** show the responses of a VPL neuron to graded distentions of the duodenum. A midline lesion of the DC at C2 had no effect (middle row of records). However, a wider lesion of the DC reduced the responses substantially (lower row). **B** shows the action potential of the neuron at different times during the experiment. **C–E** show the lesions, the cutaneous receptive field, and the recording site. (From Feng *et al.*, 1998.).

In one series of experiments, a balloon catheter was implanted into the duodenum so that distention of the duodenum could be used to evoke writhing responses in awake, behaving rats or to activate neurons in the VPL nucleus of the thalamus of anesthetized rats (Feng *et al.*, 1998). It was found that lesions of the dorsal column could dramatically reduce the writhing responses (Fig. 4.10) and the responses of VPL neurons to duodenal distention (Fig. 4.11). However, a midline dorsal column lesion (at C2) was ineffective. To be effective, the lesion had to include the boundaries between the fasciculus gracilis and the fasciculus cuneatus bilaterally (Fig. 4.11).

In a second series of experiments, pancreatic afferents were stimulated (i) by applying bradykinin to the surface of the pancreas while recording electrophysiological responses of thalamic neurons and (ii) by injection of a bile salt into and ligation of the biliary duct to study homecage behavioral activity after the development of pancreatitis.

In the first experiment (Wang *et al.*, 1998), a dorsal column lesion significantly reduced the responses of neurons in the VPL nucleus of the thalamus to bradykinin stimulation of the pancreas (Fig. 4.12A). The information was relayed through a synapse in the spinal cord since the bradykinin-induced activity of thalamic neurons was diminished by spinal administration of morphine (Fig. 4.12B). Likewise, in the second experiment (Houghton *et al.*, 1997), there was a significant increase in homecage

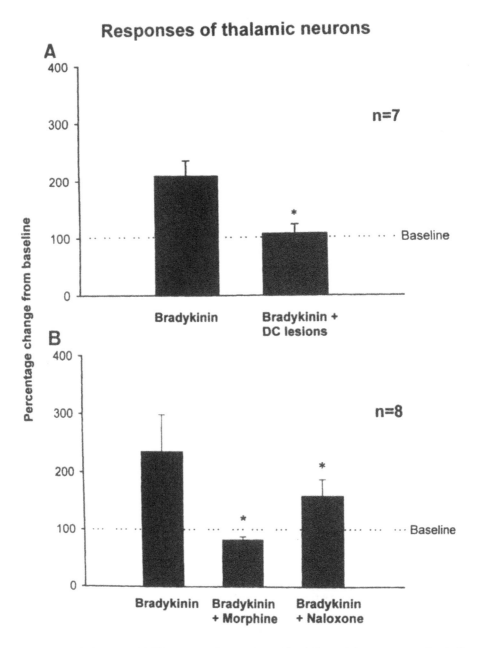

Figure 4.12 Activation of VPL neurons by noxious stimulation of the pancreas. In **A**, the mean responses (± S.E.M.) of VPL thalamic neurons to application of bradykinin to the surface of the pancreas were significantly reduced (back to baseline) after a dorsal column lesion. In **B**, the enhanced average firing rate of thalamic cells activated by bradykinin (10^{-5} M) was significantly reduced by spinal application of morphine. This effect was naloxone reversible. *denotes a significance level of $P < 0.05$ (paired t–test). (From Wang *et al.*, 1998).

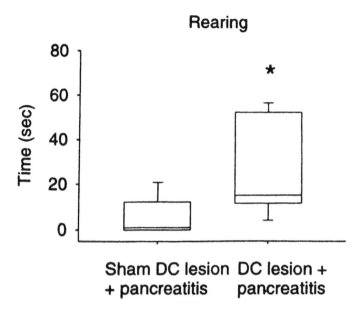

Figure 4.13 Rearing exploratory activity was increased in rats with dorsal column lesions and pancreatitis when compared to rats with sham dorsal column lesions and pancreatitis, n=9. *denotes a significance level of $P < 0.05$ (Mann–Whitney U test). (From Houghton *et al.*, 1997).

'rearing' exploratory activity in animals with an experimentally inflamed pancreas when the dorsal column was lesioned (Fig. 4.13).

Thus, from this series of studies it can be concluded that visceral input from the thoracic region of the spinal cord is also relayed by the dorsal column. The afferent input relayed may be either in response to mechanical distention or to chemically induced stimulation. Anatomical anterograde tract tracing from the central visceral processing region of the spinal cord has demonstrated a previously undescribed ascending projection through the dorsal column (Hirshberg *et al.*, 1996; Wang *et al.*, 1996, Wang *et al.*, 1999). The pathway ascends adjacent to the midline septum from sacral levels of the spinal cord and just lateral to this at the base of the dorsal intermediate septum arising from the thoracic cord (Fig. 4.14). This more lateral anatomical placement is consistent with the organization of the dorsal column and explains the necessity for larger lesions to eliminate visceral input from thoracic levels.

EVIDENCE FROM EXPERIMENTS ON MONKEYS

Partial Dependence of Responses of VPL Neurons to Colorectal Distention on Transmission Through the Dorsal Column

Investigation of the dorsal column visceral pain pathway was extended to monkeys to ensure that evidence like that obtained in rats could be obtained in a species whose nervous system closely resembles that of humans.

Visceral PSDC Pathway

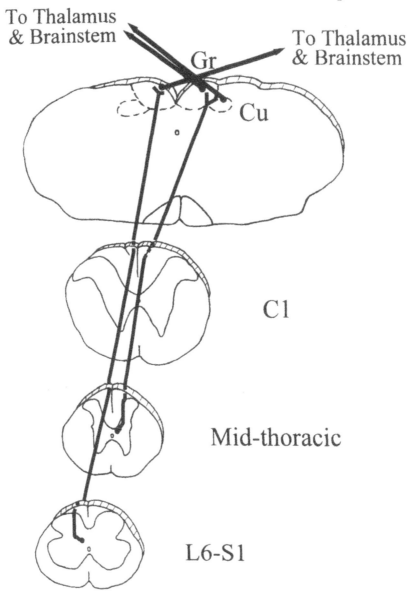

Figure 4.14 The ascending axons of cells in the central visceral processing region of the spinal cord ascend in the dorsal column near the dorsal intermediate septum from thoracic levels and near the midline septum from lumbosacral levels of the spinal cord. The central visceral processing region of the thoracic level innervates both the cuneate and the gracile nucleus while the lumbosacral level innervates only the gracile nucleus. (From Wang *et al.*, 1996).

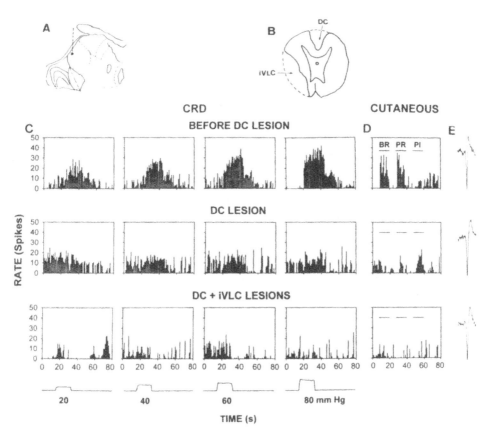

Figure 4.15 Reduction in the responses of a VPL neuron in a monkey to colorectal distention and cutanous stimuli following a DC lesion. The recording site is shown in **A** and the maximum extent of the DC and ipsilateral VLC lesions in **B**. The records in **C** and **D** show the responses to colorectal distention and to mechanical stimulation of the skin. The upper row are the control records. The middle row was recorded after a DC lesion at T10. The lower row shows the effect of additional lesion of the ipsilateral VLC. In **E** are shown action potentials at different times during the experiment. (From Al-Chaer *et al.*, 1998.).

As in rats, recordings were made from neurons of the monkey VPL nucleus of responses to colorectal distention, as well to mechanical stimulation of the skin (Al-Chaer *et al.*, 1998a). An example taken from one of the 6 experiments is illustrated in Figure 4.15. The VPL neuron was strongly activated by graded colorectal distention (Fig. 4.15C, top row of histograms) and by weak mechanical stimulation of the skin (Fig. 4.15D), but not by cutaneous pinch (Fig. 4.15D). A dorsal column lesion was then made at T10 (extent shown in Fig. 4.15B). All of the responses were dramatically reduced (Fig. 4.15C and D, middle row of histograms). A lesion of the ventrolateral quadrant of the spinal cord that would presumably have interrupted the spinothalamic tract (Fig. 4.15B) eliminated any responses that remained and reduced the background activity of the cell

(Fig. 4.15C and D, lower row of histograms). In 2 other cases, the responses of VPL neurons were reduced by dorsal column lesions but not substantially by subsequent ventrolateral quadrant lesions. Instead, bilateral lesions of the dorsal lateral funiculus eliminated the residual responses. For the sample of 5 VPL neurons, the dorsal column lesion reduced the responses to colorectal distention by more than 50%. Lesions of other tracts did not have consistent effects.

Preliminary results of experiments designed to compare the responses of postsynaptic dorsal column neurons and spinothalamic tract neurons in the sacral spinal cord to graded colorectal distention have been reported (Al-Chaer et al., 1998b). Of a total of 100 neurons whose responses were sampled, 48 were identified by antidromic activation as postsynaptic dorsal column neurons and 17% as spinothalamic tract neurons. None of the neurons in the population sampled could be identified as projecting to both the gracile nucleus and the VPL nucleus. The responses of these neurons to colorectal distention were comparable. It was concluded that the greater effectiveness of the postsynaptic dorsal column pathway than the spinothalamic tract in activating neurons of the VPL nucleus might reflect a larger number of postsynaptic dorsal column neurons that respond to colorectal input, rather than any intrinsic differences in response properties.

The responses of cuneothalamic and spinothalamic tract neurons in monkeys to stimulation of cardiopulmonary afferents have also been compared (Chandler et al., 1998). Lesions showed that the pathway that mediated the responses of cuneothalamic tract neurons to electrical stimulation of cardiopulmonary afferents traveled in the ipsilateral dorsal column and probably also in the dorsolateral funiculus. The responses of spinothalamic tract neurons had longer latency and were more prolonged than were the responses of cuneothalamic neurons. Furthermore, spinothalamic tract neurons had bilateral input from cardiopulmonary afferents, whereas the responses of cuneothalamic neurons were strictly to ipsilateral input. Cuneothalamic neurons responded chiefly to innocuous cutaneous stimuli, whereas spinothalamic tract cells responded chiefly to noxious cutaneous stimuli. It was concluded that cuneothalamic and spinothalamic neurons play different roles in signalling nociceptive cardiac information.

Functional Magnetic Resonance Imaging Studies

In order to determine the brain structures that are affected by colorectal distention and the changes that might occur following a lesion of the dorsal column, a functional magnetic resonance imaging (fMRI; Sereno, 1998) study was initiated. Relative cerebral blood flow was estimated by gradient echo bolus tracking and cerebral blood volume by steady state spin echo imaging.

Four monkeys were anesthetized with isoflurane and then imaged with a 4.7 Tesla magnet during and between innocuous or noxious distentions of the colon. The blood was labeled with a superparamagnetic iron oxide compound to cause the image intensity to decrease in proportion to increases in regional cerebral blood volume (Quast et al., 1998). Signal intensity reductions in response to colon distention were in the range of 2–7%, corresponding to increases in blood volume of 9–30%. Blood volume changes were observed in the brain stem, thalamus and cerebral cortex. Noxious colorectal distention resulted in increases in cerebral blood volume in the medial midbrain, VPL nucleus, and medial thalamus, among other structures.

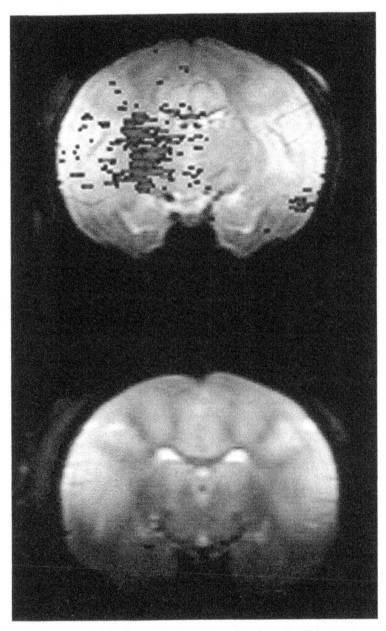

Figure 4.16 Effect of a DC lesion on changes in regional cerebral blood volume observed using fMRI. The upper image shows a cross-section through a monkey's brain at the level of the thalamus. The regional increases in blood volume induced by repeated colorectal distentions of 80 mmHg is indicated by a statistical map (colored voxels). The lower image was obtained from the same monkey 4 months after a DC lesion at T10. The section was taken through approximately the same level of the thalamus. The image shows the lack of change in blood volume in response to the same series of stimuli but following a DC lesion.

Blood volume changes did not occur symmetrically in the brain (Al-Chaer *et al.*, 1998c). Instead, they were more prominent on one side of the brain at some levels, but on the other side at other levels.

The blood volume changes produced by innocuous and noxious colorectal distention were recorded on at least two occasions before surgery was performed at the T10 spinal cord segmental level (summarized in Willis, 1998). Sham surgery was done in one animal and lesions of the dorsal column were made in three animals. After recovery from the surgery, fMRI studies were repeated at 2 weeks, 2 months and 4 months postoperatively. The sham surgery did not affect the fMRI results, but the dorsal column lesion essentially eliminated the blood volume changes that had previously resulted from colorectal distention (Fig. 4.16).

CONCLUSIONS

Clinical evidence indicates that the human dorsal column contains an important visceral pain pathway that, when interrupted, can relieve the pain of cancer affecting pelvic viscera. Interruption of this pathway accounts for the previously unexplained success of limited midline myelotomies.

Experimental evidence is consistent with the clinical results in showing the presence of a visceral nociceptive pathway in the dorsal column in both rats and monkeys. The axons that transmit visceral nociceptive information originate from neurons belonging to the postsynaptic dorsal column pathway. The cell bodies of these neurons are concentrated in the central, visceral processing area of the spinal cord that surrounds the central canal (laminae X, V and VII). Axons from visceral postsynaptic dorsal column neurons in the sacral spinal cord ascend adjacent to the midline to the gracile nucleus. Axons from comparable neurons in the mid-thoracic cord ascend near the dorsal intermediate septum and end in the gracile and cuneate nuclei. The dorsal column nuclei relay noxious visceral information to the ventral posterolateral (VPL) nucleus of the thalamus. Interruption of the dorsal column results in a profound reduction in the visceral responses of neurons in the VPL nucleus of rats and monkeys and in nociceptive behaviors in rats. Interruption of the spinothalamic tract has much less effect on visceral responses in the VPL nucleus, although it results in a substantial reduction in responses to noxious cutaneous stimuli. Spinothalamic tract neurons in the monkey sacral cord are just as responsive to visceral stimuli as are postsynaptic dorsal column cells, but they may be less numerous.

Studies in monkeys using fMRI suggest that noxious visceral stimuli can result in substantial changes in cerebral blood volume in several areas of the brain, including the posterior thalamus. Interruption of the dorsal column eliminates these changes.

REFERENCES

Aidar, O., Geohegan, W.A. and Ungewitter, L.H. (1952). Splanchnic afferent pathways in the central nervous system. *Journal of Neurophysiology*, **15**, 131–138.

Al-Chaer, E.D., Feng, Y. and Willis, W.D. (1998a). A role for the dorsal column in nociceptive visceral input into the thalamus of primates. *Journal of Neurophysiology*, **79**, 3143–3150.

Al-Chaer, E.D., Feng, Y. and Willis, W.D. (1998b). Quantitative basis for the dorsal column dominant role in visceral pain in the primate. *Digestive Diseases Week Abstract*, p. A–900.

Al-Chaer, E.D., Lawand, N.B., Westlund, K.N. and Willis, W.D. (1996a). Visceral nociceptive input into the ventral posterolateral nucleus of the thalamus: a new function for the dorsal column pathway. *Journal of Neurophysiology*, **76**, 2661–2674.

Al-Chaer, E.D., Lawand, N.B., Westlund, K.N. and Willis, W.D. (1996b). Pelvic visceral input into the nucleus gracilis is largely mediated by the postsynaptic dorsal column pathway. *Journal of Neurophysiology*, **76**, 2675–2690.

Al-Chaer, E.D., Quast, M., Feng, Y., Wei, J., Gondesen, K., Illangasekare, N., Deyo, D. and Willis, W.D. (1998c). Visceral pain: an asymmetric function of the brain? *Neuroscience Abstracts*, **24**, 1389.

Al-Chaer, E.D., Westlund, K.N. and Willis, W.D. (1997). Nucleus gracilis: an integrator for visceral and somatic information. *Journal of Neurophysiology*, **78**, 521–527.

Amassian, V.E. (1951). Fiber groups and spinal pathways of cortically represented visceral afferents. *Journal of Neurophysiology*, **14**, 445–460.

Angaut-Petit, D. (1975). The dorsal column system. II. Functional properties and bulbar relay of the postsynaptic fibres of the cat's fasciculus gracilis. *Experimental Brain Research*, **22**, 471–493.

Armour, D. (1927). On the surgery of the spinal cord and its membranes. *Lancet*, **2**, 691–697.

Barolat, G. (1998). Spinal cord stimulation for persistent pain management. In *Textbook of stereotactic and functional neurosurgery*, edited by P.L. Gildenberg and R.R. Tasker, pp. 1519–1537. New York: McGraw Hill.

Basbaum, A.I. (1973). Conduction of the effects of noxious stimulation by short-fiber multisynaptic systems of the spinal cord in the rat. *Experimental Neurology*, **40**, 699–716.

Bennett, G.J., Seltzer, Z., Lu, G.W., Nishikawa, N. and Dubner, R. (1983). The cells of origin of the dorsal column postsynaptic projection in the lumbosacral enlargements of cats and monkeys. *Somatosensory Research*, **1**, 131–149.

Berkley, K.J. and Hubscher, C.H. (1995). Are there separate central nervous system pathways for touch and pain? *Nature Medicine*, **1**, 766–773.

Berkley, K.J., Guilbaud, G., Benoist, J.M. and Gautron, M. (1993). Responses of neurons in and near the thalamic ventrobasal complex of the rat to stimulation of uterus, cervix, vagina, colon, and skin. *Journal of Neurophysiology*, **69**, 557–568.

Brown, A.G., Brown, P.B., Fyffe, R.E.W. and Pubols, L.M. (1983). Receptive field organization and response properties of spinal neurones with axons ascending the dorsal columns in the cat. *Journal of Physiology*, **377**, 575–588.

Brown-Séquard, C.E. (1860). *Course of lectures on the physiology and pathology of the central nervous system.* Philadelphia: Collins.

Brüggemann, J., Shi, T. and Apkarian, A.V. (1994). Squirrel monkey lateral thalamus. II. Viscerosomatic convergent representation of urinary bladder, colon and esophagus. *Journal of Neuroscience*, **14**, 6796–6814.

Chandler, M.J., Hobbs, S.F., Fu, Q.G., Kenshalo, D.R., Blair, R.W. and Foreman, R.D. (1992). Responses of neurons in ventroposterolateral nucleus of primate thalamus to urinary bladder distension. *Brain Research*, **571**, 26–34.

Chandler, M.J., Zhang, J. and Foreman, R.D. (1998). Cardiopulmonary sympathetic input excites primate cuneothalamic neurons: comparison with spinothalamic tract neurons. *Journal of Neurophysiology*, **80**, 628–637.

Cliffer, K.D., Hasegawa, T. and Willis, W.D. (1992). Responses of neurons in the gracile nucleus of cats to innocuous and noxious stimuli: basic characterization and antidromic activation from the thalamus. *Journal of Neurophysiology*, **68**, 818–832.

Conti. F., De Biasi, S., Giuffrida, R. and Rustioni, A. (1990). Substance P-containing projections in the dorsal columns of rats and cats. *Neuroscience*, **34**, 607–621.

Cook, A.W., Nathan, P.W. and Smith, M.C. (1984). Sensory consequences of commissural myelotomy: a challenge to traditional anatomical concepts. *Brain*, **107**, 547–568.

Dougherty, P.M., Palecek, J., Paleckova, V., Sorkin, L.S. and Willis, W.D. (1992). The role of NMDA and non-NMDA excitatory amino acid receptors in the excitation of primate spinothalamic tract neurons by mechanical, chemical, thermal, and electrical stimuli. *Journal of Neuroscience*, **12**, 3025–3041.

Fabri, M. and Conti, F. (1990). Calcitonin gene-related peptide-positive neurons and fibers in the cat dorsal column nuclei. *Neuroscience*, **35**, 167–174.

Feng, Y., Cui, M., Al-Chaer, E.D. and Willis, W.D. (1998). Epigastric antinociception by cervical dorsal column lesions in rats. *Anesthesiology*, **89**, 411–420.

Ferrington, D.G., Downie, J.W. and Willis, W.D. (1988). Primate nucleus gracilis neurons: responses to innocuous and noxious stimuli. *Journal of Neurophysiology*, **59**, 886–907.

Gammon, G.D. and Bronk, D.W. (1935). The discharge of impulses from pacinian corpuscles in the mesentery and its relation to vascular changes. *American Journal of Physiology*, **114**, 77–84.

Giesler, G.J. and Cliffer, K.D. (1985). Postsynaptic dorsal column pathway of the rat. II. Evidence against an important role in nociception. *Brain Research*, **326**, 347–356.

Gildenberg, P.L. and Hirshberg, R.M. (1984). Limited myelotomy for the treatment of intractable cancer pain. *Journal of Neurology, Neurosurgery and Psychiatry*, **47**, 94–96.

Giuffrida, R. and Rustioni, A. (1992). Dorsal root ganglion neurons projecting to the dorsal column nuclei of rats. *Journal of Comparative Neurology*, **316**, 206–220.

Gybels, J.M. and Sweet, W.H. (1989). *Neurosurgical treatment of persistent pain. Physiological and pathological mechanisms of human pain*. Basel: Karger.

Head, H. and Thompson, T. (1906). The grouping of afferent impulses within the spinal cord. *Brain*, **29**, 537–741.

Hirshberg, R.M., Al-Chaer, E.D., Lawand, N.B., Westlund, K.N. and Willis, W.D. (1996). Is there a pathway in the posterior funiculus that signals visceral pain? *Pain*, **67**, 291–305.

Hitchcock, E.R. (1970) Stereotactic cervical myelotomy. *Journal of Neurology, Neurosurgery and Psychiatry*, **33**, 224–230.

Hitchcock, E.R. (1972a). Electrophysiological exploration of the cervico-medullary region. In *Neurophysiology studied in man*, edited by G. Somjen, pp. 237–245. Amsterdam: Excerpta Medica.

Hitchcock, E.R. (1972b). Stereotaxis of the spinal cord. *Confinia Neurologica*, **34**, 229–310.

Honda, C.N. (1985). Visceral and somatic afferent convergence onto neurons near the central canal in the sacral spinal cord of the cat. *Journal of Neurophysiology*, **53**, 1059–1078.

Houghton, A.K., Kadura, S. and Westlund, K.N. (1997). Dorsal column lesions reverse the reduction of homecage activity in rats with pancreatitis. *NeuroReport*, **8**, 3795–3800.

Hu, S. and Rubly, N. (1983). Effects of morphine on ionic currents in frog node of Ranvier. *European Journal of Pharmacology*, **95**, 185–192.

Karplus, J.P. and Kreidl, A. (1914). Ein Beitrag zur Kenntnis der Schmerzleitung im Rückenmark. *Pflügers Archiv — European Journal of Physiology*, **158**, 275–287.

King, R.B. (1977). Anterior commissurotomy for intractable pain. *Journal of Neurosurgery*, **47**, 7–11.

Kuo, D.C. and De Groat, W.C. (1985). Primary afferent projections of the major splanchnic nerve to the spinal cord and the nucleus gracilis of the cat. *Journal of Comparative Neurology*, **231**, 421–434.

Leriche, R. (1936). Du traitement de la douleur dans les cancers abdominaux et pelviens inopérables ou récidivés. *Gazette des Hopitaux Paris*, **109**, 917–922.

Lu, G.W., Bennett, G.J., Nishikawa, N., Hoffert, M.J. and Dubner, R. (1983). Extra- and intracellular recordings from dorsal column postsynaptic spinomedullary neurons in the cat. *Experimental Neurology*, **82**, 456–477.

Mansuy, L., Lecuire, J. and Acassat, L. (1944). Technique de la myelotomie commissurale posterieur. *Journal de Chirurgie*, **60**, 206–213.

Melzack, R. and Wall, P.D. (1965). Pain mechanisms: a new theory. *Science*, **150**, 971–979.

Nathan, P.W., Smith, M.C. and Cook, A.W. (1986). Sensory effects in man of lesions of the posterior columns and of some other afferent pathways. *Brain*, **109**, 1003–1041.

Nauta, H.J.W., Hewitt, E., Westlund, K.N. and Willis, W.D. (1997). Surgical interruption of a midline dorsal column visceral pain pathway. *Journal of Neurosurgery*, **86**, 538–542.

Ness, T.J. and Gebhart, G.F. (1987). Characterization of neuronal responses to noxious visceral and somatic stimuli in the medial lumbosacral spinal cord of the rat. *Journal of Neurophysiology*, **57**, 1867–1892.

Ness, T.J. and Gebhart, G.F. (1988). Colorectal distension as a noxious visceral stimulus: physiologic and pharmacologic characterization of pseudoaffective reflexes in the rat. *Brain Research*, **450**, 153–169.

Ness, T.J., Metcalf, A.M. and Gebhart, G.F. (1990). A psychophysiological study in humans using phasic colonic distension as a noxious visceral stimulus. *Pain*, **43**, 377–386.

Noordenbos, W. and Wall, P.D. (1976). Diverse sensory functions with an almost totally divided spinal cord. A case of spinal cord transection with preservation of part of one anterolateral quadrant. *Pain*, **2**, 185–195.

Omote, K., Kawamata, M., Iwasaki, H. and Namiki, A. (1994). Effects of morphine on neuronal and behavioural responses to visceral and somatic nociception at the level of the spinal cord. *Acta Anesthesiologica Scandinavica*, **38**, 514–517.

Papo, I. and Luongo, A. (1976). High cervical commissural myelotomy in the treatment of pain. *Journal of Neurology, Neurosurgery and Psychiatry*, **39**, 705–710.

Patterson, J.T., Coggeshall, R.E., Lee, W.T. and Chung, K. (1990). Long ascending unmyelinated primary afferent axons in the rat dorsal column: immunohistochemical localizations. *Neuroscience Letters*, **108**, 6–10.

Patterson, J.T., Head, P.A., McNeill, D.L., Chung, K. and Coggeshall, R.E. (1989). Ascending unmyelinated primary afferent fibers in the dorsal funiculus. *Journal of Comparative Neurology*, **290**, 384–390.

Putnam, T.J. (1934). Myelotomy of the commissure: a new method of treatment for pain in the upper extremities. *Archives of Neurology and Psychiatry*, **32**, 1189–1193.

Quast, M.L., Al-Chaer, E., Wei, J., Feng, Y., Illangasekare, N., Gonzales, J.M., Deyo, D., Sell, S., Gonsesen, K.J. and Willis, W.D. (1998). High resolution fMRI in a monkey model of visceral pain. *International Society for Magnetic Resonance in Medicine*, Abstract.

Rees, H. and Roberts, M.H.T. (1993). The anterior pretectal nucleus: a proposed role in sensory processing. *Pain*, **53**, 121–135.

Rigamonti, D.D. and Hancock, M.B. (1974). Analysis of field potentials elicited in the dorsal column nuclei by splanchnic nerve A-beta afferents. *Brain Research*, **77**, 326–329.

Rigamonti, D.D. and Hancock, M.B. (1978). Viscerosomatic convergence in the dorsal column nuclei. *Experimental Neurology*, **61**, 337–348.

Sarnoff, S.J., Arrowood, J.G. and Chapman, W.P. (1948). Differential spinal block. IV. The investigation of intestinal dyskinesia, colonic atony, and visceral afferent fibers. *Surgery, Gynecology and Obstetrics*, **86**, 571–581.

Schvarcz, J.R. (1976). Stereotactic extralemniscal myelotomy. *Journal of Neurology, Neurosurgery and Psychiatry*, **39**, 53–57.

Schvarcz, J.R. (1978). Spinal cord stereotactic techniques re trigeminal nucleotomy and extralemniscal myelotomy. *Applied Neurophysiology*, **41**, 99–112.

Schvarcz, J.R. (1984). Steroeotactic high cervical extralemniscal myelotomy for pelvic cancer pain. *Acta Neurochirurgica*, **Suppl. 33**, 431–435.

Sereno, M.I. (1998). Brain mapping in animals and humans. *Current Opinion in Neurobiology*, **8**, 188–194.

Shealy, C.N., Mortimer, J.T. and Reswick, J.B. (1967). Electrical inhibition of pain by stimulation of the dorsal columns. *Anesthesia and Analgesia*, **46**, 489–491.

Tamatani, M., Senba, E. and Tohyama, M. (1989). Calcitonin gene-related peptide- and substance P-containing primary afferent fibers in the dorsal column of the rat. *Brain Research*, **495**, 122–130.

Uddenburg, N. (1968). Functional organization of long, second-order afferents in the dorsal funiculus. *Experimental Brain Research*, **4**, 377–382.

Wall, P.D. and Noordenbos, W. (1977). Sensory functions which remain in man after complete transection of dorsal columns. *Brain*, **100**, 641–653.

Wang, C.-C., Houghton, A.K. and Westlund, K.N. (1998). Pancreatic nociceptive information is transmitted by the post-synaptic dorsal column pathway. *Neuroscience Abstracts*, **24**, 394.

Wang, Chia-Chuan, Lu, Ying, Willis, W.D. and Westlund, K.N. (1996). A new visceral pain pathway in the dorsal columns: a PHA–L study of ascending projections from T7 and S1 lamina X in the rat. *American Pain Society*, **15**, A–108.

Wang, C.C., Willis, W.D. and Westlund, K.N. (1999). Ascending projection from the area around the spinal cord central canal. *Journal of Comparative Neurology*, **415**, 341–367.

White, J.C. (1943). Sensory innervation of the viscera: studies on visceral afferent neurones in man based on neurosurgical procedures for the relief of intractable pain. *Research Publications of the Association for Research in Nervous and Mental Disease*, **23**, 373–390.

White, J.C. and Sweet, W.H. (1969). *Pain and the neurosurgeon. A forty-year experience*. Springfield: Thomas.

Willis, W.D. (1998). Evidence for a visceral pain pathway in the dorsal column of the spinal cord. Abstract, National Academy of Sciences Colloquium on *The Neurobiology of Pain*.

Willis, W.D. (1985). *The pain system. The neural basis of nociceptive transmission in the mammalian nervous system*. Basel: Karger.

Willis, W.D. and Coggeshall, R.E. (1991). *Sensory mechanisms of the spinal cord*. New York: Plenum Press.

Yamamoto, S. and Sugihara, S. (1956). Microelectrode studies on sensory afferents in the posterior funiculus of cat. *Japanese Journal of Physiology*, **6**, 68–85.

Yuge, O., Matsumoto, M., Kitahata, L.M., Collins, J.G. and Senami, M. (1985). Direct opioid application to peripheral nerves does not alter compound action potentials. *Anesthesia and Analgesia*, **64**, 667–671.

CHAPTER 5
TRANSMISSION SECURITY ACROSS CENTRAL SYNAPSES FOR TACTILE AND KINAESTHETIC SIGNALS

Mark J. Rowe

School of Physiology and Pharmacology, The University of New South Wales, Sydney, Australia

INTRODUCTION: RECEPTORS AND SENSORY FIBER CLASSES THAT CONTRIBUTE TO TACTILE AND KINAESTHETIC SENSATION

Tactile and kinaesthetic sensations depend upon signals that arise from approximately eight major classes of receptors and sensory nerve fibers, the majority of which are responsible for carrying the signals for our tactile sense, while two or three classes convey the kinaesthetic signals.

The principal receptors accounting for our tactile sense in the fingertips and palms are the Pacinian corpuscle (PC) receptors, whose associated nerve fibers are usually designated the PC afferent fibers; the Meissner corpuscle, associated with the rapidly adapting (RA) afferent fibers, and the Merkel receptor complex supplied by slowly adapting type I (SAI) nerve fibers. In the hairy regions of skin the principal tactile afferent fibers include the SAI class together with an SAII class, associated with Ruffini receptors, and a major group whose endings are associated with hair follicles, the HFA afferent fibers (Jänig *et al.,* 1968; Talbot *et al.,* 1968; Chambers *et al.,* 1972; Iggo and Ogawa, 1977; Ferrington and Rowe, 1980). Similar classes appear to be present in a variety of species, including cat, macaque and human being, although there is some variation in the nomenclature for designating the different classes. For example, some of the most detailed observations on human tactile afferent fiber classes have been made by the Swedish researchers, Vallbo and Johansson, who have used an FAI and FAII designation for the RA and PC fiber classes respectively (Johansson and Vallbo, 1979; Vallbo and Johansson, 1984).

The principal sensory fiber classes subserving our kinaesthetic sense arise from muscle spindle receptors and from receptors in the joint capsules, although cutaneous mechanoreceptors, in the vicinity of joints, principally associated with the SAII fibers, are sensitive to skin stretch when flexion – extension movements occur at the joint and also contribute to our sense of limb position (Gandevia and McCloskey, 1976; McCloskey, 1978; Edin and Abbs, 1991; Edin, 1996).

Address for correspondence: Professor Mark J. Rowe, School of Physiology and Pharmacology, The University of New South Wales, Sydney, NSW 2052, Australia. Tel: 62 2 9385 1054; Fax: 62 2 9385 1059; E-mail: M.Rowe@unsw.edu.au

MICRONEUROGRAPHY AND THE STUDY OF SINGLE SENSORY FIBERS IN CONSCIOUS HUMAN SUBJECTS

Although the above afferent classes contribute to tactile or kinaesthetic experience, there is quite striking evidence, when *individual* fibers are activated in conscious human subjects, of marked differences among the fiber classes in their capacity to generate a perceptual response. This evidence has come from the use of microneurography, which was developed principally by Swedish researchers (Vallbo and Hagbarth, 1968; Vallbo *et al.*, 1979). The technique involves the insertion of a fine needle micro-electrode through the skin into a peripheral nerve in order to record from a *single* afferent or efferent nerve fiber in a conscious human subject who is attending to a particular sensory or motor task. An important extension of the procedure has been the advent of intraneural microstimulation, based upon conversion of the microneurography electrode from the recording mode to the stimulation mode (Torebjörk *et al.*, 1987). The experimenter must therefore first isolate electrophysiologically the responses of a single identified sensory fiber and characterize its receptive field before switching the electrode to stimulation mode. In this circumstance it is possible with weak electrical stimuli to selectively activate the identified sensory fiber and determine whether there are perceptual consequences for the subject. The close correspondence observed between the 'receptive' and 'perceptive' fields is shown in Figure 1 for two afferent fibers, an SAI fiber on the left, and an RA or FAI fiber on the right, in one of the major pioneering studies conducted with the intraneural microstimulation procedure by Vallbo *et al.* (1984). The stimulus train was initially delivered at very low strength ($< 1 \mu A$) and progressively increased until the subject reported a sensation, localizable to an area that corresponded to the receptive field of the SAI or RA fiber. With further increases in stimulus strength the percept extends to areas beyond those identified in Figure 5.1, presumably reflecting the recruitment of two or more afferent fibers (Vallbo *et al.*, 1984).

PERCEPTUAL CONSEQUENCES OF ACTIVATING SINGLE PC OR RA AFFERENT FIBERS

The detailed microstimulation studies carried out in Vallbo's and Torebjörk's laboratories have shown that trains of weak electrical stimuli applied to *single* RA or PC fibers in peripheral nerve fascicles of alert human subjects evoke a distinct sense of flutter or vibration depending upon the frequency of the stimuli. This perceptual response arises within the skin area identified as the receptive field for the particular PC or RA fiber under study. The studies demonstrate that inputs arising in *single* RA or PC sensory fibers can give rise to a conscious sensory experience, and that even a single impulse generated in one of these fibers (RA fibers, in particular) can give rise to a perceptual response (Ochoa and Torebjörk, 1983; Vallbo *et al.*, 1984; Macefield *et al.*, 1990).

Furthermore, this capacity for single RA or PC fibers to generate the perceptual response is apparent in almost all fibers tested, as shown in Figures 5.2 and 3 taken from Ochoa and Torebjörk (1983). Figure 5.2 shows the receptive field focus on the fingers and hand for 38 RA fibers, each activated singly by intraneural microstimulation, 35 of which elicited a perceptual response. The open circles indicate the fields for the three RA fibers that failed to produce a perceptual response. Among 14 PC fibers tested by Ochoa and Torebjörk, 12 gave rise to a perceptual response as indicated in Figure 5.3.

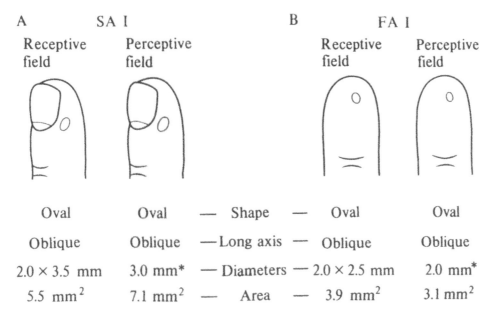

		Shape		
Oval	Oval	Shape	Oval	Oval
Oblique	Oblique	Long axis	Oblique	Oblique
2.0 × 3.5 mm	3.0 mm*	Diameters	2.0 × 2.5 mm	2.0 mm*
5.5 mm²	7.1 mm²	Area	3.9 mm²	3.1 mm²

Figure 5.1 Spatial properties of receptive fields of **(A)** an SAI and **(B)** an FAI unit compared with the spatial properties of percepts elicited by microstimulation through the recording electrode. Visual matching to circular areas (from Vallbo *et al.*, 1984).

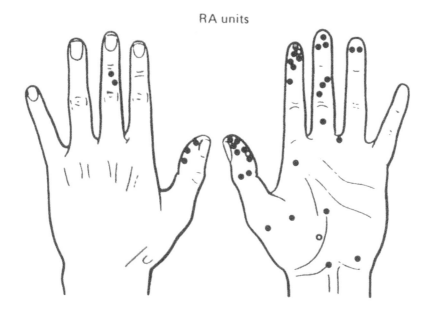

Figure 5.2 Location of receptive fields of 38 RA units. Note clustering in finger pulps. Intraneural microstimulation evoked a conscious sensation from most of the units (•): only three units had no cognitive correlate (o) (from Ochoa and Torebjörk, 1983).

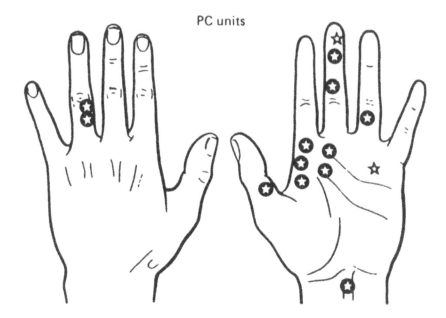

Figure 5.3 Location of receptive fields of fourteen PC units. Intraneural microstimulation evoked sensations from 12 units (encircled stars) and no sensation from two units (stars) (from Ochoa and Torebjörk, 1983).

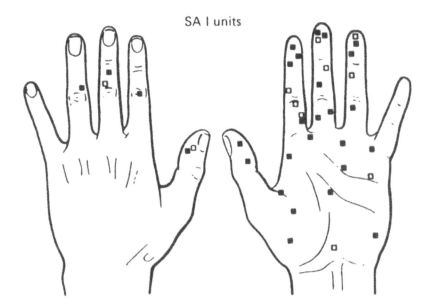

Figure 5.4 Location of receptive fields of 39 SAI units. Intraneural microstimulation evoked a sensation from 28 units (■) and no sensation from 11 units (□) (from Ochoa and Torebjörk, 1983).

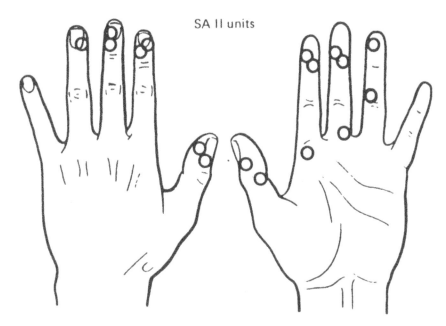

Figure 5.5 Location of receptive fields of 17 SAII units. Note clustering close to nails and joints. Intraneural microstimulation failed to evoke sensation from any of these units (from Ochoa and Torebjörk, 1983).

PERCEPTUAL CONSEQUENCES OF ACTIVATING SINGLE SAI OR SAII AFFERENT FIBERS

A very different outcome was observed when SAI and SAII fibers were stimulated singly. First, although a majority (~75%; Ochoa and Torebjörk, 1983) of single SAI fibers generate a perceptual response (Fig. 5.4), the percept described in response to trains of weak electrical stimuli at frequencies of 20–100 Hz (lasting 250–500 ms) is one of steady pressure or deformation in the region of the SAI fiber's receptive field, in contrast to the vibrotactile percept generated by the same train of stimuli delivered to individual RA or PC fibers (Ochoa and Torebjörk, 1983; Vallbo *et al.*, 1984; Macefield *et al.*, 1990). In the case of SAII fibers, all 17 studied by Ochoa and Torebjörk (1983) were unable to generate a perceptual response when activated by single stimuli or trains of stimulus (Fig. 5.5).

PERCEPTUAL CONSEQUENCES OF ACTIVATING SINGLE KINAESTHETIC AFFERENT FIBERS OF MUSCLE OR JOINT ORIGIN

The two major classes of kinaesthetic afferents, arising from muscle spindles and joint capsules, were first examined with the intraneural microstimulation procedure by Macefield *et al.* (1990) who found differences within this group (as among tactile afferents) in their capacity to generate a perceptual response. When activated singly, almost all joint afferents associated with the interphalangeal and metacarpophalangeal joints gave rise to 'deep'

Table 5.1

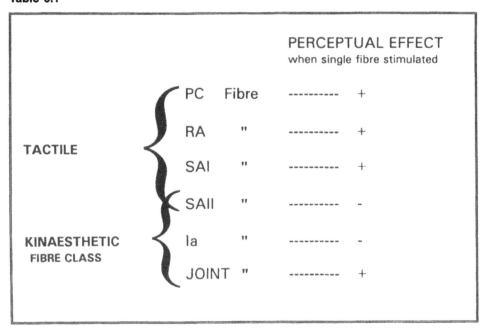

sensations associated with the joint capsule or a sense of movement in the joint from which they arose, whereas the activation of single muscle spindle afferents from a variety of muscles in the forearm, principally the intrinsic muscles of the hand, was almost always without effect. Further observations involving activation of spindle afferent fibers within the radial nerve above the elbow also revealed a failure of spindle afferents associated with more proximal muscles in the arm, to generate perceptual responses (Ni *et al.*, 1998).

Table 5.1 summarizes the results observed in the microstimulation studies for these six different classes of tactile and kinaesthetic afferents. The plus signs indicate that PC, RA, SAI and joint afferent fibers are able to elicit perceptual responses when activated singly while SAII and spindle afferent fibers are generally unable to do so. A further qualification needs to be emphasized however, that the SAI fibers fail to generate any vibrotactile sense in response to activation with a train of stimuli in contrast to the RA and PC fibers, and that a significant proportion of the SAI fibers fail to generate any percept, even of steady pressure.

HYPOTHESIS TO EXPLAIN THE DIFFERENTIAL CAPACITIES OF INDIVIDUAL FIBERS OF DIFFERENT AFFERENT CLASSES TO GENERATE PERCEPTUAL RESPONSES

One of the explanations for the marked differences observed among these kinaesthetic and tactile fiber classes in their capacity to generate perceptual responses may be based on differences in their transmission characteristics across synaptic junctions in the somatosensory pathways. A central rather than peripheral explanation for the differential effectiveness of the different fiber classes appears necessary for the following reasons. First, the electrical pulse trains delivered at 20–100 Hz in the microstimulation experiments had the same precise periodicity when testing the different fiber classes (Ochoa and Torebjörk, 1983; Vallbo *et al.*, 1984; Macefield *et al.*, 1990; Ni *et al.*, 1998). Furthermore, work in our laboratory (Gynther *et al.*, 1992; Vickery *et al.*, 1992) on SAI and SAII fibers in the cat forelimb, and in other laboratories on SAI fibers in the facial region (Gottschaldt and Vahle-Hinz, 1981) has demonstrated that in response to cutaneous sinusoidal vibration *both* SAI and SAII fibers are capable of responding at low thresholds with tightly phaselocked patterns of impulses that reflect accurately the periodicity of the vibratory stimulus, even at frequencies above 500 Hz. In this respect neither the SAI, nor SAII fibers appear to be poorer than the PC fibers in their capacity for encoding information in an *impulse pattern code*, about the frequency parameter of cutaneous vibration. One must therefore conclude that peripheral factors cannot explain why *individual* SAI and SAII fibers, in contrast to RA and PC fibers are unable to generate a subjective sense of the periodic, or temporally modulated components of peripheral stimuli such as those delivered with the intraneural microstimulation procedure. Instead, the explanation for the differential capacities of the different fiber classes must lie within the central nervous system.

A PAIRED RECORDING ARRANGEMENT FOR ANALYZING TRANSMISSION CHARACTERISTICS BETWEEN SINGLE IDENTIFIED AFFERENT FIBERS AND THEIR CENTRAL TARGET NEURONES

In order to examine and test the hypothesis that there may be differential transmission security for different fiber classes across synaptic linkages in the tactile and kinaesthetic sensory pathways we have developed an experimental paradigm in the anaesthetized cat, based on paired simultaneous recording from an *individual* afferent fiber in an *intact* peripheral nerve fascicle, and from the central target neurone of that afferent fiber. With this we were able to examine the efficacy of transmission in a one-to-one linkage between an identified primary afferent fiber and one of its central target neurones. Our initial analysis of transmission characteristics for single afferent fibers of the different classes has been conducted at the dorsal column nuclei, the first central synaptic relay interposed in the principal tactile and kinaesthetic sensory pathway that traverses the spinal dorsal columns and medial lemniscus, before further synaptic interruptions occur in the ventral posterolateral thalamus and then in somatosensory cortical areas I and II (SI and SII respectively; Ferrington and Rowe, 1980; Turman *et al.*, 1992; Zhang *et al.*, 1996).

The first class of afferent fiber for which we have been able to successfully achieve the paired simultaneous recording paradigm in our anaesthetized cat preparation was for the PC fiber class, using the hindlimb interosseous nerve, which supplies an array of PC receptors on the hindlimb interosseous membrane, as represented in Figure 5.6 (Ferrington *et al.*, 1986, 1987a,b). These Pacinian receptors show the same mechanosensitivity, including vibrational sensitivity, as cutaneous Pacinian corpuscles. Furthermore, their afferent fibers project to the dorsal column nuclei as do cutaneous PC fibers (Fig. 5.6). The interosseous nerve is sufficiently fine that if one carefully dissects it free of the flexor hallucis longus muscle through which it passes, and places it over a simple platinum hook

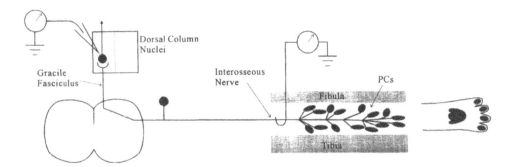

Figure 5.6 Experimental arrangement for paired simultaneous recording showing the recording sites within the gracile nucleus and on the hind-limb interosseous nerve proximal to the clusters of PC receptors associated with the nerve and the interosseous membrane (from Ferrington, Rowe and Tarvin, 1987a).

electrode, it is then possible to identify, in the intact nerve, the spike activity of each active PC fiber.

Individual PC fibers were activated in these experiments by means of vibration applied with a 200 μm diameter probe to individual corpuscles that could be visualized along the nerve or the interosseous membrane. A second electrode, a microelectrode was inserted into the gracile division of the dorsal column nuclei to record simultaneously from the central target neurones of the afferent fiber being selectively activated and monitored in the interosseous nerve (Fig. 5.6).

TRANSMISSION SECURITY FOR SINGLE PC AFFERENT FIBERS AT SYNAPTIC JUNCTIONS IN THE DORSAL COLUMN NUCLEI

In our analysis of transmission characteristics for PC afferent fibers we isolated 35 PC fiber-gracile neurone pairs in which activity in a *single* identified PC fiber, in fact a single impulse in the single fiber, was capable of evoking suprathreshold responses, that is, spike output in the gracile neurone (Ferrington *et al.*, 1986, 1987a). This means of course that unitary EPSPs generated by PC afferent fibers commonly exceed threshold in the target gracile neurones.

Among the 35 gracile neurones studied that were driven by single PC afferent fibers the location in the nucleus was verified in 24 cases and found to be between 2.5 and 4 mm caudal to the obex and at mediolateral positions between 0.13 and 0.35 mm from the midline. These locations conform to the 'cluster' region of dorsal column nuclei (Berkley, 1975; Hand and Van Winkle, 1977; Ellis and Rustioni, 1979) where most neurones have projections to the contralateral VPL thalamus and are therefore involved in conveying sensory signals rostrally to the thalamo-cortical centres crucial for perceptual experience. The mean latency of response for these gracile neurones to their single PC fiber inputs was 10.3 ± 1.5 ms ($n = 35$; Ferrington *et al.*, 1987a) which for a hindlimb source of group II afferent input is consistent with a monosynaptic action of the PC fibers on the gracile neurones.

For all 35 PC fiber-gracile neurone pairs studied the synaptic linkage is clearly very

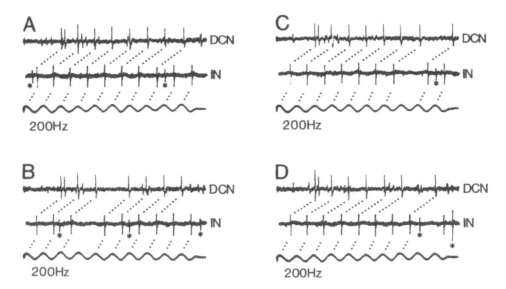

Figure 5.7 Single sensory nerve impulses in PC fibres of the interosseous nerve (IN) evoke spike output from gracile neurone of dorsal column nuclei (DCN; uppermost traces) and a PC sensory fibre of the i.n. (middle traces) with direct evidence for the high security of the linkage in this sensory relay. The lowest trace shows a train of ten cycles of 200 Hz vibration at an amplitude of < 2 μm where, in **B**, **C** and **D** the PC fibre failed occasionally to respond on each cycle of the vibration. On those cycles of vibration that failed to elicit a PC fibre spike (the fourth cycle in B, the eighth in C and the ninth in D) the target neurone also failed to respond. (from Ferrington, Rowe and Tarvin, 1987a).

secure as reflected in the spike output as a consequence of the action of a single sensory fiber. However, for a great many of the pairs the potency of the one-to-one synaptic linkage was quite remarkable as may be seen in Figure 5.7, where the paired traces in A–D show gracile neurone responses in the uppermost traces, (d.c.n.), PC fiber activity recorded from the interosseous nerve (i.n.) in the middle traces, and the 200 Hz vibration stimulus wave form in the lowest traces. The high security of the linkage for this PC fiber-gracile neurone pair is reflected first, by the presence in each set, A–D, of a pair of spikes in the gracile neurone in response to the first PC fiber spike and secondly, by the presence of a gracile neurone spike in response to almost all PC fiber spikes in each of the traces. The only instances where the fiber spike failed to elicit a gracile neurone response were on the ninth cycle of vibration in A and B and on the seventh cycle in D.

We believe that the observed PC fiber spikes in records such as Figure 5.7 are uniquely responsible for the associated gracile neurone responses on several grounds. First, the stimulus parameters and extensive leg denervation preclude inputs other than interosseous PC fibers. Secondly, the interosseous nerve recording was only considered satisfactory if there was a clear discontinuity between the noise level on the recording trace and the height of each PC fiber spike. In most experiments this involved a signal-to-noise ratio of 5–10:1 for PC fiber activity and ensured that the gracile responses could not be attributed to PC fibers whose incoming spike activity was buried in the noise of the recording trace. Thirdly,

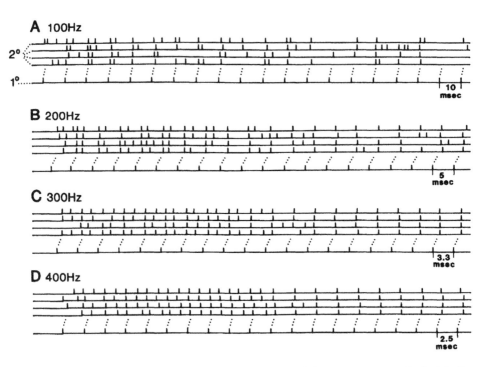

Figure 5.8 Amplification of output across the synaptic linkage between a single PC fibre and a gracile neurone. In each set (**A–D**) of impulse trace replicas the upper four traces represent separate responses of the gracile neurone to different PC fibre spike trains. Only one PC fibre impulse trace is represented in each set (lowest traces) as successive response traces for a given PC fibre were essentially identical. The potency of the linkage for this PC fibre-gracile neurone pair is seen at all four vibration frequencies (from 100 to 400 Hz) where the gracile neurone responds with pairs or triplets of spikes to each PC fibre spike evoked by the first five to 15 cycles of vibration. The gracile neurone traces have been displaced to the left by approximately 8 ms in order to align the PC fibre-target neurone responses more easily (from Ferrington, Rowe and Tarvin, 1987a).

as seen in Figure 5.7, gracile neurone responses were dependent on the prior occurrence of the *observed* PC fiber spike recorded from the interosseous nerve and followed it at a steady latency. When the PC fiber failed to respond on some vibration cycles, for example on the fourth cycle in Figure 5.7B, the eight in C, and the ninth in D, there was correlated absence of response in the gracile neurone.

AMPLIFICATION OF INPUT IN THE PC FIBER–GRACILE NEURONE SYNAPTIC LINKAGE

The generation of pairs or triplets of spikes in the gracile neurone by single PC input spikes, in particular at the onset of a train of input spikes, was a common occurrence. For example in twenty out of twenty-eight PC fiber-gracile neurone pairs studied using 200 Hz vibration

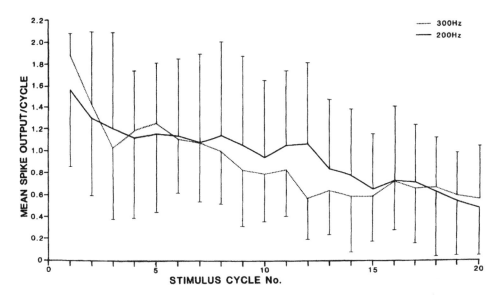

Figure 5.9 Security of linkage in PC fibre-gracile neurone pairs expressed as the averaged spike output on successive cycles of vibration for all pairs studied quantitatively at 200 and 300 Hz. The measure of gracile neurone output was obtained for each pair when the PC fibre was responding to the vibration with a regular 1:1 pattern of activity. The mean spike output on each cycle was then calculated, up to cycle number 20, for all pairs for which data were available and plotted along with S.D. (bars). At 200 Hz the number of pairs (*n*) contributing to the plot fell from *n* = 28 in each of the first six cycles to *n* = 25 (tenth cycle), *n* = 13 (fifteenth cycle) and *n* = 9 (twentieth cycle). At 300 Hz, for each of the first sixteen cycles *n* = 8, and for the seventeenth to twentieth cycles, *n* = 7 (from Ferrington, Rowe and Tarvin, 1987a).

it was found, as seen in Figure 5.7, that the first spike arriving over the single PC fiber elicited pairs or triplets of spikes in the gracile neurone. This is one index of the extraordinarily potent linkage between this class of tactile afferent fiber and its central target neurone, and of course, represents an amplification mechanism at the synapse that may operate in some cases for perhaps the first 10–15 impulses arriving over the PC fiber (Fig. 5.8). Furthermore, the high security of the linkage was apparent over a range of frequencies in the input spike train as seen in the impulse-trace replicas of Figure 8 obtained for one PC fiber-gracile neurone pair at four different vibration frequencies between 100 and 400 Hz. Indeed, there is a higher level of response at the higher frequencies that probably reflects temporal summation of the unitary synaptic potentials induced by successive PC fiber spikes.

The security of the linkage for different PC fiber-gracile neurone pairs studied varied from the high levels seen in Figures 5.7 and 8 to somewhat weaker suprathreshold actions for other pairs. An average measure of the security of linkage was plotted in Figure 5.9 for all PC fiber-gracile neurone pairs in which quantitative measures of spike output were made. The measure obtained was the average spike output for the gracile neurone on each cycle of vibration, up to cycle number 20, when the single PC fiber providing the input was responding with one spike on each cycle of vibration, that is, a 1:1 level of response. This was done at 200 and 300 Hz and the mean spike output (± S.D.) then obtained for the whole sample of gracile neurones. For 28 PC fiber-gracile neurone pairs on which quantitative data were obtained at 200 Hz, the mean spike output was 1.56 on the first cycle and declined gradually on subsequent cycles (Fig. 5.9). At 300 Hz the mean response was 1.86 spikes in the first cycle and fell thereafter reaching an approximately steady level of about 0.6 spikes per cycle after the first ten to eleven cycles of vibration (Ferrington *et al.*, 1987a).

Our analysis has revealed a second mechanism of amplification operating for PC inputs at this first synaptic processing site in the central pathways, as individual PC fibers diverge and may exert suprathreshold actions on multiple target neurones (Ferrington *et al.*, 1986). Thus two potent mechanisms are available for amplification of minimal sensory inputs arriving over PC sensory fibers. At the extreme, a *single* impulse in *one* sensory fiber can be directed, by divergence of the input fiber, to several target neurones within which pairs or bursts of spikes can be generated. With the PC input sustained over a full one second by means of a maintained train of (100–400 Hz) vibration activating the single PC fiber, the central target neurone in many cases was driven to respond at impulse rates of 100–200/s, or even in one case, at 400 impulses/s (Ferrington *et al.*, 1987a).

SECURITY PLUS TEMPORAL FIDELITY IN THE TRANSMISSION OF SIGNALS ACROSS THE LINKAGE BETWEEN SINGLE PC FIBERS AND CENTRAL TARGET NEURONES

The remarkable security displayed by the linkage between PC sensory fibers and their target gracile neurones ensures a very high safety or security, and is in stark contrast to the dependence, at the Ia-motoneurone synapse, upon elaborate convergence and summation before motoneurone output can be generated.

What we have observed in the transmission characteristics for individual PC fibers is entirely consistent with their capacity to generate a perceptual response in the microneurography experiments. One aspect of the percept described by human subjects in response to single PC fiber inputs generated by a train of electrical pulses at 20–100 Hz was a sense of skin vibration. If one examines the impulse traces of the PC fiber and its target central neurone in Figure 5.7 one can see first, that the PC fiber responds to the mechanical vibratory stimulus with a precise, temporally patterned sequence of spikes which reflect the periodicity of the vibratory stimulus. Furthermore, the gracile target neurone displays tightly phaselocked patterned activity that is also able to reflect the periodicity of the vibration, as the interspike intervals approximate the 5 ms cycle period of the 200 Hz vibration stimulus. As the neural code for the frequency parameter of vibrotactile stimuli appears to depend upon an impulse pattern code of this type (Mountcastle *et al.*, 1969; Ferrington and Rowe, 1980), it appears from response traces such as those of Figure 5.7 that, in the process of synaptic transmission within the dorsal column nuclei, the central neurone retains reliable information about the frequency parameter of the stimulus even in response to a selective input from a single PC sensory fiber. This behaviour of neurones at this first central synapse is at least consistent with the ability of human subjects to recognize the vibrotactile character of the input when single PC fibers are driven

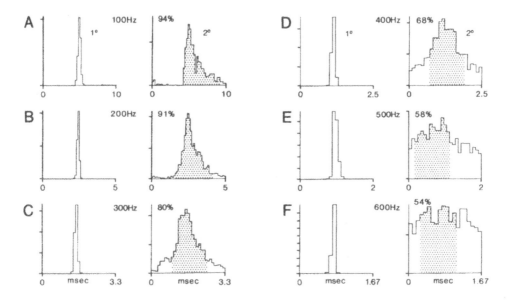

Figure 5.10 Cycle histogram pairs showing the distribution of impulse activity within the vibration cycle period for the responses of a PC fibre-gracile neurone pair at six frequencies of vibration (from 100 to 600 Hz). The analysis period indicated on the abscissa of each histogram corresponds to the cycle period at the vibration frequency used for each set in **A–F**. The left-hand cycle histogram of each pair was constructed from the responses of the primary PC fibre. The ordinate scale division for the left-hand series of histograms represents 500 impulse counts except in A where it represents 400. The right-hand histograms in A-F were constructed from the responses of the gracile neurone. The quantitative measure of the tightness of phase locking (percentage entrainment) is indicated on the right-hand histograms. The continuous half-cycle segment of the histogram in which this percentage of impulses occurred is designated by the stippling. Each ordinate division on the gracile neurone's histograms represents 50 impulse counts. (from Ferrington, Rowe and Tarvin, 1987a).

at frequencies of 20–100 Hz with the intraneural microstimulation procedure (Ochoa and Torebjörk, 1983; Vallbo *et al.* 1984). In fact, when quantitative analysis is conducted on the impulse patterning in gracile responses to vibratory inputs carried over single PC fibers we find that tightly phaselocked patterns can be retained over a vibration frequency bandwidth extending up to ~ 400 Hz (Ferrington *et al.*, 1987a). This is demonstrated in Figure 5.10 in the paired cycle histograms constructed from responses of a representative PC fiber-gracile neurone pair to six different vibration frequencies in the range 100–600 Hz. The cycle histograms, which have an analysis time on the abscissa corresponding to the cycle period of the vibration (10,5,3.3 ms etc) display the probability of impulse occurrence throughout the period of the vibration cycle. Those on the left hand side show that the impulse occurrences in the PC fiber are tightly confined within little more than ~10% of the vibration cycle, even at the highest frequencies of 400, 500 and 600 Hz, reflecting the metronome-like precision in the impulse patterning of the PC fibers in response to vibration. Although there is a greater dispersion of impulse occurrences in the gracile responses in the right hand histograms, these central neurone responses remain well phaselocked even at vibration frequencies up to 400 Hz. The percentage values on each of these histograms provide a quantitative measure of phaselocking and represent the maximum percentage of total impulse occurrences confined within a continuous half-cycle segment (indicated in each histogram by the stippling). Beyond 400 Hz the percentage entrainment values approach the lower limit of 50% consistent with the approximately rectangular shape of the distributions indicating an absence of phaselocking at these highest frequencies. Nevertheless, the observations indicate that transmission across this synapse for PC fiber inputs preserves temporal information that could permit the vibratory character of the input to be signalled for frequencies beyond 400 Hz.

SECURITY OF TRANSMISSION FOR THE ONE-TO-ONE SYNAPTIC LINKAGE BETWEEN SINGLE SLOWLY ADAPTING TACTILE AFFERENT FIBERS AND THEIR CUNEATE TARGET NEURONES

Analysis of transmission characteristics for single fibers of the SAI and SAII classes of tactile afferents was undertaken with a paired recording paradigm in which the peripheral recording was made from fine fascicles of the lateral branch of the superficial radial nerve in the cat forearm, and the central recording from single target neurones of the cuneate nucleus (Vickery *et al.*, 1994; Gynther *et al.*, 1995). When activated in the microneurography experiments single SAII fibers were ineffective in generating a perceptual response and single SAI fibers were either ineffective or unable with trains of stimuli to generate any vibrotactile sense or percept. Each of these classes of SA fibers can be easily distinguished from the purely dynamically sensitive tactile afferents such as PC, RA and HFA fibers, as they respond in a maintained way to static skin displacement. Furthermore, the distinction between SAI and SAII fibers can be made according to a concatenation of structural and functional criteria (Burgess *et al.*, 1968; Chambers *et al.*, 1972). The SAI fibers innervating hairy skin have punctate receptive fields comprising one or a few points each corresponding to the 200–400 µm diameter *touch domes* beneath which Merkel receptor complexes are aggregated (Iggo and Muir, 1969), whereas SAII fibers have a single spot-like field without skin specialization. The SAII fibers are very sensitive to skin stretch and have a very regular impulse pattern in response to static skin displacement, whereas SAI fibers are insensitive to skin stretch and have irregular discharge patterns.

TRANSMISSION CHARACTERISTICS FOR SA FIBER-CUNEATE NEURONE PAIRS FOR INPUTS GENERATED BY STATIC SKIN DISPLACEMENT

The paired recording arrangement for the analysis of transmission characteristics for *single* SAI and SAII fibers is represented in Figure 5.11 along with paired recording traces of activity for an SAII fiber (1° in Fig. 5.11) that was capable of driving a cuneate target neurone (2° in Fig. 5.11) with high security. Two spikes are apparent in the peripheral nerve recording in Figure 5.11, one predominantly upgoing which was spontaneously active, unaffected by the static, 300 µm amplitude skin indentation indicated by the upper waveform, and had no effect on the recorded cuneate neurone. The other, downgoing spike,

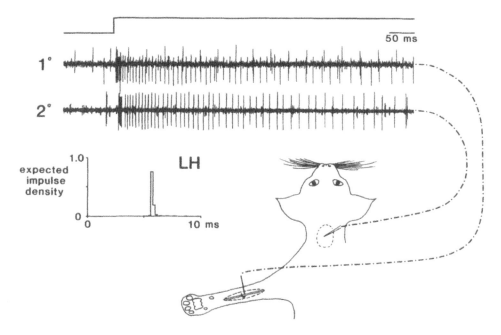

Figure 5.11 Paired recording format for analysis of transmission characteristics between individual SA afferent fibres and cuneate neurones, and evidence for a secure synaptic linkage between a slowly adapting type II (SAII) peripheral fiber (1°, downwards impulses) and a cuneate neuron (2°). Simultaneously recorded impulse records (700 ms duration) are shown with a schematic of the recording sites. The *top trace* is a replica of the 300 μm step indentation stimulus. The latency histogram (LH) was constructed from activity accumulated over the illustrated 700 ms during six stimulus repetitions. The time window over which cuneate impulses were accumulated in the LH was from 5 to 10 ms after each SAII spike and SAII instantaneous frequencies greater than 200 Hz were not analyzed. Each column in the LH is 0.2 ms in width (from Gynther, Vickery and Rowe, 1995).

associated with the SAII fiber, was highly responsive to the static skin stimulus and almost unfailingly elicited a response in the cuneate neurone at a fixed latency of 5.7 ms as shown in the illustrated *latency* histogram which is a modified cross-correlation histogram (Vickery *et al.*, 1994; Gynther *et al.*, 1995). The transmission security for this SAII-cuneate neurone pair (defined as the percentage of SAII impulses that evoked spike output from the cuneate neurone) was almost 100%. In three SAII fiber-cuneate pairs in which transmission security could be quantified in association with such static skin displacements, the security was ≥84% (Gynther *et al.*, 1995).

It was also found that individual SAI fibers had extremely tight synaptic linkages with their cuneate target neurones when activated by steady skin displacement as shown in the paired response traces of Figure 5.12 during 50 ms segments of a static 1.5s skin displacement at the start (A column), in the middle (B), and at the end (C) when the displacement amplitude was 200 μm, 300 μm, and 400 μm in the three rows. The dotted lines link cuneate

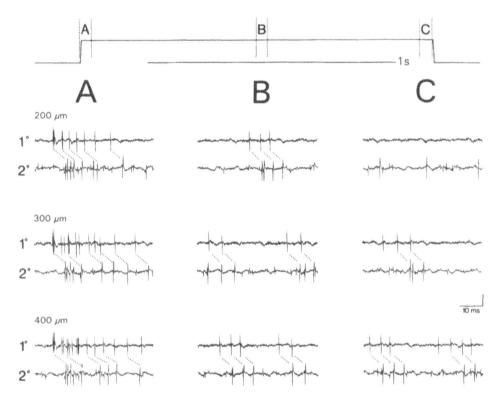

Figure 5.12 Impulse records of a synaptically linked SAI fibre (primary) and cuneate neurone (secondary) responding to a 1.5 s skin indentation. The responses to three indentation amplitudes (200, 300 and 400 µm) are shown in the three rows, while the columns labelled **A**, **B** and **C** contain responses drawn from the first, middle, and last 50 ms of indentation (indicated on schematic of stimulus). Dashed lines link primary fibre impulses and the secondary neuronal impulses they evoke. At stimulus onset **(A)** the high impulse rates make correct assignment of primary and evoked secondary impulses difficult. The vertical scale bar is 43 µV for the primary fibre, and 173 µV for the secondary neurone (from Vickery, Gynther and Rowe, 1994).

spikes with the causal SAI spike and show that the vast majority of SAI spikes elicit cuneate impulse activity. Occasional exceptions are seen in A (200 µm) and C (300 and 400 µm). The security of the linkage ranged from 80–95% in seven out of nine SAI-cuneate pairs studied quantitatively for their responses to static skin displacement. As was observed for PC fiber linkages, the SAI-cuneate synaptic linkage shows examples of amplification, with two or more spikes being generated on occasion in the cuneate neurone by a single SAI spike. This high security of transmission for the SAI fibers is, of course, consistent with the capacity of this class of fibers to generate a perceptual response, — that of steady skin pressure following microstimulation in conscious human subjects. However, as *single* SAII fibers can display a similar high security of transmission in their synaptic linkage to cuneate neurones it appears that the failure of single SAII fibers to elicit perceptual responses in the human microneurography experiments cannot be explained in terms of a less secure synaptic linkage, at least at the level of the dorsal column nuclei (Gynther *et al.*, 1995).

TRANSMISSION SECURITY FOR SA FIBER-CUNEATE NEURONE SYNAPTIC LINKAGES IN RESPONSE TO VIBROTACTILE STIMULI

As stimuli based on static skin indentation generally elicit lower rates of impulse activity in the afferent fiber (< 50–100 impulses/s) than do dynamic forms of stimuli such as cutaneous vibration, we have also examined the cuneate transmission characteristics for SAI and SAII fibers when each was driven at the high impulse rates (many hundreds/s) that are attained in PC fibers in response to vibration (Figs. 5.7 and 8). It was of particular importance to examine this issue for the SAI fibers because the human microstimulation experiments showed no capacity of individual SAI fibers, when stimulated with a train of repetitive electrical pulses (20–100 Hz), to generate a sense of flutter or vibration as occurs in association with comparable stimulation of RA and PC fibers. Instead, they are reported to generate merely a sense of steady pressure. A possible explanation for this discrepancy in the perceptual consequences of activating individual PC and SAI fibers in an identical patterned way could be that, in contrast to the PC fiber-central neurone linkage, the synaptic linkage between SAI fibers and their target cuneate neurones is unable to retain the temporally precise, phaselocked pattern of spike activity in the transmission process to the output neurone. With such a degradation of temporal patterning in the central neurone's response there would be a loss of information about the vibrotactile character of the original peripheral input. However, even in the case of the SAII fiber-cuneate neurone linkage it was important to examine transmission security in association with the higher levels of peripheral drive associated with vibrational inputs as this may reveal a breakdown in transmission security not apparent at the relatively low impulse rates generated by the static skin displacements (Fig. 5.11).

The use of such vibrotactile stimuli has provided an unequivocal result in testing the above hypotheses. First, in the case of the SAI fiber-cuneate neurone linkage, Figure 5.13 shows 100 ms paired recording segments at the start (in A) and after 500 ms (in B) for responses to four frequencies (50, 100, 200 and 400 Hz) of vibration delivered to the SAI fiber's touch dome receptive field on the cat forearm. If we concentrate on the 100 Hz responses, the SAI fiber is firing in a regular 1:1 pattern to the vibration, i.e. an impulse on each cycle of the vibration (as it does at 50 (A) 200 (B) and 400 Hz (B) as well). However, the point to be emphasized is that the cuneate neurone is responding with a faithful and precise impulse pattern reflecting that in the incoming SAI spikes in a way that allows the cuneate neurone to retain a precise signal of the vibratory character of the stimulus. This was observed for other SAI-cuneate neurone pairs as shown in Figure 5.14 illustrating paired responses at four amplitudes of a 100 Hz vibration stimulus. Once again, the regularity of the cuneate neurone's activity is remarkable and, when quantified by means of the cycle histogram analyses described above, the phaselocking in the cuneate responses was no poorer than that of the dorsal column nuclei neurones driven by single PC fibers (Vickery *et al.*, 1994).

We may conclude confidently that the failure of SAI afferent fibers to generate a vibrotactile percept in the microneurography experiments cannot be explained by poor

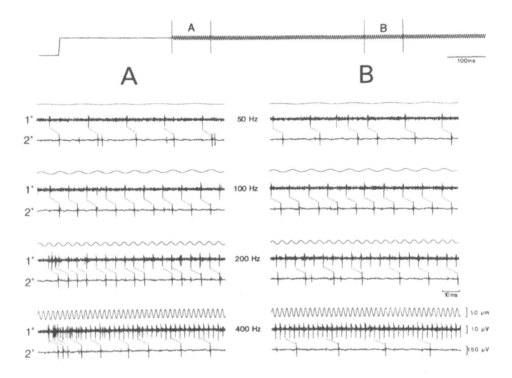

Figure 5.13 Impulse patterning in the cuneate response to a 1:1 response level in its SAI input is shown by the impulse records for the vibration-induced responses of the SAI (primary)-cuneate (secondary) pair. Segments (100 ms) of responses to vibration, at its onset and after 500 ms (see schematic), are shown in columns **A** and **B** respectively for four vibration frequencies. Dashed lines link the cuneate impulses to the SAI impulses presumed to have caused them. The top trace in each record is a replica of the stimulus waveform showing the relative timing and amplitude (250 μm probe, pre-indentation 200 μm). (from Vickery, Gynther and Rowe, 1994).

transmission security, and degradation of the temporal pattern of incoming activity within this first central relay nucleus in the dorsal column nuclei.

When the cuneate transmission characteristics for individual SAII fibers were examined in association with vibrotactile stimuli we found behaviour that was indistinguishable from that observed in the linkages for individual PC and SAI fibers (Gynther *et al.*, 1995).

As with SAI and PC fibers it was possible with sinusoidal vibrotactile stimuli to activate the SAII fibers in a regular, metronome-like pattern of response at high discharge rates (> 500 impulses/s). Furthermore, in response to just the *one* activated SAII fiber, the target cuneate neurone could be driven to respond at impulse rates in excess of 100–200 impulses/s. Figure 5.15 shows the capacity of cuneate neurones to respond to the regular one-to-one impulse pattern arriving over the SAII fiber. The dotted lines between each pair of impulse traces indicate the link between the cuneate impulse activity (lower trace, 2°) and the SAII impulses (upper trace, 1°) responsible for the cuneate activity. The central neuron

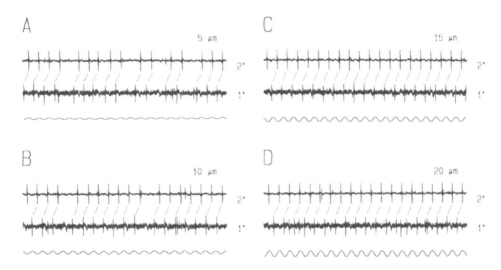

Figure 5.14 Impulse traces recorded from an SAI fibre-cuneate neurone pair (1° and 2° respectively) responding to a 100 Hz vibration train at four amplitudes (5–20 μm) delivered to the receptive field focus on the skin of the cat forearm.

followed, at a fixed modal latency of 5.6 ms (Fig. 5.15, 100 Hz *latency histogram*, LH) almost all (93%) of the incoming SAII impulses in response to 100 Hz vibration.

Although *transmission security* declined somewhat when the SAII afferent fiber was driven by the vibration stimulus to respond at the higher rates of 200 Hz and 400 Hz, this type of decline was also observed in the central response at high input driving frequencies in PC and SAI fibers. Even at these high driving frequencies, the cuneate neurone response was a much better match to that the SAII fiber over earlier segments of the response to the usual 1s long stimulus. For example, over the initial 200 ms segment of the response to the 200 Hz and 400 Hz input from the SAII fiber, there was a much higher transmission security for the linkage (Gynther *et al.*, 1995). However, it should be emphasized that these frequencies (200 and 400 Hz) were well above those examined in the human intraneural microstimulation studies. At the frequencies (≤100 Hz) used to activate single SAII fibers in the human studies it can be seen from Figure 5.15 first, that at the lower input driving frequency of 30 Hz, individual spikes in the SAII fiber generate almost always a paired spike output from this cuneate neurone, reflecting an amplification in the synaptic linkage reminiscent of that observed for both PC and SAI fibers. The doublet spike response from the cuneate neurone observable in the impulse trace record is also apparent in the double peaks (of approximately the same areas) separated by ~1.5ms in the *latency histogram*. A second point that is apparent, in particular in the cuneate responses to the 100 Hz input in Figure 5.15, is that the cuneate response displays a precision of patterning and tightness of phaselocking that is little different from that apparent in the metronome-like spike pattern in the SAII fiber. Quantitative measures of phaselocking in the two elements of the pair confirmed this point (Gynther *et al.*, 1995) and indicate that the linkages formed by single SAII fibers and dorsal column nuclei neurones appear little different from those

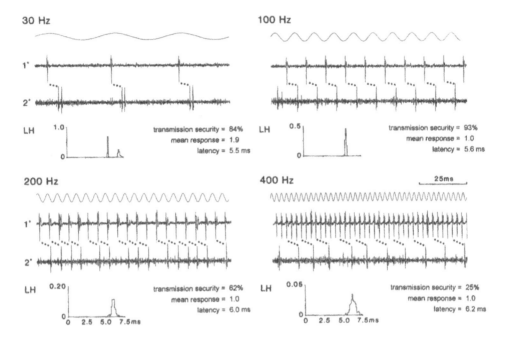

Figure 5.15 Transmission of impulse activity in an SAII-cuneate pair at a 1:1 response level in the SAII fiber for four vibration frequencies. The impulse records for the SAII fiber (1°) and cuneate neuron (2°) and the stimulus replicas are all 100 ms segments, taken 250 ms after the onset of vibration. The latency histograms (LHs) were constructed from cuneate impulses accumulated over five repetitions of the 1s stimulus and have 0.1ms columns and a 5.2 to 7.5 ms time window. SAII instantaneous frequencies of greater than 435 Hz were not analyzed. The preindentation amplitude was 100 μm, the probe diameter was 250 μm, and the vibration amplitudes at 30, 100, 200 and 400 Hz were, respectively, 1, 10, 50, and 50 μm. These amplitudes were chosen to achieve a regular 1:1 pattern of response in the SAII fiber at each of these vibration frequencies (from Gynther, Vickery and Rowe, 1995).

formed by PC and SAI fibers in their capacity for retaining temporal fidelity in impulse transmission.

TRANSMISSION SECURITY ACROSS THE CUNEATE SYNAPTIC LINKAGE FOR SINGLE KINAESTHETIC AFFERENT FIBERS

In order to extend these analyses to the two principal kinaesthetic afferent classes we once again required a preparation that enabled recordings to be made from single fibers in an intact peripheral nerve or nerve fascicle. This peripheral nerve preparation has been achieved in the cat for two joint nerves; one, the medial articular nerve, that supplies the medial surface of the knee joint (Mackie *et al.*, 1995), the other, the wrist joint nerve which is a fine branch of the deep radial nerve (Coleman *et al.*, 1998). If either of these fine joint nerves is freed over a distance of ~3–4 cm from the joint and remains in continuity

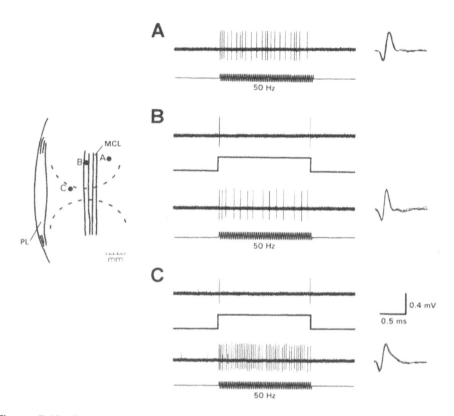

Figure 5.16 Responses of intact joint afferent fibres of the medial articular nerve to mechanical stimuli applied to the knee joint capsule at the point of peak sensitivity for each fibre. Locations of the punctate receptive fields for the 3 fibres **(A, B, C)** are shown on the sketch of the medial surface of the knee joint (PL = patellar ligament; MCL = medial collateral ligament). The 2-s duration impulse traces show responses to 1s trains of 50 Hz (100 μm) sinusoidal vibration or 1s long steady indentation (200 μm) delivered by means of 1–2 mm diameter circular probes. Superimposed expanded spike waveforms are shown on the right-hand side with the indicated time and voltage scales. For all spikes negativity is shown upwards (from Mackie *et al.*, 1995).

with the central nervous system, one can monitor all active group II joint afferent fibers with an excellent signal-to-noise ratio, as illustrated in Figure 5.16, and examine the central actions of individual fibers on their target neurones. Selective activation of single joint afferent fibers may be accomplished most effectively by punctate, focal stimulation of the joint capsule with a fine (usually 250 μm diameter) mechanical stimulator probe used to apply either static displacements or vibration as effective stimuli to the receptive field of individual joint afferents. In order to evaluate transmission characteristics for individual muscle spindle afferent fibers we have made use of a very fine nerve in the cat forearm, the nerve that supplies the indicis proprius muscle, a forearm extensor muscle (Mackie and Rowe, 1997). Once again, selective activation of individual afferents has been achieved most effectively by focal stimulation of the muscle over the presumed site of individual spindles.

Preliminary reports have been published which present evidence for a secure transmission between individual afferents of *both* these kinaesthetic fiber classes and their target cuneate neurones (Coleman *et al.*, 1997, 1998; Mackie and Rowe, 1997, 1998) and full papers are in preparation.

SUMMARY AND CONCLUSIONS

In summary, the differential capacity of the various tactile and kinaesthetic fiber classes to generate a perceptual response when activated singly in human microneurography experiments does not appear to be explicable in terms of differential transmission characteristics across the first synaptic relay site in the principal tactile and kinaesthetic sensory system that traverses the dorsal column-lemniscal pathway. In particular, the failure of individual SAII fibers and muscle spindle afferent fibers to generate sensation upon activation with microneurography, does not appear to be attributable to a breakdown in synaptic transmission security at the level of the dorsal column nuclei. Both these classes almost certainly contribute to sensory and perceptual experience, spindle afferents contributing to the kinaesthetic sense, in particular, our sense of limb and body position (Goodwin *et al.*, 1972; Gandevia and McCloskey, 1976; McCloskey, 1978; Gandevia, 1985), while SAII fibers appear to contribute to sensations of static displacement and stretch of the skin (Johansson and Vallbo, 1979; Vallbo and Johansson, 1976, 1984) which, when it occurs around joints, may contribute to the kinaesthetic sense (Edin and Abbs, 1991; Edin, 1996) as well as to tactile experience. However, presumably for sensations to be generated by inputs arising in either of these fiber classes requires the concurrent activation of at least a small population of each afferent class before a percept will ensue. Whether this is related to a need for adequate spatial summation to take place at thalamic or cortical levels before transmission of the SAII or spindle inputs is effective at these levels, remains uncertain. An alternative hypothesis is that individual spindle afferents and SAII afferents (and, in a proportion of cases, at least, some SAI fibers which fail to generate a perceptual response) do not diverge in their ascending projection paths to the same extent that single PC, RA, joint afferent and the majority of SAI fibers do, and that for the 'critical mass' of cortical tissue to be activated by these inputs for the perceptual threshold to be crossed, may require peripheral activation of a 'ensemble' of each of these classes.

A further issue that remains unclear is why single SAI fibers can, on microneurographic activation with a train of pulses, generate a pressure sensation but are unable to contribute any sense of flutter or vibration as can the individual RA and PC fibers when stimulated in this way. Our hypothesis that this failure of SAI single fiber inputs might be based on degradation in the fidelity of the temporal patterning of responses across the SAI-cuneate synaptic linkage was not supported by our observations as this linkage proved just as secure as that formed by single PC afferent fibers, and just as well able to retain the precision of impulse patterning. Current investigations at the thalamic and cortical levels are addressing this paradox; first, in terms of whether the temporal signalling from SAI inputs is markedly degraded at these higher synaptic relays, and second, in terms of analyzing the locus and extent of the somatosensory area I and/or II (SI and SII) involvement in processing these signals from single SAI fibers.

ACKNOWLEDGMENTS

Work from the author's laboratory has been supported by grants from the National Health and Medical Research Council of Australia and the Australian Research Council. Invaluable contributions have been made by a number of PhD students and other colleagues to the work from the author's laboratory reported in this chapter; in particular, those of D.G. Ferrington, R.P. Tarvin, R.M. Vickery, B.D. Gynther, H.Q. Zhang, P.D. Mackie and G.C. Coleman.

REFERENCES

Burgess, P.R., Petit, D. and Warren, R.M. (1968). Receptor types in cat hairy skin supplied by myelinated fibers. *Journal of Neurophysiology*, **31**, 833–848.

Berkley, K.J. (1975). Different targets of different neurons in nucleus gracilis of the cat. *Journal of Comparative Neurology*, **163**, 285–304.

Chambers, M.R., Andres, K.H. Duering, M.V. and Iggo, A. (1972). The structure and function of the slowly adapting type II mechanoreceptor in hairy skin. *Quarterly Journal of Experimental Physiology*, **57**, 417–445.

Coleman, G.T., Zhang, H.Q., Mackie, P.D. and Rowe, M.J. (1998). An intact peripheral nerve preparation for examining the central actions of single kinaesthetic afferent fibres arising in the wrist joint. *Primary Sensory Neuron*, **3**, 61–70.

Coleman,, G.T., Zhang, H.Q. and Rowe, M.J. (1998). Transmission security for sustained inputs in the linkage between single joint afferent fibres and cuneate neurones. *Proceedings of the Australian Neuroscience Society*, **9**, 27.

Edin B.B. (1996). Strain-sensitive mechanoreceptors in the human skin provide kinaesthetic information. In *Somesthesis and the Neurobiology of the Somatosensory Cortex*, edited by O. Franzén, R. Johansson and L. Terenius, pp. 283–294. Basel: Birkäuser Verlag.

Edin, B.B. and Abbs, J.H. (1991). Finger movement responses of cutaneous mechanoreceptors in the dorsal skin of the human hand. *Journal of Neurophysiology*, **65**, 657–670.

Ellis Jr., L.C and Rustioni, A. (1979). Thalamic relay neurons and interneurons in the feline dorsal column nuclei: a Golgi and HRP study. *Neuroscience Letters, suppl.*, **3**, S273.

Ferrington, D.G. and Rowe, M. (1980). Differential contributions to coding of cutaneous vibratory information by cortical somatosensory areas I and II. *Journal of Neurophysiology*, **43**, 310–331.

Ferrington, D.G., Rowe, M.J. and Tarvin, R.P. (1986). High gain transmission of single impulses through dorsal column nuclei of the cat. *Neuroscience Letters*, **65**, 277–282.

Ferrington, D.G., Rowe, M.J. and Tarvin, R.P. (1987a). Actions of single sensory fibres on cat dorsal column nuclei neurones: vibratory signalling in a one-to-one linkage. *Journal of Physiology*, **386**, 293–309.

Ferrington, D.G., Rowe, M.J. and Tarvin, R.P. (1987b). Integrative processing of vibratory information in cat dorsal column nuclei neurones driven by identified sensory fibres. *Journal of Physiology*, **386**, 311–331.

Gandevia, S.C. (1985). Illusory movements produced by electrical stimulation of low-threshold muscle afferents from the hand. *Brain*, **108**, 965–981.

Gandevia, S.C. and McCloskey, D.I. (1976). Joint sense, muscle sense, and their combination as position sense, measured at the distal interphalangeal joint of the middle finger. *Journal of Physiology*, **260**, 387–407.

Goodwin, G.M., McCloskey, D.I. and Matthews, P.B.C. (1972). The contribution of muscle afferents to kinaesthesia shown by vibration induced illusions of movement and by the effects of paralysing joint afferents. *Brain*, **95**, 705–748.

Gottschaldt, K.M. and Vahle-Hinz, C. (1981). Merkel cell receptors: structure and transducer function. *Science*, **214**, 183–186.

Gynther, B.D., Vickery, R.M. and Rowe, M.J. (1992). Responses of slowly adapting type II afferent fibres in cat hairy skin to vibrotactile stimuli. *Journal of Physiology*, **458**, 151–169.

Gynther, B.D., Vickery, R.M. and Rowe, M.J. (1995). Transmission characteristics for the 1:1 linkage between slowly adapting type II fibers and their cuneate target neurons in cat. *Experimental Brain Research*, **105**, 67–75.

Hand, P.J. and Van Winkle .T. (1977). The efferent connections of the feline nucleus cuneatus. *Journal of Comparative Neurology*, **171**, 83–109.

Iggo, A. and Muir, A.R. (1969). The structure and function of a slowly adapting touch corpuscle in hairy skin. *Journal of Physiology*, **200**, 763–796.

Iggo, A. and Ogawa, H. (1977). Correlative physiological and morphological studies of rapidly adapting mechanoreceptors in cat's glabrous skin. *Journal of Physiology*, **266**, 275–296.

Jänig, W., Schmidt, R.F. and Zimmermann, M. (1968). Single unit responses and the total afferent outflow from the cat's foot pad upon mechanical stimulation. *Experimental Brain Research*, **6**, 100–115.

Johansson, R.S. and Vallbo, Å.B. (1979). Detection of tactile stimuli. Thresholds of afferent units related to psychophysical thresholds in the human hand. *Journal of Physiology*, **297**, 405–422.

Macefield, G., Gandevia, S.C. and Burke, D. (1990). Perceptual responses to microstimulation of single afferents innervating joints, muscles and skin of the human hand. *Journal of Physiology*, **429**, 113–129.

Mackie, P.D. and Rowe, M.J. (1997). An intact peripheral nerve preparation for monitoring inputs from single muscle afferent fibres. *Experimental Brain Research*, **113**, 186–188.

Mackie, P.D. and Rowe, M.J. (1998). Central projection of proprioceptive information from the wrist joint via a forearm 'muscle' nerve in the cat. *Journal of Physiology*, **510**, 261–267.

Mackie, P.D., Zhang, H.Q., Schmidt, R.F. and Rowe, M.J. (1995). An intact nerve preparation for monitoring inputs from single joint afferent fibres. *Journal of Neuroscience Methods,* **56**, 31–35.

McCloskey, D.I. (1978). Kinesthetic sensibility. *Physiological Reviews*, **58**, 763–820.

Mountcastle, V.B., Talbot, W.H., Sakata, H. and Hyvarinen, J. (1969). Cortical neuronal mechanisms in flutter-vibration studied in unanesthetized monkeys. Neuronal periodicity and frequency discrimination. *Journal of Neurophysiology*, **32**, 452–484.

Ni, S., Wilson, L.R., Burke, D. and Gandevia S. (1998). Microneurographic stimulation of muscle spindle afferents in the radial nerve fails to elicit proprioceptive sensation. *Proceedings of the Australian Neuroscience Society*, **9**, 183.

Ochoa, J. and Torebjörk, E. (1983). Sensations evoked by intraneural microstimulation of single mechanoreceptor units innervating the human hand. *Journal of Physiology*, **342**, 633–654.

Talbot, W.H., Darian-Smith, I., Kornhuber, H.H. and Mountcastle, V.B. (1968). The sense of flutter-vibration: comparison of the human capacity with response patterns of mechanoreceptive afferents from the monkey hand. *Journal of Neurophysiology*, **31**, 301–334.

Torebjörk, H.E., Vallbo, Å.B. and Ochoa, J.L. (1987). Intraneural microstimulation in man: Its relation to specificity of tactile sensations. *Brain*, **110**, 1509–1529.

Turman, A.B., Ferrington, D.G., Ghosh, S., Morley, J.W. and Rowe, M.J. (1992). Parallel processing of tactile information in the cerebral cortex of the cat: effect of reversible inactivation of SI on responsiveness of SII neurons. *Journal of Neurophysiology*, **67**, 411–429.

Vallbo, Å.B. and Hagbarth, K.E. (1968). Activity from skin mechanoreceptors recorded percutaneously in awake human subjects. *Experimental Neurology*, **21**, 270–289.

Vallbo, Å.B., Hagbarth, K.E., Torebjörk, H.E. and Wallin, B.G. (1979). Somatosensory, proprioceptive, and sympathetic activity in human peripheral nerves. *Physiological Reviews*, **59**, 919–957.

Vallbo, Å.B. and Johansson, R.S. (1976). Skin mechanoreceptors in the human hand: Neural and psychophysical thresholds. In *Sensory Functions of the Skin in Primates*, edited by Y. Zotterman, pp.185–199. Oxford: Pergamon.

Vallbo, Å.B. and Johansson, R.S. (1984). Properties of cutaneous mechanoreceptors in the human hand related to touch sensation. *Human Neurobiology*, **3**, 3–14.

Vallbo, Å.B., Olsson, K.Å., Westberg, K.-G. and Clark, F.J. (1984). Microstimulation of single tactile afferents from the human hand: Sensory attributes related to unit type and properties of receptive fields. *Brain*, **107**, 727–749.

Vickery, R.M., Gynther, B.D. and Rowe, M.J. (1992). Vibrotactile sensitivity of slowly adapting type I sensory fibres associated with touch domes in cat hairy skin. *Journal of Physiology*, **453**, 609–626.

Vickery, R.M., Gynther, B.D. and Rowe, M.J. (1994). Synaptic transmission between single slowly adapting type I fibres and their cuneate target neurones in cat. *Journal of Physiology*, **474**, 379–392.

Zhang, H.Q. Murray, G.M. Turman, A.B. Mackie, P.D. Coleman, G.T. and Rowe, M.J. (1996). Parallel processing in cerebral cortex of the marmoset monkey: effect of reversible SI inactivation on tactile responses in SII. *Journal of Neurophysiology*, **76**, 3633–3655.

CHAPTER 6

PROCESSING OF HIGHER ORDER SOMATOSENSORY AND VISUAL INFORMATION IN THE INTRAPARIETAL REGION OF THE POSTCENTRAL GYRUS

Yoshiaki Iwamura, Atusi Iriki, Michio Tanaka, Miki Taoka and Takashi Toda

Department of Physiology, Toho University School of Medicine, Otaku, Tokyo, Japan

INTRODUCTION

In the postcentral gyrus of awake monkeys, there is a gradual and systematic increase in the receptive field (RF) size and in the complexity of response characteristics of single neurons along the rostro-caudal axis (Hyvarinen and Poranen, 1978; Iwamura *et al.*, 1980, 1983a,b, 1985a,b, 1993). This tendency continues further into the anterior bank of the intraparietal sulcus (IPS), the caudalmost part of the gyrus. In recent years we have paid much attention to this region, and have found several unique groups of neurons such as those responding to unique objects (Iwamura and Tanaka, 1978, 1996; Iwamura *et al.*, 1996a), those with bilateral RFs (Iwamura *et al.*, 1994, 1996b), those under the influence of attention (Iriki *et al.*, 1996a), and those with both somatosensory and visual inputs (Iriki *et al.*, 1996b). In this article we review some of these results.

ANATOMICAL BACKGROUND

In macaque monkeys, cutaneous inputs reach areas 3b and 1, and deep inputs reach areas 3a and 2 via the thalamic ventrobasal nuclear complex (Jones and Friedmann, 1982; Pons and Kaas, 1985). Rich intrinsic cortico-cortical connections exist within the postcentral gyrus and convey information from area 3b to areas 1 and 2, and from area 1 to area 2 (Jones *et al.*, 1978; Kunzle, 1978; Vogt and Pandya, 1978; Shanks and Powell, 1981). There are massive anatomical inflows into the anterior bank of IPS from the more rostral parts of the postcentral gyrus (Seltzer and Pandya, 1986). Thus the IPS region is regarded as the final stage of the hierarchical organization within the postcentral gyrus. Additional inputs from thalamic association nuclei such as anterior pulvinar and the posterior lateral nucleus reach area 2 as well as areas 5 and 7 (Pons and Kaas, 1985; Burton and Sinclair, 1996; Darian-Smith *et al.*, 1996). Their presence suggests that a part of area 2 is already a transitional zone to the parietal association cortex. In the anterior bank of IPS, area 2 shifts to either area 5 (in the medial part) or area 7 (in the lateral part) (Mountcastle *et al.*, 1975). It was difficult sometimes to determine the border based on the anatomical criteria described by Powell and Mountcastle (1959). The transitional border between areas 2 and 5 could be determined by several physiological criteria (Iwamura *et al.*, 1993).

It has been known that neurons at the bottom of the IPS (VIP) as well as those in area 7b responded to both somatosensory and visual stimuli (Leinonen *et al.*, 1979; Duhamel *et al.*, 1991), but how visual signals reach there has not yet been clarified.

BILATERAL REPRESENTATION OF THE HAND

It has long been thought that the representation of digits in the postcentral somatosensory cortex is for the contralateral side only and that the integration of input from bilateral digits takes place in the second somatosensory area (SII) (Ridley and Ettlinger, 1978; Manzoni *et al.*, 1989; Berlucchi, 1990). However, we recently discovered that there are substantial numbers of neurons with bilateral or ipsilateral RFs in the caudalmost part (areas 2 and 5) of the postcentral digit region (Iwamura *et al.*, 1994, 1995, 1996a). The bilateral RFs were large and symmetrical, the largest and the most complex types in each hand, and thus they fit with the concept of the highest hierarchical processing taking place within this gyrus (Fig. 6.1). The bilateral RFs disappeared after the destruction of the opposite hemisphere indicating their dependence on the callosal connections. Previous anatomical studies indicated that the caudal part of the postcentral digit region is not a-callosal in contrast to the more rostral part of the digit region (Jones and Powell, 1969; Pandya and Vigonolo, 1969; Karol and Pandya, 1971; Jones *et al.*, 1979; Jones and Hendry, 1980; Killackey *et al.*, 1983; Shanks *et al.*, 1985; Caminiti and Sbriccoli, 1985).

BILATERAL REPRESENTATION OF THE UPPER ARMS, SHOULDERS AND THE TRUNK

We explored the more medial part of the postcentral gyrus for the arm and trunk representation. Previous studies had shown the existence of neurons with bilateral RFs, but only those that had RFs on the trunk included the body midline (Conti *et al.*, 1986; Manzoni *et al.*, 1989). We found bilateral RF neurons of different types there (Fig. 6.2) (Iwamura *et al.*, 1996a; Taoka *et al.*, 1997, 1998). One class was based on skin-submodality types. Their RFs were on either the trunk, the forelimb or the combination of the two. The trunk type and combined type included the body midline in their RF. The other was a deep-submodality type activated by the manipulation of joints at the shoulder, elbows, shoulders and elbows (proximal type), wrists and digits (distal type), and those which responded to both the proximal and distal joint manipulation (combined type).

The distribution of neurons with ipsilateral RFs overlapped with that of the bilateral neurons. The majority of the bilateral or ipsilateral RF neurons were found in the anterior bank of IPS, and some in the more rostral crown of the postcentral gyrus. The combined type, in either skin or deep submodality, were found in the anterior bank of IPS (the majority in area 5) only. Thus there is a hierarchy in the distribution of bilateral RF neurons: the more caudal, the more complex the neuron's characteristics. Bilateral or ipsilateral RF neurons were rare in the cortical region for the forearm representation. Thus the distribution of bilateral RF neurons was discontinuous between the hand and the shoulder-trunk regions. In the trunk and proximal arm regions diffuse callosal connections were found in areas 3, 1 and 2 (Manzoni *et al.*, 1989). We assume that our bilateral input neurons receive ipsilateral inputs via callosal connections. However, the bilateral RF may not be attributed solely to callosal connections as bilateral integration may occur at the subcortical levels as well (Schwartz and Fredrickson, 1971; Manzoni *et al.*, 1989).

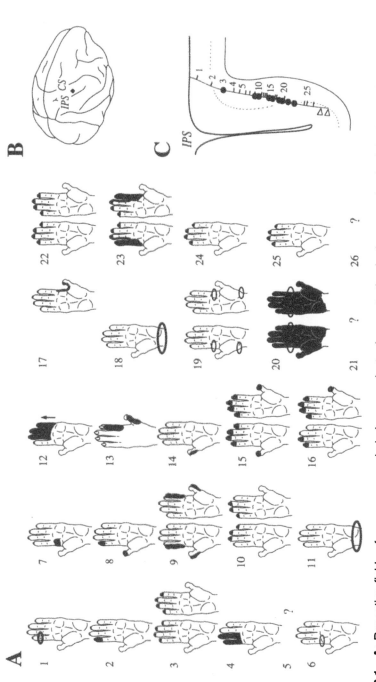

Figure 6.1 A: Receptive fields of neurons recorded along an electrode penetration in the upper bank of the intraparietal sulcus in the finger region. The extent of skin receptive fields is shaded dark in each illustration. Effective positions of joint manipulation neurons are encircled. Neurons with no identifiable receptive fields are shown as '?'. The arrow at receptive field 12 indicates the preferred direction of a moving stimulus across the receptive field. A total of 9 neurons had bilateral receptive fields along this penetration. Their recording sites are indicated by filled circles in **C**. Note the symmetry in the ipsilateral and contralateral receptive field configuration in most of these pairs. **B:** A dot on the postcentral gyrus indicates the entry of the electrode penetration in the same animal from which receptive fields were determined in A. Sites of neuronal recordings are shown by either horizontal bars (for unilateral receptive fields) or filled circles (for bilateral receptive fields). **C:** A section perpendicular to CS including the electrode track described in A. CS: central sulcus. IPS: intraparietal sulcus. IPS: intraparietal sulcus. Numbers represent the locations of neurons corresponding to receptive fields depicted in A. Open triangles at the end of the electrode track indicate sites of lesions used as landmarks for later reconstruction of recording sites. A dotted line indicates layer IV (Iwamura *et al.*, 1994).

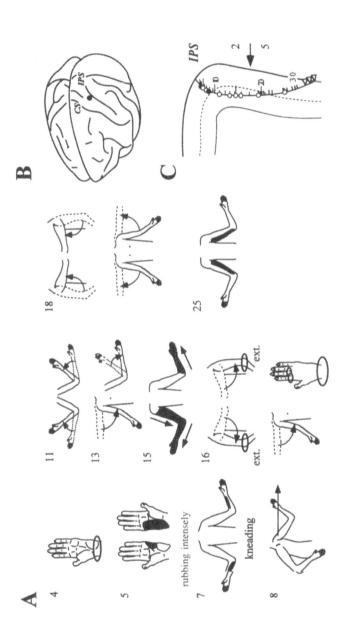

Figure 6.2 **A:** Ipsilateral or bilateral receptive fields of neurons recorded along an electrode penetration in the upper bank of the intraparietal sulcus in the arm-trunk region. A total of 8 neurons had bilateral receptive fields and an additional two neurons (#4 and 8) had ipsilateral receptive fields along this penetration. The extent of skin receptive fields is shown by dark shading in each illustration. The straight arrows near receptive fields indicate the preferred direction of a moving stimulus across the receptive field. Effective positions of joint manipulation neurons are encircled. Curved arrows indicate the effective direction of the joint movement. Neurons with no identifiable receptive fields are shown as "?". **B:** A dot on the postcentral gyrus indicates the entry of the electrode penetration in the same animal from which receptive fields were determined in A (Anterior or rostral is to the left in this case in contrast to the illustration in Figure 1). CS: central sulcus. IPS: intraparietal sulcus. **C:** A section perpendicular to CS including the electrode track described in A. Sites of neuronal recordings are shown by either horizontal bars (for contralateral receptive fields) or filled circles (for ipsilateral receptive fields) or open circles (bilateral receptive fields). Numbers represent the locations of neurons corresponding to receptive fields depicted in A. Open triangles at the end of the electrode track indicate sites of lesions used as landmarks for later reconstruction of recording sites. A dotted line indicates layer IV.

MIDLINE FUSION THEORY

In the visual cortex, callosal fibers are absent in the primary receiving area. They appear at the border between areas 17 and 18 or V1 and V2 (Chouldhury *et al.*, 1965; Berlucchi *et al.*, 1967; Berlucchi and Rizzolatti, 1968; Hubel and Wiesel, 1967; Zeki, 1978). There, the vertical meridian, the midline between two visual hemi-fields is represented, and neurons there tend to have receptive fields representing the midline. Thus the midline fusion theory was proposed. In the primary somatosensory cortex the presence of neurons with bilateral RFs has been known for the representation of face, oral cavity (Schwartz and Fredrickson, 1971; Dreyer *et al.*, 1975; Ogawa *et al.*, 1989) or trunk (Conti *et al.*, 1986; Manzoni *et al.*, 1989). RFs of these neurons were often across the midline of the body. Thus as an analogy with the visual cortex, the midline fusion theory was claimed to be applicable to the somatosensory cortex. We confirmed the presence of the midline neurons and further found the bilateral representation of the hand and shoulders in the caudalmost part of the postcentral somatosensory cortex. The hands are functionally analogous to the fovea which has the highest spatial resolution power and is involved in the vertical meridian. Thus without considering the hands the midline fusion theory was defective. Our findings thus reinforce the idea that the midline fusion theory is applicable to the somatosensory cortex. Even though the hands are apparently remote from the body midline, they often meet in the midline when they perform actions. However shoulders are the structures to work in cooperation of both sides together with the hands. Moreover, the anatomical distribution pattern of the callosal connections in the hand and digit region resembles that of the visual cortex, in that callosal inputs are believed to be practically absent in the hand region except for the caudalmost part of it.

BIMODAL NEURONS IN THE ANTERIOR BANK OF IPS: CONVERGENCE OF SOMATOSENSORY AND VISUAL INPUTS

We found that substantial numbers of neurons in the anterior bank of the IPS were activated by both somatosensory and visual stimulation. The presence of somatosensory-visual bimodal neurons has been reported in area 7b (Leinonen *et al.*, 1979) and VIP, the fundus of IPS (Duhamel *et al.*, 1991), but not previously in the anterior bank of the IPS. The effective visual stimulus was to move an object or the experimenter's hand in the space over or near the somatosensory RF of the neuron. In those neurons with a discontinuous somatosensory RF, the visual stimulus was effective in the space created between the two body parts represented in the somatosensory RF. About one-third of the bimodal neurons had bilateral somatosensory RFs. Thus, the space or regions represented by these neurons could be those between two hands, hands and mouth, or those over the two arms, trunk, neck and face etc. It should be pointed out that neurons with RFs including face were found in the IPS region far medial to the usual face representation.

TOOL USE AND THE ACTIVITY OF BIMODAL NEURONS

In bimodal neurons found in the IPS region, visual responses were often dependent on the monkey's arm position; strong responses were evoked by visual stimuli presented within reaching distance of the hand. We expected that some of these neurons code the

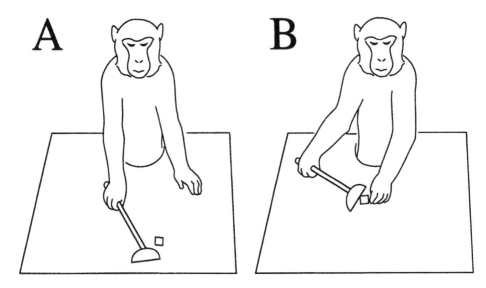

Figure 6.3 Sketch of a monkey trained to use a rake to retrieve an apple piece beyond the reach of the arm. The monkey wielded the tool **(A)** and pulled the food closer to retrieve it with the other hand **(B)**. (from Iriki *et al.*, 1996b).

schema of the hand in space. It has been known that monkeys use a stick to reach objects beyond the limit of the arm. The stick becomes an extension of the hand in both a physical and perceptual sense. Modification of body schemata is postulated as the basis of such perceptual assimilation of a tool held in the hand, but the neural mechanism has been unknown. We trained macaque monkeys to use a rake as a tool to retrieve distant objects, and studied neural correlates for the assimilation of the tool and the hand. A food pellet was dispensed beyond the reach of the monkey's hand but accessible using a rake as a tool to extend its reach. The monkey seized the food pellet with the tip of the rake and pulled it closer to retrieve it by the other hand (Fig. 6.3) (Iriki *et al.*, 1996b).

EXPANSION OF THE VISUAL RECEPTIVE FIELD AFTER TOOL USE

There were two types of neurons of which visual RFs were modified after tool use, the distal type and the proximal type. In the distal type (Fig. 6.4), the somatosensory RF was on the glabrous skin of the hand or the forearm skin. The visual RF was defined as the territory in space in which visual stimuli evoked action potentials. The most effective visual stimulus was a piece of food held in the experimenter's hand moving towards the soma- tosensory RF. The location of the visual target was measured by a 3-dimensional electromagnetic tracking device with a spatial resolution of 0.8 mm. The visual RF before tool use encompassed the hand area where the somatosensory RF was located. After the monkey held a rake and repeated food-retrieving actions for 5 min, the visual RF was elongated while the somatosensory RF remained unchanged. Its excitatory area became elongated along the axis of the tool, as if the image of the tool was incorporated into that

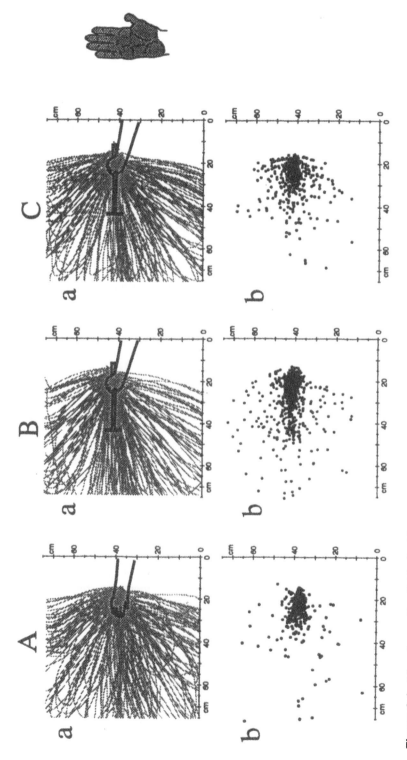

Figure 6.4 Visual receptive field of a 'distal type' bimodal neuron. The somatosensory receptive field of this neuron was on the glabrous skin of the right hand. **A:** before tool use; **B:** immediately after tool use; **C:** 3 minutes after retrieving food without the tool, but holding it in the hand. **a:** Trajectories (shaded lines) of a scanning object projected to a horizontal plane. The scanning object was visible to the monkey; it was moved toward or away from the monkey's hand equally in all directions at a speed of about 0.5 m/s. **b:** The locations (mediolateral on ordinate and anteroposterior on abscissa) of the scanning object in the horizontal plane caused the neurons to fire; each dot represents 1 spike discharged at an instantaneous frequency higher than 3.0 Hz (the spontaneous firing level of this neuron).

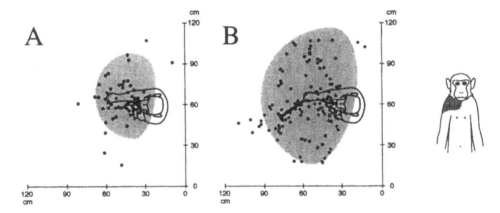

Figure 6.5   Visual receptive field of a 'proximal type' neuron before **(A)** and immediately after **(B)** tool use. Locations of the scanning object in the horizontal plane effective to make neurons fire are shown by dots (each dot represents a 1 spike discharge; this neuron exhibited no spontaneous firing). The areas reachable by either a free arm (A) or with a rake (B) are shaded. Tactile receptive field of this neuron was on the skin covering the right shoulder.

of the hand. After the monkey retrieved food without using the tool for 3 min, the expanded visual RF shrunk back, even though the monkey kept holding the tool during the recording period. Responses to somatosensory stimulation were not modified after the tool-use.

In the proximal type (Fig. 6.5), somatosensory RFs were on the skin of the face, neck, trunk, or arm. They were activated when the experimenter's hand with a food pellet moving toward the RF entered the space accessible to the hand. After the tool use, they became responsive to the stimuli presented over a wider region, namely the space accessible to the hand-held rake. Both in the distal and proximal type neurons, shrinking of the expanded visual RF was observed while the tool was being held but not used. Thus, the expansion of the visual RF was not the result of physical changes in the configuration of the body, but it appeared to be associated with the monkey's immediate intention to use it. None of these neurons was activated in association with the monkey's action to retrieve food, suggesting that activities of these neurons are not coding any aspect of tool-using arm movements. These neurons were found most frequently in the arm/hand region of the postcentral gyrus, but not in the digit region (Fig. 6.6). Some of them were found in the crown of the gyrus, in area 2, but the majority were in the anterior bank up to the very fundus of the IPS. The distribution of the distal and proximal type neurons overlapped each other.

SUMMARY

In the anterior bank of the intraparietal sulcus (IPS), we found neurons with receptive fields (RF) on bilateral digits, arms, shoulders or trunk. The size and configuration of the RFs were generally large and complex, and some of them were considered to be at the highest

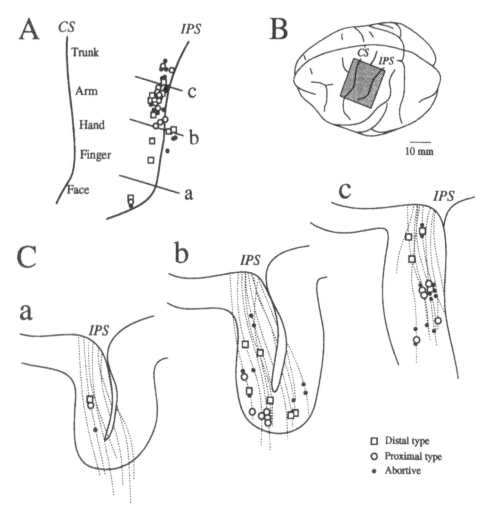

Figure 6.6 **A:** Recording sites of the bimodal neurons (open squares – the distal type; open circles – the proximal type; filled circles – no modification of visual receptive fields during tool use) plotted on the dorsal surface of the left postcentral gyrus (hatched area in **B**). CS: central sulcus; IPS: intraparietal sulcus. The approximate sites of representation of the body parts in area 3b are as indicated. **C:** Off-sagittal sections orthogonal to IPS (indicated by lines a-c in A), on which recording sites are projected. Broken lines indicate electrode tracks which covered the whole extent of the anterior bank and the fundus of the IPS.

level of the hierarchical processing in this gyrus. We found another group of neurons which receive visual as well as somatosensory inputs. The somatosensory RFs were on the hand or shoulders, arms, neck or face. Many of these bimodal neurons had bilateral RFs. The visual RF was in a limited region in the visual space just over the somatic RF and moved, keeping a close spatial relationship with the somatosensory counterpart. When the monkey used a tool held by hand, the depth and shape of the visual RF was transformed so that it covered the whole extent of the tool. The phenomenon was interpreted as a neural correlate of the modified schema of the hand in which the tool was incorporated. These observations are in line with the hierarchical scheme of information processing within the postcentral gyrus and suggest that the anterior bank of the IPS region plays a unique role in the performance of hand function.

REFERENCES

Berlucchi, G. (1990). Commissurotomy studies in animals. In *Handbook of Neuropsychology*, edited by F. Boler and J. Grafman, 4, pp. 9–46. Amsterdam: Elsevier.

Berlucchi, G., Gazzaniga, M.S. and Rizzolatti, G. (1967). Microelectrode analysis of transfer of visual information by the corpus callosum. *Archives Italiennes de Biologie*, **105**, 583–596.

Berlucchi, G. and Rizzolatti, G. (1968). Binocularly driven neurons in visual cortex of split-chiasm cats. *Science*, **159**, 308–310.

Burton, H. and Sinclair, R. (1996). Somatosensory cortex and tactile perceptions, In *Touch and Pain*, edited by L. Kruger, pp. 105–177. London: Academic.

Caminiti, R. and Sbriccoli, A. (1985). The callosal system of the superior parietal lobule in the monkey. *Journal of Comparative Neurology*, **237**, 85–99.

Choudhury, B.P., Whitteridge, D. and Wilson, M.E. (1965). The function of the callosal connections of the visual cortex. *Quarterly Journal of Experimental Physiology*, **50**, 214–219.

Conti, F., Fabri, M. and Manzoni, T. (1986). Bilateral receptive fields and callosal connectivity of the body midline representation in the first somatosensory area of primates. *Somatosensory Research*, **3**, 273–289.

Darian-Smith, I., Goodwin, A., Sugitani, M. and Heywood, J. (1984). The tangible features of textured surfaces: their representation in the monkey's somatosensory cortex. In: *Dynamic aspects of neocortical function*, edited by G.M. Edelman, W.E. Gall, W.M. Cowan, pp. 475–500. New York: John Wiley.

Darian-Smith, I., Galea, M.P., Darian-Smith, C., Sugitani, M., Tan, A. and Burman, K. (1996). The anatomy of manual dexterity: the new connectivity of the primate sensorimotor thalamus and cerebral cortex. *Advances in Anatomy, Embryology and Cell Biology*, **133**, 1–142.

Dreyer, D.A., Loe, P.R., Metz, C.B. and Whitsel, B.L. (1975). Representation of head and face in postcentral gyrus of the macaque. *Journal of Neurophysiology*, **38**, 714–733.

Duhamel, J.-R. Colby, C.L. and Goldberg, M.E. (1991). Congruent representations of visual and somatosensory space in single neurons of monkey ventral intra-parietal cortex (area VIP) In Brain and Space. Edited by J. Paillard, pp. 223–236. Oxford: Oxford University Press

Hubel, D.H. and Wiesel, T.N. (1967). Cortical and callosal connections concerned with the vertical meridian of visual fields in the cat. *Journal of Neurophysiology*, **30**, 1561–1573.

Hyvarinen, J. and Poranen, A. (1978). Receptive field integration and submodality convergence in the hand area of the postcentral gyrus of the alert monkey. *Journal of Physiology (London)*, **257**, 199–227.

Iriki, A., Tanaka, M. and Iwamura, Y. (1996a). Attention-induced neuronal activity in the monkey somatosensory cortex revealed by pupillometrics. *Neuroscience Research*, **25**, 173–181.

Iriki, A., Tanaka, M. and Iwamura, Y. (1996b). Coding of modified body schema during tool use by macaque postcentral neurons. *Neuroreport*, **7**, 2325–2330.

Iwamura, Y., Iriki, A. and Tanaka, M. (1994). Bilateral hand representation in the postcentral somatosensory cortex. *Nature*, **369**, 554–556.

Iwamura, Y., Iriki, A., Tanaka, M., Taoka, M. and Toda, T. (1996a). Bilateral receptive field neurons in the postcentral gyrus: two hands meet at the midline. In *Perception, Memory, and Emotion*: Frontier in Neuroscience, edited by T. Ono, B.L. McNaughton, S. Molotchnikoff, E.T. Rolls, and H. Nishijo, pp. 33–44. Oxford, New York: Pergamon.

Iwamura, Y. and Tanaka, M. (1978a). Postcentral neurons in hand region of area 2: their possible role in the form discrimination of tactile objects. *Brain Research*, **150**, 662–666.

Iwamura, Y. and Tanaka, M. (1996). Representation of reaching and grasping in the monkey postcentral gyrus. *Neuroscience Letters*, **214**, 147–150.

Iwamura, Y., Tanaka, M. and Hikosaka, O. (1980). Overlapping representation of fingers in the somatosensory cortex (area 2) of the conscious monkey. *Brain Research*, **197**, 516–520.

Iwamura, Y., Tanaka, M., Sakamoto, M. and Hikosaka, O. (1983a). Functional subdivisions representing different finger regions in area 3 of the first somatosensory cortex of the conscious monkey. *Experimental Brain Research*, **51**, 315–326.

Iwamura, Y., Tanaka, M., Sakamoto, M. and Hikosaka, O. (1983b). Converging patterns of finger representation and complex response properties of neurons in area 1 of the first somatosensory cortex of the conscious monkey. *Experimental Brain Research*, **51**, 327–337.

Iwamura, Y., Tanaka, M., Sakamoto, M. and Hikosaka, O. (1985a). Diversity in receptive field properties of vertical neuronal arrays in the crown of the postcentral gyrus of the conscious monkey. *Experimental Brain Research*, **58**, 400–411.

Iwamura, Y., Tanaka, M., Sakamoto, M. and Hikosaka, O. (1985b). Vertical neuronal arrays in the postcentral gyrus signaling active touch: a receptive field study in the conscious monkey. *Experimental Brain Research*, **58**, 412–420.

Iwamura, Y., Tanaka, M., Sakamoto, M. and Hikosaka, O. (1993). Rostrocaudal gradients in neuronal receptive field complexity in the finger region of alert monkey's postcentral gyrus. *Experimental Brain Research*, **92**, 360–368.

Iwamura, Y., Tanaka, M., Sakamoto, M. and Hikosaka, O. (1996b). Postcentral neurons of alert monkeys activated by the contact of the hand with objects other than the monkey's own body. *Neuroscience Letters*, **186**, 127–130.

Jones, E.G., Coulter, J.D. and Hendry, S.H.C. (1978). Intracortical connectivity of architectonic fields in the somatic sensory, motor and parietal cortex of monkeys. *Journal of Comparative Neurology*, **181**, 291–348.

Jones, E.G., Coulter, J.D. and Wise, S.P. (1979). Commissural columns in the sensory-motor cortex of monkeys. *Journal of Comparative Neurology*, **188**, 113–136.

Jones, E.G. and Friedman, D.P. (1982) Projection pattern of functional components of thalamic ventrobasal complex on monkey somatosensory cortex. *Journal of Neurophysiology*, **48**, 521–44.

Jones, E.G. and Hendry, S.H.C. (1980). Distribution of callosal fibers around the hand representations in monkey somatic sensory cortex. *Neuroscience Letters*, **19**, 167–172.

Jones, E.G. and Powell, T.P.S. (1969). Connections of the somatic sensory cortex of the rhesus monkey.II. Contralateral cortical connections. *Brain*, **92**, 717–730.

Karol, E.A. and Pandya, D.N. (1971). The distribution of the corpus callosum in the rhesus monkey. *Brain*, **94**, 471–486.

Killackey, H.P., Gould, H.J.III, Cusick, C.G., Pons, T.P. and Kaas, J.H. (1983). The relation of corpus callosum connections to architectonic fields and body surface maps in sensorimotor cortex of new and old world monkeys. *Journal of Comparative Neurology*, **219**, 384–419.

Kunzle, H. (1978). Cortico-cortical efferents of primary motor and somatosensory regions of the cerebral cortex in macaca fascicularis. *Neuroscience*, **3**, 25–39.

Leinonen, L., Hyvarinen, J., Nyman, G. and Linnankoski, I. (1979). Functional properties of neurons in lateral part of associative area 7 in awake monkeys. *Experimental Brain Research*, **34**, 299–320.

Manzoni, T., Barbaresi, F., Conti, P. and Fabri, M. (1989). The callosal connections of the primary somatosensory cortex and the neural bases of midline fusion. *Experimental Brain Research*, **76**, 251–266.

Mountcastle, V.B., Lynch, J.C., Georgopoulos, A., Sakata, H. and Acuna, C. (1975). Posterior parietal association cortex of the monkey: command functions for operations within extrapersonal space. *Journal of Neurophysiology*, **38**, 871–908.

Ogawa, H., Ito, S. and Nomura, T. (1989). Oral cavity representation at the frontal operculum of macaque monkeys. *Neuroscience Research*, **6**, 283–298.

Pandya, D.N. and Vignolo, L.A. (1969). Interhemispheric projections of the parietal lobe in the rhesus monkey. *Brain Research*, **15**, 49–65.

Pons, T.P. and Kaas, J.H. (1985). Connections of area 2 of somatosensory cortex with the anterior pulvinar and subdivisions of the ventroposterior complex in macaque monkeys. *Journal of Comparative Neurology*, **240**, 16–36.

Powell, T.P.S and Mountcastle, V.B. (1959). The cytoarchitecture of the postcentral gyrus of the monkey macaca mulatta. *Bulletin of Johns Hopkins Hospital*, **105**, 108–131.

Schwartz, D.W.F. and Frederickson, J.M. (1971). Tactile direction sensitivity of area 2 oral neurons in the rhesus monkey cortex. *Brain Research*, **27**, 397–401.

Seltzer, B. and Pandya, D.N. (1986). Posterior parietal projections to intraparietal sulcus of the rhesus monkey. *Experimental Brain Research*, **62**, 459–469.

Shanks, M.F. and Powell, T.P.S. (1981). An electron microscopic study of the termination of thalamocortical fibres in areas 3b, 1 and 2 of the somatic sensory cortex in the monkey. *Brain Research*, **218**, 35–47.

Shanks, M.F., Pearson, R.C.A. and Powell T.P.S. (1985). The callosal connections of the primary somatic sensory cortex in the monkey. *Brain Research Reviews*, **9**, 43–65.

Taoka, M., Toda, T. and Iwamura, Y. (1997). Bilateral representation of the arm and trunk in the monkey postcentral somatosensory cortex. *Society for Neuroscience Abstract*, pp. 1007, #398.4.

Taoka, M., Toda, T. and Iwamura, Y. (1998). Representation of the midline trunk, bilateral arms, and shoulders in the monkey postcentral somatosensory cortex. *Experimental Brain Research*, **123**, 315–322.

Vogt, B.A. and Pandya, D.N. (1978). Corticocortical connections of somatic sensory cortex (areas 3,1, and 2) in the rhesus monkey. *Journal of Comparative Neurology*, **177**, 179–192.

Zeki, S. (1978). Functional specialization in the visual cortex of rhesus monkey. *Nature*, **274**, 423–428.

CHAPTER 7

SENSORY AND MOTOR FUNCTIONS OF FACE PRIMARY SOMATOSENSORY CORTEX IN THE PRIMATE

Gregory M. Murray[1], Li-Deh Lin[2], Dongyuan Yao[3] and Barry J. Sessle[3]

[1]Faculty of Dentistry, University of Sydney, Sydney, Australia; [2]School of Dentistry, National Taiwan University, Taipei, Taiwan; [3]Faculty of Dentistry, University of Toronto, Toronto, Canada

INTRODUCTION

The primary somatosensory cortex (SI), comprising cytoarchitectonic areas 3a, 3b, 1 and 2 of the cerebral cortex, plays a critical role in primates in the processing of somatosensory information generated by passive stimulation of peripheral tissues or elicited by active movements. In the case of spinally innervated tissues, numerous studies employing stimulation, ablation/inactivation, imaging or neurone recording of limb SI have clearly established its importance in tactile acuity, detection and discrimination (for review, see Penfield and Rasmussen, 1950; Mountcastle, 1984; Whitsel *et al.*, 1989; Carlson, 1990; Kaas, 1996). In addition, analogous studies point to a role for SI in perceptual processes related to pain and possibly temperature sensations (Roland, 1987; Kenshalo and Willis, 1991; Backonja, 1996; Casey and Minoshima, 1997). There is however, evidence that SI may also play an important role in limb motor control in primates. This evidence includes the abundant interconnections between motor cortex (MI) and SI (Jones *et al.*, 1978; Mountcastle, 1984), the corticofugal modulation exerted by SI on somatosensory inputs during movements (for review, see Wiesendanger, 1981; Chapin, 1987; Chapman 1994), and the motor effects of surface electrical stimulation (Penfield and Rasmussen, 1950; Woolsey, 1958) or lesions of SI (Freund, 1987; Asanuma, 1989) in monkeys and humans. Moreover, single neurone recordings in behaving monkeys have revealed that the activity patterns of limb SI neurones may correlate with position, force, direction or velocity, during trained forelimb movements, although the extent to which SI contributes to the acquisition versus the execution of motor skills is still unclear (for review, see Evarts, 1986; Asanuma, 1989; Keller, 1996; Nelson, 1996).

In contrast, there is much less information available of the role of face SI in primate orofacial somatosensation and motor control. The limited number of studies using SI stimulation, ablation/inactivation or imaging do, however, indicate that face SI is important

---

Address for Correspondence: Barry J. Sessle, Faculty of Dentistry, University of Toronto, Toronto M5G 1G6, Canada. Tel: 416 979-4921; Fax: 416 979-4936; E-mail: barry.sessle@totonto.ca

for somatosensory perceptual processing (see Penfield and Rasmussen, 1950; Dubner *et al.*, 1978; Mountcastle, 1984), consistent with findings of the functional properties of face SI neurones activated by tactile or noxious stimulation of orofacial tissues in awake or anaesthetised monkeys (Schwarz and Frederickson, 1971; Dreyer *et al.*, 1975; Cusick *et al.*, 1986; Huang *et al.*, 1988, 1989a,b; Kenshalo and Willis, 1991; Iwata *et al.*, 1998). Until our own recent investigations, the few studies of the possible role of face SI in primate motor control had focussed on the effects of surface electrical stimulation and optical imaging (Penfield and Rasmussen, 1950; Woolsey, 1958; Haglund *et al.*, 1992). In view, therefore, of the limited information available of the role of face SI in primate orofacial somatosensation and especially orofacial motor control, we carried out a series of investigations utilising, in particular, reversible inactivation of face SI and recordings of the activity of single neurones in awake monkeys to clarify the possible role of face SI in orofacial somatosensory processing and motor control.

METHODS

The studies were carried out on female monkeys (*Macaca fascicularis*) cared for in accordance with the Guiding Principles of the American Physiological Society and the guidelines of the Canadian Council on Animal Care. The methods have been extensively detailed elsewhere (Huang *et al.*, 1988, 1989a,b: Murray *et al.*, 1991; Murray and Sessle, 1992a,b; Lin *et al.*, 1993, 1994; Moustafa *et al.*, 1994; Martin *et al.*, 1997) and will therefore only be briefly outlined. Each animal was trained to protrude its tongue to engage a force transducer placed at a set distance from its mouth (see Murray *et al.*, 1991; Murray and Sessle, 1992b, and Fig. 7.1). The transducer output moved a cursor (on a video screen in front of the animal), and the animal's task was to move the cursor into a target window that reflected a force level of 1.0 N (Fig. 7.2). The animal was rewarded with 0.4 ml of juice for a successful task trial if the cursor remained within the baseline window during the pre-trial period, exited the baseline window within 3 s of target appearance, and remained within the target window for at least 0.5 s. Some animals were also similarly trained in a biting task requiring the animal to bite with the incisor teeth onto a force transducer placed between the teeth and to produce bite forces of around 20 N. The electromyographic (EMG) activities of several muscles, including genioglossus, masseter and anterior digastric, were recorded with chronically implanted EMG electrodes. The EMG activities as well as neuronal activity (see below) were also recorded in some instances when the animal carried out semi-automatic movements associated with the mastication and swallowing of standard-sized pieces of apple or sultanas or swallowed fruit juice. Orofacial movements were also sometimes recorded on video tape.

After a task success rate of 50–70% was achieved, a stainless-steel cylinder was implanted over one or both cerebral hemispheres to allow transdural access to the sensorimotor cortex by glass-coated tungsten microelectrodes. A series of microelectrode penetrations was made in face SI, MI or more lateral cortical regions that included the classical cortical masticatory area (CMA) and swallow cortex. In each penetration, intracortical microstimulation (ICMS) was systematically applied at neuronal recording sites to allow for functional identification of the face MI, CMA and swallow cortex based on characteristically evoked movements and associated EMG patterns (Huang *et al.*, 1989a,b; Murray and Sessle, 1992a; Martin *et al.*, 1997).

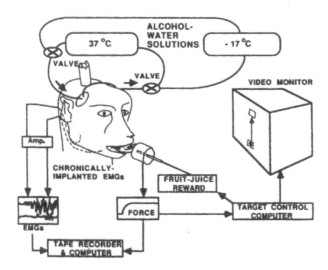

Figure 7.1 Experimental arrangement with monkey performing the tongue-protrusion task and the experimental set-up used to deliver alcohol-water solution at −17° or 37°C to two thermodes for bilateral cold block-induced inactivation of face primary somatosensory cortex (SI). The flow of solution from one or the other bath was controlled by the valves; the arrows indicate the direction of flow. The ellipse on the side of the monkey's head represents one of the bilaterally implanted chambers through which thermodes were placed on SI (see Fig. 5.3). (from Lin *et al.*, 1993).

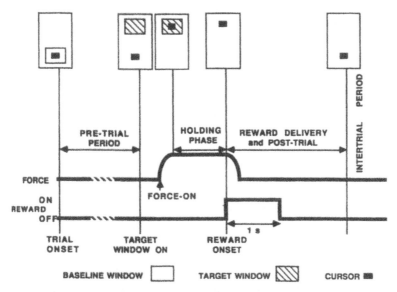

Figure 7.2 Paradigm for tongue-protrusion task or for biting task; long, thin vertical lines indicate digital markers. The rectangular squares at the top of the vertical lines represent the monitor that presents to the monkey the baseline window or target window at different stages of the task and a cursor controlled by the force output from the transducer. (from Lin *et al.*, 1993).

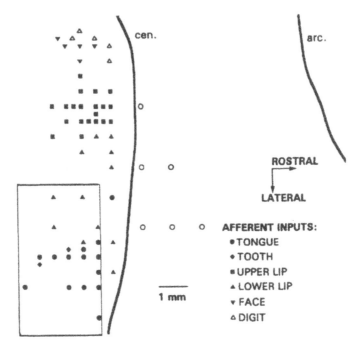

Figure 7.3 Surface view of right lateral pericentral cortices and adjacent regions showing the location of face SI and its relation to a cooling thermode. Symbols show the sensory inputs in SI for each penetration. Open circles indicate those penetrations in which a tongue or facial movement could be evoked in face primary motor cortex (MI). arc., arcuate sulcus; cen., central sulcus. (from Lin *et al.*, 1993).

In addition to characterising subsequently the movement-related activity and afferent input properties of single neurones in face SI or MI (see below), the orofacial representations within the face SI of each hemisphere were initially mapped in five monkeys by defining the mechanoreceptive field (RF) of single neurones (Lin *et al.*, 1993) so that a thermode (2.5×5.5 mm or 2.5×7.5 mm) could be positioned on the dura or the pia over the lateral half of the face SI in one or both hemispheres, that is, over the SI representations for the tongue and other intraoral regions and part of the perioral area (Figs. 7.1, 3). This part of the face SI is also the only SI region from which rhythmical jaw movements or swallowing can be evoked by ICMS (Huang *et al.*, 1989b; Martin *et al.*, 1993; Sessle *et al.*, 1995a; Martin *et al.*, unpublished observations). The thermode provided an experimental means by which the effects of reversible cold block-induced inactivation of SI could be tested on orofacial motor behaviour and on MI neuronal activity (see below) in the awake monkey. An alcohol-water solution maintained the thermode temperature either at 37°C during control (precool and postcool) periods, or at 2–4°C during the test (cool) period. A series of trials of the tongue and biting tasks, as well as sequences of masticatory trials, were studied during these three periods. For each masticatory trial, the total masticatory time was defined as the period from the end of anterior digastric EMG activity associated with mouth opening to obtain the food to be masticated to the beginning of anterior digastric EMG activity associated with the swallow that concluded the masticatory movements for that trial. The chewing time was defined as the time from the beginning of the total masticatory time to the end of the rhythmic bursts of masseter, anterior digastric, and genioglossus EMG activity associated with the masticatory movements. The oral transport time was defined as the period from the end of the chewing time to the end of the total masticatory time.

After a precool control sequence of trials, each thermode was cooled to 2–4°C and 4 min later a sequence of test trials was conducted over a 5–6 min period. This represents ample time for cortical inactivation, since our previous studies in these same animals have shown that within 4 min both evoked and spontaneous activity in the cortex beneath the thermode could be abolished (Murray *et al.*, 1991). We have also shown with isotherms and neurone recordings that thermode cooling blocks synaptic transmission within a considerable portion of the lateral half of face SI, and that the activity of the face MI would not have been directly affected by cooling face SI (Murray *et al.*, 1991; Lin *et al.*, 1993; and see below). After the last test (cool) trial, the thermodes were rewarmed to 37°C, and 4 min later, a final series of trials constituted the postcool control condition. Analysis procedures have been previously described in detail (Lin *et al.*, 1993; Moustafa *et al.*, 1994; Lin *et al.*, 1998) and included assessments of successful task performance as well as visual observations and off-line analyses from the EMG data of several variables of the animal's task, masticatory or swallowing behaviour.

In addition to testing the efficacy of unilateral or bilateral face SI cold block on task or semi-automatic motor behaviours, in three of the animals we also tested the effect of unilateral face SI cold block on the activity of single face MI neurones. The spontaneous or chewing-related activity of the neurone, or responses evoked from the neuronal orofacial mechanoreceptive field, were assessed before, during and after cold block to determine if MI neuronal activity is dependent upon an intact face SI.

In two other monkeys, single neurone recordings were instead made in face SI to determine if they showed activity patterns associated with orofacial motor behaviour. Superficial and deep sensory inputs to each neurone and the neuronal RF were assessed by the application of non-noxious mechanical or electrical stimuli to the orofacial tissues. Their extracellularly recorded evoked or ongoing activity was then studied in relation to the force-development and force-holding phases of the trained tongue or biting task (Fig. 7.2), and in some cases to semi-automatic movements.

A neurone in face SI or face MI was considered to have task, mastication, or swallow-related activity if its firing frequency during the relevant period was significantly different from that in the pre-trial period (repeated-measures ANOVA and post-hoc Duncan comparisons, $P < 0.05$). Other statistical tests used for the analyses of SI or MI neuronal properties are detailed elsewhere (Murray and Sessle, 1992b; Lin and Sessle, 1994; Martin *et al.*, 1997).

Recording and ICMS sites in each animal were histologically reconstructed and correlated with cytoarchitectonic areas of the cortex (Huang *et al.*, 1989a; Murray and Sessle, 1992a; Lin *et al.*, 1994; Martin *et al.*, 1997).

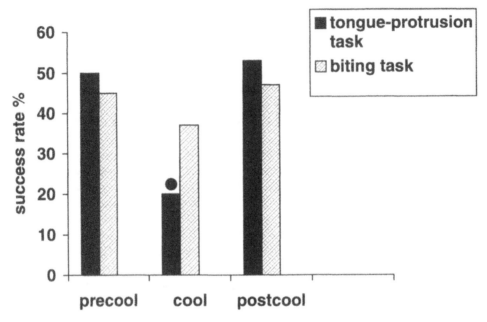

Figure 7.4 Success rate (%) for performance of the tongue-protrusion task and the biting task for precool, cool, and postcool conditions during bilateral cooling of face SI. Filled dot indicates a significant difference in the success rates under the cool condition in comparison with those under the precool and postcool conditions (Chi square test, $P < 0.05$). (from Lin *et al.*, 1993).

RESULTS AND DISCUSSION

Effects of SI Cold Block on Orofacial Motor Behaviour

(a) Task performance

During bilateral cooling of face SI in two monkeys tested, there was a statistically significant ($P < 0.05$) reduction in the success rates for the performance of the tongue-protrusion task in comparison with the control (precool and postcool) series of trials (Fig. 7.4). Detailed analyses of force and EMG activity showed that the principal deficit was the significant ($P < 0.05$) reduction in the monkey's ability to maintain a steady tongue-protrusive force in the force-holding period during each trial and to exert a consistent tongue-protrusion force across different trials (see Lin *et al.*, 1993). Unilateral face SI cold block did not interfere with tongue-task performance (Yao *et al.*, 1996), indicating that the remaining intact contralateral SI may be sufficient to allow for successful performance of tongue protrusion. However, preliminary MI neuronal data acquired during unilateral cold block of face SI raise the possibility that some features of face MI neuronal activity may be influenced by face SI (see below).

Identical cooling conditions involving bilateral face SI cold block did not significantly affect the success rates for the performance of the biting task. The lack of any significant effect on the monkeys' success rate in the performance of the biting task is consistent with a study reporting that the performance of a biting task was 'essentially' normal in one monkey with bilateral cortical lesions limited to the postcentral gyrus (Luschei and Goodwin, 1975). However, cooling did significantly ($P < 0.05$) affect the ability of the monkeys to maintain a steady force in the holding period during each trial and to exert a consistent bite force across different trials.

We attribute the effects of face SI cold block on tongue-task performance to a disturbance in synaptic transmission and not to an interruption in fibre-tract conduction (see also Murray *et al.*, 1991). We have previously shown that transdural cooling results in a rapid and powerful reduction of evoked and spontaneous neuronal activity directly beneath the thermode when the thermode temperature has cooled to 28°C (Murray *et al.*, 1991), and minimal interference in fibre-tract conduction occurs during the cooling of fibre tracts to temperatures between 20° and 10°C (Brooks, 1983). These single-neurone data, together with isotherm data showing that a considerable portion of the face SI cortex was at a temperature < 20°C during cooling of face SI, suggest that synaptic transmission indeed would have been blocked in most of lateral face SI (Murray *et al.*, 1991).

Our data providing the first documentation that selective inactivation of face SI significantly disrupts tongue protrusion, are in accord with our recent data on bilateral cooling of face MI which also results in a profound effect on the success rate for the performance of the tongue-protrusion task but, as for face SI cooling, not for performance of the biting task (Murray *et al.*, 1991). Nevertheless, differences do exist between face SI and MI in the effects of cooling these regions. For example, reversible inactivation of face MI by cooling significantly affects the monkey's ability to exert enough tongue-protrusive force to maintain the required force (Murray *et al.*, 1991), whereas bilateral cooling of SI mainly affects the animal's ability to maintain a steady tongue-protrusion force during each trial and to exert a consistent force between different trials. These data suggest that face MI is important in the generation and fine control of voluntary tongue movements while face SI is important especially in the fine control of these movements.

The significant increase in the variation with which tongue-protrusive force could be maintained indicates that the effect of bilateral cooling of face SI on the success rates for the performance of the tongue-protrusion task was likely due to an impairment in sensorimotor integration and in the fine control of force maintenance and, as noted above, to a lesser extent in the generation of adequate force with the tongue. Our data are consistent with the observations that digital tactile inputs are important in precision grip to adjust the ratio between grip force and load force and that grip force regulation is impaired under digital cutaneous anaesthesia (Johansson *et al.*, 1992). They also are consistent with clinical neurological observations that many patients with cortical lesions in the postcentral gyrus have deficient digital force control, although some of them do develop initial paresis of the limbs contralateral to the lesion (Pause *et al.*, 1989). The poor performance of the monkeys in the tongue-protrusion task also could be due to cold block-induced impairment of kinaesthesia of the tongue and other orofacial structures associated with the tongue-protrusion task. Removal of the postcentral gyrus does not result in motor paralysis but does produce a loss of the sense of movement and the sense of position in space in the opposite extremities (Penfield and Rasmussen, 1950), and this defect is most apparent in the distal parts of the limbs. Indeed, clumsiness in use of the contralateral hand is the general finding after ablation or reversible inactivation of the primate forelimb SI (e.g.

Table 7.1 Median and range of the total masticatory time, oral transport time, and chewing time in the control and cool conditions tested in one monkey.

	control (n=14)	cool (n=11)
Total masticatory time(s)	10 (7.8–13.2)	13 (7.5–15.8)
Oral transport time(s)	2.7 (1.1–5.4)	5.2 (1.1–7.2)
Chewing time(s)	7.6 (5.2–10.2)	7.8 (5.1–9.7)

Asanuma and Arissian, 1984; Iwamura and Tanaka, 1991) and in clinical neurological observations (Pause *et al.*, 1989; Roland, 1987; for review see Freund, 1987).

Although the monkeys showed deficits in maintaining a steady force and exerting a consistent force for both tasks during bilateral cooling of face SI, the mean maximum force and mean force for the holding phase during the cool condition were not significantly different from those during the precool and postcool conditions except in one monkey which showed a slight reduction in the mean tongue-protrusion force during the cortical cooling. These data suggest that the deficit observed after cooling of face SI may be due to deficient sensorimotor integration instead of deficient force generation or movement execution and that face SI may be important in the fine control of tongue movements and, to a lesser extent, jaw-closing movements.

(b) Masticatory and swallowing performance

Further evidence that face SI may influence orofacial motor behaviour is provided by our novel findings that bilateral cooling of SI impaired rhythmical jaw and tongue movements and EMG activity associated with mastication in one of the two monkeys tested, and modified masticatory and swallowing activity in the other (Table 7.1). In this second monkey, the total masticatory time was significantly increased by SI cold block, due principally to an increase in the oral transport time (control 2.7 s, cool 5.2 s; $P < 0.05$); the bolus was manipulated by the tongue during this period before it was swallowed. Within the chewing time, cold block produced an increase in the duration of anterior digastric EMG activity, a delay in the onset of masseter EMG activity, and an increase in the variance of genioglossus EMG duration; these significant ($P < 0.05$) changes occurred in the absence of any effect on the amplitude of EMG activity in these same three muscles. The data suggest that while the face SI is not itself the generator of the basic magnitude and temporal relations of orofacial muscles during these semi-automatic movements, which appear to be dependent on central pattern generators in the brainstem (Dubner *et al.*, 1978; Luschei and Goldberg, 1981; Sessle and Henry, 1989; Lund, 1991), it may nevertheless modulate the timing and duration of anterior digastric, masseter, and genioglossus EMG activity.

It is possible that the impairments in mastication and swallowing during bilateral cold block of face SI may have been related to some disturbance in sensory perception from the tongue dorsum, along the lines we noted above for the impaired performance in the tongue task. However, part of the reason may relate to the demonstrated role of the lateral face SI in the modulation of the central pattern generators for mastication (Huang *et al.*, 1989b) and for swallowing (Martin *et al.*, 1993; Sessle *et al.*, 1995a; Martin *et al.*, unpublished observations). It is most likely that such a role of the lateral face SI also requires an extensive, low-threshold tactile input from intraoral structures, and modulation of this input could play a part in selecting the appropriate movement pattern during mastication and swallowing. The thermodes used for cold block of SI indeed covered the lateral part of the face SI, which contains an extensive cortical representation for the tongue and other intraoral regions and part of the perioral area (Cusick *et al.*, 1986; Huang *et al.*, 1989b; Lin *et al.*, 1993). This lateral part of face SI also is the only SI region from which rhythmical jaw movements or swallowing can be evoked by ICMS (Huang *et al.*, 1989a; Martin *et al.*, 1993; Sessle *et al.*, 1995a; Martin *et al.*, unpublished observations). Our ICMS and cold block data of the involvement of face SI in orofacial motor control are also supported by the effects of surface stimulation of lateral SI which evokes facial, jaw, tongue and swallowing movements in humans and monkeys (Penfield and Rasmussen, 1950; Woolsey, 1958), by recent optical imaging findings of face SI activity during tongue movements in humans (Haglund *et al.*, 1992), and by our single neurone recordings in face SI (see below). As noted above, there are also reports that lesions of the limb SI in monkeys and man may lead to marked motor disturbance and, furthermore, extensive bilateral cortical lesions, including areas 44, 6b, 4, 3, 1 and 2, do result in the inability of monkeys to feed themselves and to swallow immediately after the ablation (Kirzinger and Jürgens, 1982). Parallel observations come from the effects of lesions or cold block of the face MI or the far-lateral pericentral cortex (i.e. cortical masticatory area) in non-human primates, which do result in highly uncoordinated tongue and chewing movements (Luschei and Goodwin, 1975; Larson *et al.*, 1980; Enomoto *et al.*, 1987; Narita *et al.*, 1995; Sessle *et al.*, 1995b). In one of these studies, we found that reversible bilateral cold block of the lateral pericentral region from which chewing or swallowing could be evoked by ICMS results in a marked and statistically significant reduction in the incidence of swallowing and slowing or interruption of chewing, and significant changes in the EMG patterns associated with chewing and swallowing (Narita *et al.*, 1995; Sessle *et al.*, 1995b).

The present data thus suggest that the changes after cold block of the lateral face SI are more subtle than the changes in mastication and swallowing after cold block of the ICMS-defined masticatory and swallow areas in the lateral pericentral region and the lateral part of the face MI (Narita *et al.*, 1995). Our observations support the view that these different cortical regions may be differentially involved in the selection or control of chewing patterns during mastication and swallowing, and provide further evidence in support of a role for the lateral part of the face SI cortex in the control of both operantly conditioned (i.e. task) and semi-automatic orofacial movements. This latter conclusion is analogous to one we have recently drawn from similar studies of face MI, namely that the role of MI in motor control is not restricted to operantly conditioned motor behaviour but also includes semi-automatic movements (Martin *et al.*, 1995, 1997; Sessle *et al.*, 1995a).

SI Neuronal Activity — Afferent Input Properties and Activity Patterns Related to Orofacial Motor Behaviour

(a) Afferent input properties

The face SI was mapped in five monkeys, and a detailed assessment of the neuronal properties made in two of these monkeys. The representation of the orofacial region was found immediately lateral to that of the hand, and there was a clear somatotopic pattern of organisation within face SI: the periorbital or nose region was located most medially, then followed laterally in sequence the representation of the upper lip, lower lip, and intraoral area in the face SI. A mechanoreceptive field (RF) was identified for 253 face SI neurones, which included 162 'lip RF' neurones receiving mechanosensitive afferent inputs from the upper lip, lower lip, or both; 72 'tongue RF' neurones that received mechanosensitive afferent inputs from the tongue; 11 'periodontium RF' neurones receiving periodontal inputs; and 8 neurones that received inputs from other orofacial regions. Nearly all (249/253) the neurones responded to light tactile stimulation of a localised orofacial RF and with a short latency response (mean±SD: 9.0±2.1 ms; n=40) to electrical stimulation of the RF. Most (78%) of the neurones received contralateral orofacial inputs and most (82%) showed a rapidly adapting response to tactile stimulation. These findings confirm previous observations in awake monkeys (Dreyer *et al.*, 1975; Huang *et al.*, 1989b), and are comparable with those in anaesthetised monkeys (Pons *et al.*, 1985; Cusick *et al.*, 1986; Huang *et al.*, 1988) and also parallel our observations of somatosensory inputs to face MI (Huang *et al.*, 1988, 1989b; Murray and Sessle, 1992a; Martin *et al.*, 1997). While there appears to be no previous documentation of the proportion of rapidly versus slowly adapting neurones in primate face SI, our findings are consistent with previous findings that 23% of neurones receiving mechanosensitive afferent inputs in primate face MI are slowly adapting neurones (Murray and Sessle, 1992a). The properties of single neurones in face SI were thus consistent with the involvement of face SI in tactile perceptual processes that has been well documented in cortical stimulation, imaging and ablation/inactivation studies (see references cited in the introduction to this chapter).

(b) Movement-related SI neuronal activity

Other properties of the face SI neurones indicated the involvement of face SI also in motor control. The activity of most tongue RF (79% of 56) neurones and lip RF (60% of 93) neurones tested was significantly altered during the tongue-protrusion task (Fig. 5). In addition, 27% of face SI neurones related to the tongue-protrusion task exhibited statistically significant changes in firing rate in advance (40±32 ms) of onset of genioglossus EMG activity. Tongue and lip RF neurones also exhibited a variety of different neuronal activity patterns during the tongue-protrusion task, and in accordance with the patterns described for face MI neurones (Murray and Sessle, 1992b), five patterns of face SI neuronal activities could be classified in relation to the task period of the tongue-protrusion task. An example of one of the activity patterns, the phasic-tonic pattern, is shown in Figure 5C. Directional relations were also found in 25 (58%) of 43 neurones studied. The 25 neurones exhibited a single preferred direction of firing, that is, there were statistically significant changes in firing rate with changes in the direction of the tongue-protrusive movement.

Although many face SI neurones, including those exhibiting tongue task-related activity, showed rhythmic activity associated with rhythmic jaw movements, few showed activity related to the biting task. For example, 14% of 36 tongue RF neurones and 34% of the 92 lip RF neurones tested showed a significant change in firing rate during the biting task. Most of the 116 face SI neurones studied during both tasks exhibited a preferential relation to the tongue-protrusion task as distinct from the biting task, and none showed task-related activity during the biting task only. The significantly larger population of face SI neurones

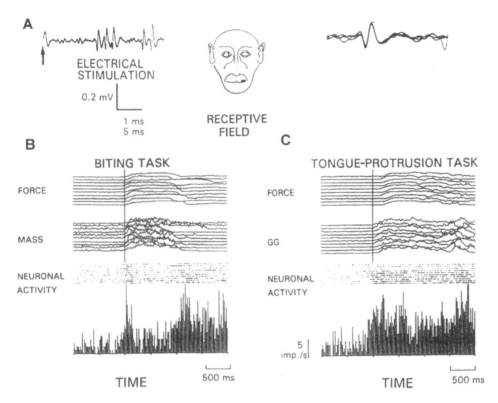

Figure 7.5 Examples of the evoked and task-related activity of a lip mechanoreceptive field (RF) neurone. **A:** Single trace of activity evoked by electrical stimulation of the RF is shown (arrow: stimulus onset) together with the location of the neurone's RF, and 4 superimposed traces of the single neurone's waveform. **B** and **C:** in the absence of the electrical stimulation to the RF, the activity of this neurone is shown during the biting task and tongue-protrusion task, respectively. The neurone showed a phasic-tonic activity pattern in the tongue-prostrusion task and a phasic activity pattern during the biting task. All the trials were aligned at the EMG onset in genioglossus (GG) for the tongue-protrusion task and at the force-on for the biting task. MASS, masseter.

related to the tongue-protrusion task in comparison with that related to the biting task thus establishes a neuronal correlate for our findings that bilateral cold block of face SI results in a marked impairment in the performance of the tongue-protrusion task but only minor effects on the biting task (see above).

These original neuronal data indicate an important role for face SI in operantly conditioned orofacial movements. However, our earlier ICMS findings and our cold block data from face SI, together with findings of changes in face SI neuronal firing during rhythmic jaw movements (see above), suggest that face SI may also be involved in semi-automatic (i.e. mastication, swallowing) as well as operantly conditioned (i.e. motor task) orofacial movements. This conclusion is analogous to one reached for MI from our ICMS and cold block data (see above) and from our recent findings that face MI neurones also show activity in relation to both task and semi-automatic movements (Sessle *et al.*, 1995a; Martin *et al.*, 1997).

Superficial mechanosensory inputs appear to be important in the control of orofacial movements during speech and of the fingers during fine digital manipulation (Asanuma, 1981; for review, see Johansson *et al.*, 1988; Smith, 1992). Further, optical imaging of functional activity in human cerebral cortex reveals that activity during side-to-side tongue movements is greatest at SI sites where electrical stimulation can evoke intraoral sensations (Haglund *et al.*, 1992). Our data indicate that most face SI neurones receive inputs from superficial mechanoreceptors and that the activity of most of them is related to the tongue-protrusion task. The activity patterns of these neurones might reflect partially the pattern of stimulation of the neurones' RF during the tongue-protrusion task and some of this information may be used in the guidance of the movement (see below). Similarly, neurones at tongue-MI sites (i.e. sites located in face MI from which ICMS ≤ 20 μA, could evoke tongue movement) also are characterised by a prominent input from intraoral superficial mechanosensitive afferents especially, and the activity of most of these face MI neurones also is related to the tongue-protrusion task (Murray and Sessle, 1992a, b). These findings raise the possibility that face SI contributes to the activity of face MI neurones, and indeed we have recent evidence that unilateral cold block of face SI reduces the spontaneous activity of six out of 25 single neurones recorded in the ipsilateral face MI and decreases the chewing-related activity in five of the 25 neurones tested (Yao *et al.*, 1997). In addition, preliminary data suggest that some peripherally evoked activity in face MI neurones may also be reduced during unilateral face SI cold block (Yao, Cairns and Sessle, unpublished observations).

On the other hand, part of the task-related activity of some face SI neurones in the tongue-protrusion task might originate from MI or other motor centres, i.e. 'corollary discharge' or 'efference copy' (for review, see McCloskey, 1981; Matthews, 1988). This concept suggests that the afferent signals elicited by movements, in conjunction with the motor command, informs the analysing centres (e.g. SI) that the sensory signals are self-generated rather than externally generated. Indeed, as noted above, we found that 27% of face SI neurones that were related to the tongue-protrusion task showed significant changes in firing rate in advance of onset of genioglossus EMG activity (40 ± 32 ms) associated with tongue-protrusion, and analogous premovement activity also has been observed by some investigators in forelimb SI (for review see Nelson, 1996) where it has been suggested to be of central origin rather than a reflection of reafference. We have observed previously (Murray and Sessle, 1992b) in monkeys performing the tongue-protrusion task that 24% of tongue MI neurones show significant changes in firing rate in advance of onset of EMG activity in GG (74 ± 49 ms). It therefore appears that tongue MI neurones may alter their firing rates in advance (34 ms on average) of face SI neurones related to the tongue-protrusion task, and that tongue MI neurones potentially could contribute to the premovement activity observed in some face SI neurones. Although it is difficult to delineate the precise role and origin of the task-related activities of the face SI neurones (see above), we believe that our data nonetheless suggest that these activities are important in the fine control of tongue movements. This view is based on the variety of different SI neuronal activity patterns during tongue protrusion, and our observation that cold block of face SI induces disruption of the tongue-protrusion task and alters the peripherally evoked, spontaneous or movement-related activity of some face MI neurones.

Further, we noted above that many face SI neurones also exhibited directional relations; such directional selectivity has also been identified for face MI neurones (Murray and Sessle, 1992c). Many neurones in forelimb SI as well as MI also exhibit a single preferred direction of firing and show systematic changes in firing frequency with changes in arm-reaching direction in two- and three-dimensional space (Crammond and Kalaska, 1989; Georgopoulos, 1994). The directional properties that we have documented in many face SI neurones may serve to regulate somatosensory inputs into the central nervous system or provide the necessary sensory information to efferent zones in face MI to effect the appropriate changes in tongue shape and position and the appropriate orofacial movements associated with the different directions of tongue-protrusion movements.

Although the proportion of face SI neurones altering their firing rate during the biting tasks was significantly less than that during the tongue-protrusion task, a significant number of lip RF and periodontium RF neurones did relate to the biting task. This is consistent with our observation that cold block of face SI may affect the fine control of jaw movement during the biting task (see above). The speculations we have made above for the function of the activity of face SI neurones during the tongue-protrusion task could equally well apply to the biting task.

(c) Movement-related modulation of evoked activity

Having established that many face SI neurones with an orofacial RF show movement-related activity, we then tested if the neuronal activity evoked by low-threshold stimulation of the RF was modulated during the tongue-protrusion and/or biting tasks. A total of 44 face SI neurones with a tongue or lip or lateral face RF was tested. For face SI neurones tested during the task period, the evoked activity (i.e. the number of evoked spikes in 50 ms after the onset of stimulation) was decreased for the majority (90%) of 31 neurones studied during the tongue-protrusion task and 61% of 23 studied during the biting task (Fig. 7.6). No neurones tested showed a clear facilitation of evoked activity during the task period of either task. For example, nine tongue RF and 17 lip RF neurones were tested during the tongue-protrusion task: all tongue RF and 11 of the 17 lip RF neurones were task-related. All but one of the 26 neurones showed a decrease in evoked activity during the task period; this suggests a generalised suppression of evoked neuronal activity, for tongue RF and lip RF neurones, during tongue-protrusion movements.

The evoked activity of 10 face SI neurones was also studied in the premovement period (i.e. the period between visual go-cue and movement onset) of the tongue-protrusion task; five showed a decrease in evoked activity up to 100-200 ms before the onset of EMG activity in the geniglossus muscle. The decrease in evoked neuronal activity before the movement onset suggests that modulation, at least in the premovement period, may be central in origin.

Our findings provide the first documentation of modulation of somatosensory responses of face SI neurones during trained orofacial movements. The findings are consistent with recent observations that cutaneously evoked activity in forelimb SI neurones is decreased during forelimb movements in awake rats and monkeys (see Chapin, 1987; Chapman, 1994). They also in accord with observations in the medial lemniscus, ventral posterolateral thalamus, and limb SI that the amplitude of evoked potentials decreases during digit or forelimb movements (Wiesendanger, 1981; Chapin, 1987; Chapman, 1994).

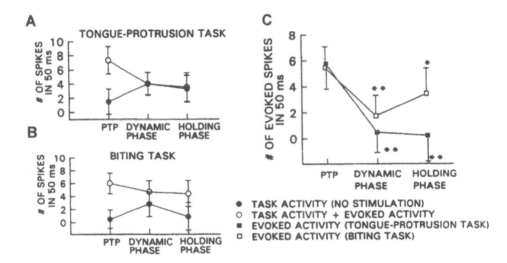

Figure 7.6 Modulation of evoked activity of the task-related neurone reported in Fig 5, during the dynamic phase and holding phase of the tongue-protrusion task and the biting task. **A** and **B** show the task-related activity as well as evoked activity plus task activity for this neurone during the tongue-protrusion task and the biting task, respectively. **C:** Number of evoked spikes in 50 ms period immediately following electrical stimulus and calculated for the pretrial period (PTP), force dynamic phase, and force holding phase for the tongue-protrusion and biting task. Vertical bars: 1 SD of the mean number of spikes indicated; the evoked activity value is significantly different from the corresponding PTP value (*P<0.05; **P<0.01). (From Lin *et al.*, 1994).

The reduction of short-latency evoked neuronal activity that we found to be a general feature of somatosensory responses evoked in neurones in face SI during the trained orofacial movements further suggest that differences exist between tongue-protrusion and biting tasks in the RF properties of the neurones affected in each task. A suppression of evoked activity occurred during the tongue-protrusion task in the majority of the face SI neurones, except those with an RF distant from the intraoral or perioral tissues. During the biting task, most neurones with a lip RF but none with a tongue RF or a distant RF on lateral face, showed a decrease of evoked neuronal activity. These data suggest that the modulation of low-threshold somatosensory responses during an orofacial movement is not generalised or nonspecific but depends on the location of a neurone's RF and the type of movement performed (see also Lin and Sessle, 1994, for a fuller account of the arguments).

Our data are compatible with the view that the motor centres can adjust the gain of transmitting elements in the somatosensory afferent pathways to compensate for the increased sensory input produced during or as a result of the movement (Chapin, 1987; MacKay and Crammond, 1989; Chapman, 1994). Our findings are also not inconsistent with the theory of corollary discharge or efference copy that is thought to modulate the transmission of sensory information and thus perception (see above); this modified sensory information also may be used by other motor centres in the control of movement. Many of the face SI neurones that we recorded likely had connections to orofacial sensorimotor centres, such as face MI and brainstem sensory and motor nuclei, e.g. there is both electrophysiological and anatomical evidence of direct projections from primate face SI to the brainstem (Darian-Smith, 1973; Dubner *et al.*, 1978; Kuypers, 1981; Sirisko and Sessle, 1983; Bushnell *et al.*, 1987). Thus, some of these corticofugal projection neurones might be concerned with the subcortical modulation of somatosensory responses during movements and may explain our findings of somatotopically organised and movement-specific modulation of the low-threshold somatosensory responses of face SI neurones during tongue-protrusion and jaw movements. These corticofugal activities may modulate the somatosensory responses in such a way as to adjust the gain of somatosensory inputs evoked during the movements or generated by the movements, and thus maintain the sensorimotor system's sensitivity to externally applied disturbances.

SUMMARY

This paper reports on a series of studies utilising reversible cold block-induced inactivation of face SI during trained and semi-automatic orofacial movements in awake monkeys and electrophysiological recordings of the orofacial-evoked responses and activity patterns of single face SI neurones during the movements. Bilateral cold block of face SI disrupts the awake monkeys' performance of trained motor behaviour as well as the semi-automatic movements of chewing and swallowing, and this may in part be related to our demonstration that SI inactivation may disrupt the activity of some face MI neurones. Furthermore, most face SI neurones show activity related to the trained or semi-automatic movements, and in the case of a trained tongue-protrusion task, the activity of many face SI neurones is preferentially related to the direction of the task. In addition, somatosensory inputs to the neurones evoked by stimulation of the neuronal mechanoreceptive field (RF) are modulated during the animals' performance of the trained motor behaviour; this gating of face SI somatosensory responses may be related to gain adjustment in somatosensory pathways during or following movements and serve an important function in orofacial motor control. We have thus provided novel data which, together with our earlier findings that intracortical microstimulation (ICMS) in face SI evokes chewing and swallowing, strongly point to a role for face SI not only in cortical processes related to somatosensory perception but also in orofacial motor control of both operantly conditioned and semi-automatic orofacial movements.

ACKNOWLEDGMENTS

We gratefully acknowledge the secretarial assistance of Mrs Tracey Bowerman. The studies of the authors were supported by grant MT-4918 to B.J. Sessle from the Medical Research Council of Canada; D. Yao also was supported by a MRC studentship.

REFERENCES

Asanuma, H. (1981). Functional role of sensory inputs to the motor cortex. *Progress in Neurobiology*, Oxford, **16**, 241–262.
Asanuma, H. (1989). *The Motor Cortex*. New York: Raven Press.

Asanuma, H. and Arissian, K. (1984). Experiments on functional role of peripheral input to motor cortex during voluntary movements in the monkey. *Journal of Neurophysiology*, **52**, 212–227.

Backonja, M.-M. (1996). Primary somatosensory cortex and pain perception. *Pain Forum*, **5**, 174–180.

Brooks, V.B. (1983). Study of brain function by local, reversible cooling. *Reviews in Physiology,Biochemistry,and Pharmacology*, **95**, 1–109.

Bushnell, M.C., Duncan, G.H. and Lund, J.P. (1987). Gating of sensory transmission in the trigeminal system. In *Higher Brain Functions: Recent Explorations of the Brain's Emergent Properties*, edited by S.P. Wise, pp. 211–237. John Wiley & Sons, Inc.

Carlson, M. (1990). The role of the somatic sensory cortex in tactile discrimination in primates. In *Cerebral Cortex*, edited by E.G. Jones and A. Peters, pp. 251–486. Amsterdam: Plenum Press.

Casey, K.L. and Minoshima, S. (1997). Can pain be imaged? In *Proceedings of the 8th World Congress on Pain, Progress in Pain Research and Management*, edited by T.S. Jensen, J.A. Turner and Z. Wiesenfeld-Hallin, 8th edn, pp. 855–866. Seattle: IASP Press.

Chapin, J.K. (1987). Modulation of cutaneous sensory transmission during movement: possible mechanisms and biological significance. In *Higher Brain Functions*, edited by S.P. Wise, pp. 181–209. New York: Wiley.

Chapman, C.E. (1994). Active versus passive touch: factors influencing the transmission of somatosensory signals to primary somatosensory cortex. *Canadian Journal of Physiology and Pharmacology*, **72**, 558–570.

Crammond, D.J. and Kalaska, J.F. (1989). Neuronal activity in primate parietal cortex area 5 varies with intended movement direction during an instructed-delay period. *Experimental Brain Research*, **76**, 458–462.

Cusick, C.G., Wall, J.T. and Kaas, J.H. (1986). Representations of the face, teeth, and oral cavity in areas 3b and 1 of somatosensory cortex in squirrel monkeys. *Brain Research*, **370**, 359–364.

Darian-Smith, I. (1973). The trigeminal system. In *Handbook of Sensory Physiology. Somatosensory System*, edited by A. Iggo, pp. 271–314. Berlin: Springer-Verlag.

Dreyer, D.A., Loe, P.R., Metz, C.B. and Whitsel, B.L. (1975). Representation of head and face in postcentral gyrus of the macaque. *Journal of Neurophysiology*, **38**, 714–733.

Dubner, R., Sessle, B.J. and Storey, A.T. (1978). *The Neural Basis of Oral and Facial Function*, New York: Plenum Press.

Enomoto, S., Schwartz, G. and Lund, J.P. (1987). The effects of cortical ablation on mastication in the rabbit. *Neuroscience Letters*, **82**, 162–166.

Evarts, E.V. (1986). Motor cortex output in primates. In *Cerebral Cortex. Vol. 5. Sensory-Motor Areas and Aspects of Cortical Connectivity*, edited by E.G. Jones and A. Peters, pp. 217–241. New York: Plenum Press.

Freund, H.-J. (1987). Abnormalities of motor behaviour after cortical lesions in humans. In *Handbook of Physiology. The Nervous System. Higher Functions of the Brain*, pp. 763–810. Bethesda: American Physiological Society.

Georgopoulos, A.P. (1994). New concepts in generation of movement. *Neurology*, **13**, 257–268.

Haglund, M.M., Ojemann, G.A. and Hochman, D.W. (1992). Optical imaging of epileptiform and functional activity in human cerebral cortex. *Neurology*, **358**, 668–671.

Huang, C.-S., Hiraba, H., Murray, G.M. and Sessle, B.J. (1989a). Topographical distribution and functional properties of cortically induced rhythmical jaw movements in the monkey (Macaca fascicularis). *Journal of Neurophysiology*, **61**, 635–650.

Huang, C.-S., Hiraba, H. and Sessle, B.J. (1989b). Input-output relationships of the primary face motor cortex in the monkey (Macaca fascicularis). *Journal of Neurophysiology*, **61**, 350–362.

Huang, C.-S., Sirisko, M.A., Hiraba, H., Murray, G.M. and Sessle, B.J. (1988). Organization of the primate face motor cortex as revealed by intracortical microstimulation and electrophysiological identification of afferent inputs and corticobulbar projections. *Journal of Neurophysiology*, **59**, 796–818.

Iwamura, Y. and Tanaka, M. (1991). Organization of the first somatosensory cortex for manipulation of objects: an analysis of behavioral changes induced by muscimol injection into identified cortical loci of awake monkeys. In *Information Processing in the Somatosensory System*, edited by O. Franzén and J. Westman, pp. 371–380. Hampshire, UK: MacMillan.

Iwata, K., Tsuboi, Y. and Sumino, R. (1998). Primary somatosensory cortical neuronal activity during monkey's detection of the perceived change in tooth-pulp stimulus intensity. *Journal of Neurophysiology*, **79**, 1717–1725.

Johansson, R.S., Häger, C. and Bäckström, L. (1992). Somatosensory control of precision grip during unpredictable pulling loads. III. Impairments during digital anesthesia. *Experimental Brain Research*, **89**, 204–213.

Johansson, R.S., Trulsson, M., Olsson, K.A. and Abbs, J.H. (1988). Mechanoreceptive afferent activity in the infraorbital nerve in man during speech and chewing movements. *Experimental Brain Research*, **72**, 209–214.

Jones, E.G., Coulter, J.D. and Hendry, S.H.C. (1978). Intracortical connectivity of architectonic fields in the somatic sensory, motor, and parietal cortex of monkeys. *Journal of Comparative Neurology*, **181**, 291–348.

Kaas, J.H. (1996). The somatosensory cortex. In *Somesthesis and the Neurobiology of the Somatosensory Cortex*, edited by O. Franzén, R. Johansson and L. Terenius, pp. 163–171. Basel: Birkhäuser Verlag.

Keller, A. (1996). Exploring the functions of the motor cortex: Hiroshi Asanuma's legacy. *NeuroReport*, **7**, 2253–2260.

Kenshalo, D.R., Jr and Willis, W.D. (1991). The role of the cerebral cortex in pain sensation. In *Cerebral Cortex*, Vol. 9, edited by A. Peters and S.P. Wise, pp. 153–212. Amsterdam: Plenum Press.

Kirzinger, A. and Jürgens, U. (1982). Cortical lesion effects and vocalization in the squirrel monkey. *Brain Research*, **233**, 299–315.

Kuypers, H.G.J.M. (1981). Anatomy of the descending pathways. In *Handbook of Physiology, The Nervous System, Vol. II, Motor Control*, edited by V.B. Brooks, pp. 597–666. Bethesda: American Physiological Society.

Larson, C.R., Byrd, K.E., Garthwaite, C.R. and Luschei, E.S. (1980). Alterations in the pattern of mastication after ablations of the lateral precentral cortex in rhesus monkeys. *Experimental Neurology*, **70**, 638–651.

Lin, L.-D., Murray, G.M. and Sessle, B.J. (1993). The effect of bilateral cold block of the primate face primary somatosensory cortex on the performance of trained tongue-protrusion task and biting tasks. *Journal of Neurophysiology*, **70**, 985–996.

Lin, L.-D., Murray, G.M. and Sessle, B.J. (1994). Functional properties of single neurons in the primate face primary somatosensory cortex. I. Relations with trained orofacial motor behaviors. *Journal of Neurophysiology*, **71**, 2377–2390.

Lin, L.-D., Murray, G.M. and Sessle, B.J. (1998). Effects on non-human primate mastication of reversible inactivation by cooling of the face primary somatosensory cortex. *Archives of Oral Biology*, **43**, 133–141.

Lin, L.-D. and Sessle, B.J. (1994). Functional properties of single neurons in the primate face primary somatosensory cortex. III. Modulation of responses to peripheral stimuli during trained orofacial motor behaviors. *Journal of Neurophysiology*, **71**, 2401–2413.

Lund, J.P. (1991). Mastication and its control by the brain stem. *Critical Reviews in Oral Biology and Medicine*, **2**, 33–64.

Luschei, E.S. and Goldberg, L.J. (1981). Neural mechanisms of mandibular control: mastication and voluntary biting. In *Handbook of Physiology, The Nervous System, Vol. II., Motor Control*, edited by V.B. Brooks, pp. 1237–1274. Bethesda: American Physiological Society.

Luschei, E.S. and Goodwin, G.M. (1975). Role of monkey precentral cortex in control of voluntary jaw movements. *Journal of Neurophysiology*, **38**, 146–157.

MacKay, W.A. and Crammond, D.J. (1989). Cortical modification of sensorimotor linkages in relation to intended action. In *Volitional Action*, edited by W.A. Hershberger, pp. 169–193. Amsterdam: Elsevier.

Martin, R.E., Kemppainen, P., Masuda, Y., Sunakawa, M. and Sessle, B.J. (1993). Swallowing evoked by intracortical microstimulation (ICMS) of lateral pericentral cerebral cortex of awake monkeys. *Society for Neuroscience Abstracts*, **19**, 777.

Martin, R.E., Murray, G.M., Kemppainen, P., Masuda, Y. and Sessle, B.J. (1997). Functional properties of neurons in the primate tongue primary motor cortex during swallowing. *Journal of Neurophysiology*, **78**, 1516–1530.

Martin, R.E., Murray, G.M. and Sessle, B.J. (1995). Cerebral cortical control of primate orofacial movements: role of face motor cortex in trained and semi-automatic motor behaviours. In *Alpha and Gamma Motor Systems*, edited by A. Taylor, M.H. Gladden and R. Durbaba, pp. 350–355. New York: Plenum Press.

Matthews, P.B.C. (1988). Proprioceptors and their contribution to somatosensory mapping: complex messages require complex processing. *Canadian Journal of Physiology and Pharmacology*, **66**, 430–438.

McCloskey, D.I. (1981). Corollary discharges: motor commands and perception. In *Handbook of Physiology, The Nervous System, Vol. II, Motor Control*, edited by V.B. Brooks, pp. 1415–1480. Bethesda: American Physiological Society.

Mountcastle, V.B. (1984). Central mechanisms in mechanoreceptive sensibility. In *Handbook of Physiology. The Nervous System. Sensory Processes*, pp. 789–878. Bethesda: American Physiological Society.

Moustafa, E.M., Lin, L.-D., Murray, G.M. and Sessle, B.J. (1994). An electromyographic analysis of orofacial motor activities during trained tongue-protrusion and biting tasks in monkeys. *Archives of Oral Biology*, **39**, 955–965.

Murray, G.M., Lin, L.-D., Moustafa, E. and Sessle, B.J. (1991). The effects of reversible inactivation by cooling of the primate face motor cortex on the performance of a trained tongue-protrusion task and a trained biting task. *Journal of Neurophysiology*, **65**, 511–530.

Murray, G.M. and Sessle, B.J. (1992a). Functional properties of single neurons in the face primary motor cortex of th primate. II. Relations with trained orofacial behavior. *Journal of Neurophysiology*, **67**, 759–774.

Murray, G.M. and Sessle, B.J. (1992b). Functional properties of single neurons in the face primary motor cortex of the primate. I. Input and output features of tongue motor cortex. *Journal of Neurophysiology*, **67**, 747–758.

Murray, G.M. and Sessle, B.J. (1992c). Functional properties of single neurons in the face primary motor cortex of the primate. III. Relations with different directions of trained tongue protrusion. *Journal of Neurophysiology*, **67**, 775–785.

Narita, N., Sessle, B.J., Raouf, R. and Huang, C.-S. (1995). Effects on mastication of reversible cold block of lateral pericentral cortex of awake monkey. *Society for Neuroscience Abstracts*, **21**, 1900.

Nelson, R.J. (1996). Interactions between motor commands and somatic perception in sensorimotor cortex. *Current Opinion in Neurobiology*, **6**, 801–810.

Pause, M., Kunesch, E., Binkofski, F. and Freund, H.-J. (1989). Sensorimotor disturbances in patients with lesions of the parietal cortex. *Brain*, **112**, 1599–1625.

Penfield, W. and Rasmussen, T. (1950). *The Cerebral Cortex of Man*. New York: Macmillan.

Pons, T.P., Garraghty, P.E., Cusick, C.G. and Kaas, J.H. (1985). The somatotopic organization of area 2 in macaque monkeys. *Journal of Comparative Neurology*, **241**, 445–466.

Roland, P.E. (1987). Somatosensory detection of microgeometry, macrogeometry and kinesthesia after localized lesions of the cerebral hemispheres in man. *Brain Research Reviews*, **12**, 43–94.

Schwarz, D.W.F. and Fredrickson, J.M. (1971). Tactile direction sensitivity of area 2 oral neurons in the rhesus monkey cortex. *Brain Research*, **27**, 397–401.

Sessle, B.J. and Henry, J.L. (1989). Neural mechanisms of swallowing: neurophysiological and neurochemical studies on brain stem neurons in the solitary tract region. *Dysphagia*, **4**, 61–75.

Sessle, B.J., Martin, R.E., Murray, G.M., Masuda, Y., Kemppainen, P., Narita, N., Seo, K. and Raouf, R. (1995a). Cortical mechanisms controlling mastication and swallowing in the awake monkey. In *Brain and Oral Functions*, edited by T. Morimoto, T. Matsuya and K. Takada, pp. 181–189. Amsterdam: Elsevier Science B.V.

Sessle, B.J., Narita, N. and Martin, R.E. (1995b). Effects of reversible cold block of lateral pericentral cortex on swallowing in awake monkey. *Society for Neuroscience Abstracts*, **21**, 1900.

Sirisko, M.A. and Sessle, B.J. (1983). Corticobulbar projections and orofacial and muscle afferent inputs of neurons in primate sensorimotor cerebral cortex. *Experimental Neurology*, **82**, 716–720.

Smith, A. (1992). The control of orofacial movements in speech. *Critical Reviews in Oral Biology and Medicine*, **3**, 233–267.

Whitsel, B.L., Favorov, O.V., Tommerdahl, M., Diamond, M.E., Juliano, S.L. and Kelly, D.G. (1989). Dynamic processes governing the somatosensory cortical response to natural stimulation. In *Sensory Processing in the Mammalian Brain*, edited by J.S. Lund, pp. 84–116. New York: Oxford University Press.

Wiesendanger, M. (1981). The pyramidal tract: its structure and function. In *Handbook of Behavioral Neurobiology. Vol. 5. Motor Coordination*, edited by A.L. Towe and E.S. Luschei, pp. 401–491. New York: Plenum Press.

Woolsey, C.N. (1958). Organization of somatic sensory and motor areas of the cerebral cortex. In *Symposium on Interdisciplinary Research*, edited by H.F. Harlow and C.N. Woolsey, pp. 63–81. Madison: University of Wisconsin.

Yao, D.Y., Narita, N., Murray, G.M., Cairns, B.E. and Sessle, B.J. (1997). Effects of reversible cold block of primate face primary somatosensory cortex (SI) on face motor cortex (MI) chewing-related and peripherally evoked neuronal activity. *Society for Neuroscience Abstracts*, **23**, 2081.

Yao, D.Y., Narita, N. and Sessle, B.J. (1996). Effects of reversible cold block of primate face primary somatosensory cortex (SI) on performance of trained motor tasks and face motor cortex (MI) neurones. *Society for Neuroscience Abstracts*, **22**, 2024.

CHAPTER 8

MECHANISMS OF SOMATOSENSORY PLASTICITY

P.J. Snow[1] and P. Wilson[2]

[1]*Cerebral and Sensory Functions Unit, Department of Anatomical Sciences, University of Queensland, Q1 QLD, 4072 Australia*
[2]*Division of Human Biology, School of Biological Sciences, University of New England, Armidale, NSW, Australia*

As often used in relation to the central nervous system (CNS), the term plasticity refers to the malleability of neuronal connections. The degree to which the nervous system shows plasticity has been one of the seminal questions that has driven man's curiosity about the brain. On the one hand, it is clear that the CNS pathways in mammals do not fully recover from injury, suggesting that, once formed, the brain and spinal cord must be structurally inert. Yet it is obvious that our physical and mental abilities can change and that, at least to some degree, we can regulate these phenomena in relation to environmental conditions.

Following widespread acceptance of the neuron doctrine, the initial challenge to neurobiologists in this century was to map out the pathways within the central nervous system and thereby to draw, wherever possible, correlations between structure and function in the CNS. In order to put this strategy into practice it was necessary to focus attention on the *rigidity* of nervous organization — a viewpoint that naturally tends to suppress the possibility that neuronal connections may be in a state of constant flux. Today it is true to say that neither of these views of the CNS can be wholly supported but that, instead, some middle path seems to hold the greater truth. The somatic sensory system has provided an admirable model for studies of plasticity in the mature brain. In this article we will explore some of the mechanisms underlying the malleability of connections in the somatic sensory system and show some new data that indicate how the brain itself can exploit this unstable substrate. In this paper we describe experiments that have revealed connectional changes (plasticity) at the level of (a) the spinal cord, and (b) the cerebral cortex, and we outline some recent results that indicate the potential of major regulatory systems to alter functionally connectivity in the pathway between the skin and the seat of consciousness — the cerebral cortex.

(A) PLASTICITY IN THE SPINAL DORSAL HORN

Within the lumbosacral enlargement of the spinal cord in the anaesthetized cat the receptive fields (RFs) of neurons of laminae IV and V of the dorsal horn form a stable and stereotyped map of the distribution of tactile receptors across the skin of the hindlimb (Fig. 8.1, Wilson *et al.*, 1986). Transganglionic tracing studies (Koerber and Brown, 1982) and intra-axonal injection of horseradish peroxidase (HRP) into single, group II, cutaneous afferent fibers (Brown *et al.*, 1977) showed that afferents from proximal skin (the thigh) project to the lateral part of the dorsal horn and that afferents from successively more distal skin terminate

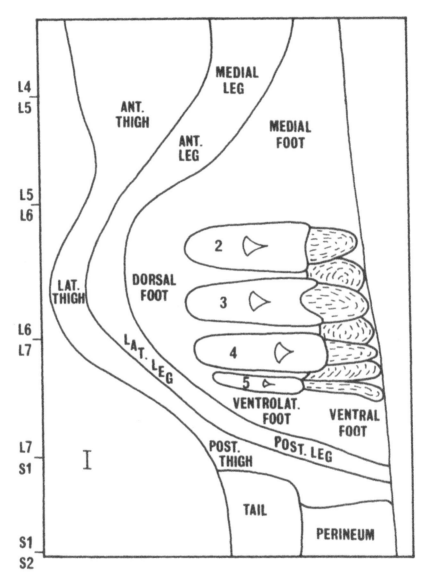

Figure 8.1 Somatotopic organization of the dorsal horn in the lumbosacral enlargement of the cat spinal cord. The illustration shows a map of the location of the centres of receptive fields of neurons in a horizontal section in the plane of lamina IV of the left dorsal horn. Rostral is to the top of the figure, and caudal to the bottom. Segmental boundaries are shown to the left, and the dorsal column is to the right. The representation of the thigh in the L6 and L7 segments has, for clarity, been shown lateral to the leg representation, though in reality it lies ventrolateral to the latter, due to the curvature of the lateral edge of lamina IV. Note the rostrocaudal sequence of the representations of the digits in segments L6–L7, with the representations of digits 2–4 each occupying approximately 3 mm in the rostrocaudal axis. Scale bar represents 1 mm in the rostrocaudal axis, and approximately 0.2 mm in the mediolateral axis. (Redrawn from Wilson *et al.*, 1986).

progressively more medially. Thus, across the medio-lateral axis of the dorsal horn there is a tight correspondence between the termination of primary afferent fibers and the somatotopic organization of dorsal horn neurons (DHNs).

Attempts to demonstrate plasticity of this somatotopic map have utilized acute and chronic transection of peripheral nerves to determine whether deafferentation of specific parts of the somatotopic map results in a reorganization of the RFs of DHNs. These studies have shown that following nerve section there is a reorganization of the RFs of DHNs such that cells that have been deprived of their normal input develop new RFs on adjacent areas of skin that have intact innervation. Early studies, involving massive denervation of the entire lower part of the limb, by transection of the sciatic nerve, indicated that neurons in the medial part of the dorsal horn could develop new RFs on proximal skin (Devor and Wall, 1978). In contrast, our own studies (Wilson, 1987), as well as those of others (Brown *et al.*, 1983) were unable to confirm such an effect of widespread deafferentation. In contrast, we found that a more limited denervation, of only one or two digits, could lead to the development of new RFs on the adjacent digits, represented rostrally or caudally to the deprived region of the dorsal horn. In chloralose anaesthetized cats, this reorganization is not observed immediately following nerve transection, but requires weeks or months to become fully apparent (Wilson and Snow, 1987, 1991).

Meyers and Snow (1984) had earlier used intra-axonal injection of HRP to demonstrate that HFAs give rise to collaterals in regions of dorsal horn immediately rostral and caudal to the region where they strongly excite DHNs. However, these collaterals appeared to make very few or no synapses, indicating that they might be normally functionless, yet constitute a pool of potentially functional projections in the event of partial deafferentation of the dorsal horn. Clearly, it was important to determine whether the rostro-caudal reorganization was the result of sprouting of such collaterals. Our approach (Wilson and Snow, 1993) to this question was to attempt to intra-axonally stain single HFAs in cats in which the reorganization of digit somatotopy had been demonstrated. Owing to the limited rostro-caudal distance over which HRP completely stains the collaterals of injected primary afferent fibers (approximately 2 mm either side of the point of impalement), we could not rely on HRP injection at random sites along primary afferents to reveal such collaterals, and so we sought to impale, in the dorsal funiculus overlying the center of the reorganized region, afferents with RFs on the adjacent digits. These technically difficult experiments revealed that collaterals of such afferents indeed gave rise to significant numbers of synaptic boutons in the reorganized region, suggesting that there had been sprouting and synaptogenesis of those collaterals of intact HFAs that innervated digits 2 and 5 — skin that is normally represented only caudally and rostrally to the denervated digits 4 and 3, respectively (Fig. 8.2).

Using intra-axonal injection of the tracer Neurobiotin, which more completely fills axons and their branches than HRP, we have since demonstrated what we believe to be an anatomical substrate for the rostro-caudal pattern of somatotopic reorganization in response to a limited denervation of the digits (Wilson *et al.*, 1996). Thus in normal cats, a single HFA, with an RF confined to the skin covering a small part of one digit, was found to project collaterals to a region of the medial part of the dorsal horn approximately 20 mm in length, containing the representation of not only all the digits, but also parts of the foot (Figs. 8.3 and 4). Single primary afferent fibres with RFs on the digits therefore project their collaterals to a strip of dorsal horn that is at least 10 times as extensive as that region

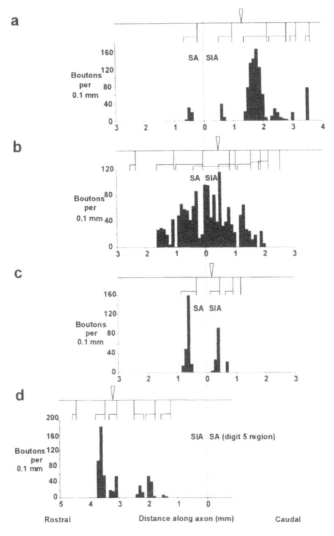

Figure 8.2 Direct morphological evidence for collateral sprouting of group II hair follicle afferent fibres (HFAs) in the dorsal horn of the spinal cord in the cat. This animal had undergone transection and ligation of the digital nerves of digits 3 and 4 on one side approximately one year earlier, and mapping the cutaneous responses of spinocervical tract neurons in the deafferented region of dorsal horn revealed that many had acquired new receptive fields on digit 2 and/or digit 5. Hair follicle afferents with receptive fields on the adjacent digits were sought and impaled with horseradish peroxidase (HRP)-filled micropipettes as near as possible to the centre of the deafferented region. Histograms show the rostrocaudal distribution of the arborizations of HRP-filled collaterals and their synaptic boutons through the chronically deprived digit 3 and digit 4 representations for three HFAs with RFs on the tip of digit 2 **(a–c)**, and a single HFA **(d)** with a RF on the tip of digit 5. All histograms share the same rostrocaudal axis and are aligned on the boundary between the digit 2 and digit 3 representations (vertical dashed line). Arrowheads indicate the point of HRP injection. SA-somatotopically appropriate region, SIA-somatotopically inappropriate region of the dorsal horn, with respect to the receptive field of the afferent. (Redrawn from Wilson and Snow, 1993).

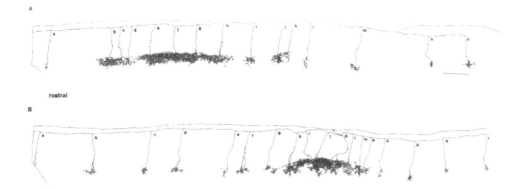

Figure 8.3 Drawings of the terminal arborizations of two cat group II hair follicle afferent fibers (**A** and **B**), superimposed on outlines of parasagittal sections of the spinal cord. The receptive fields and bouton distributions of these afferents are shown in Figure 8.4. Afferents were reconstructed from camera lucida drawings of 100 μm-thick parasagittal sections of spinal cord containing single afferents injected intraaxonally with Neurobiotin. A and B were injected on different sides of the cord in the same animal. Scale bar =1 mm. Note the large number, and extensive rostrocaudal spread (up to approximately 17 mm) of the collateral arborizations (small letters) of each afferent. Note also the relatively simple branching pattern of the collateral arborizations emitted rostral and caudal to the relatively short complex region. The caudal main axon branch of both afferents continued caudally beyond the point where the staining faded. The afferent in (A) had an RF on the dorsum of digit 3 (see Fig. 8.4A), while that in (B) had an RF on digit 5 (see Fig. 8.4B). Note that the most complex parts of the arborizations are offset rostrocaudally by an amount that corresponds to the distance between the digit 3 and digit 5 representations (6 mm, see Fig. 8.1). (Redrawn from Wilson *et al.*, 1995).

in which they strongly excite DHNs. Similar results of Neurobiotin injection have been obtained for a variety of different types of cutaneous primary afferent fibres in both the cat (Koerber and Mirnics, 1995) and the rat (Wilson and Miller, 1994).

Intracellular recording from SCT cells with small RFs on the digits confirmed the presence of previously undetected subthreshold excitatory postsynaptic potentials (EPSPs) ranging in amplitude down to approximately 100 μV, which could be evoked monosynaptically by focal electrical stimulation of the skin as far as two digits distant from the firing zone of the SCT cell's RF (Wilson *et al.*, 1994; see Fig. 8.5). Such EPSPs may be evoked by neurotransmitter secretion at a single release site, or synaptic bouton. Thus subliminal monosynaptic excitatory inputs to SCT cells, and most likely other DHNs, arises from a much wider region of the skin than previously thought (cf. Brown *et al.*, 1987). In the event of peripheral denervation, plasticity of these small inputs may provide stronger excitatory drive to DHNs. The reported blockade of somatotopic reorganization in the rat dorsal horn by chronic infusion of the specific NMDA receptor antagonist AP5 (Lewin *et al.*, 1994) suggests a role for NMDA receptors of dorsal horn neurons in this plasticity, but the prolonged time course suggests that anatomical changes, such as axonal sprouting and synaptogenesis, also might be involved.

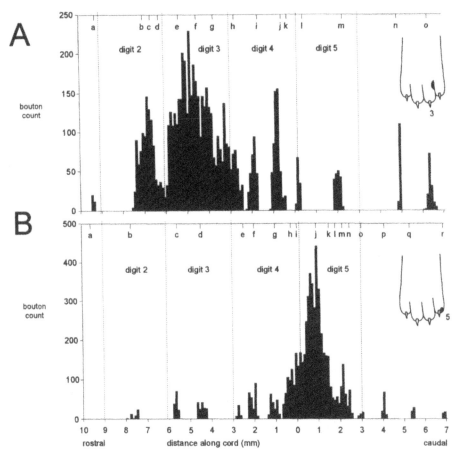

Figure 8.4 Rostrocaudal distribution of stained main axon and collateral branches (small letters at top), and synaptic boutons (100 μm bins), for the two Neurobiotin-injected cat hair follicle afferents (**A** and **B**) whose morphology is illustrated in Figure 8.3. The receptive fields of the afferents are shown in the illustrations at right. The boundaries of the digit 3 and digit 4 somatotopic representations (dotted lines) were determined by mapping the receptive fields of spinocervical tract neurons. Note that collaterals arising over many millimeters of the dorsal horn can give rise to large numbers of synaptic boutons, and that *all* collaterals give rise to at least a few boutons (in B the terminal portion of collateral *a* was not recovered). Note also that both afferents, though possessing small RFs on part of a single digit, provide collaterals with boutons to the dorsal horn throughout the entire rostrocaudal extent of the digit representation. (Wilson *et al.*, 1994).

In conclusion, we suggest that:

(a) The tactile afferent projections to the dorsal horn are much greater in their rostrocaudal distribution than in their representation in the responses of dorsal horn neurons.

(b) Following nerve section, the terminations of collaterals of intact afferents in the region of dorsal horn deprived of normal afferent input undergo an enhancement of synaptic efficacy, which may initially involve NMDA receptor activation and later possible sprouting and synaptogensis, thereby increasing their excitatory effect on DHNs.

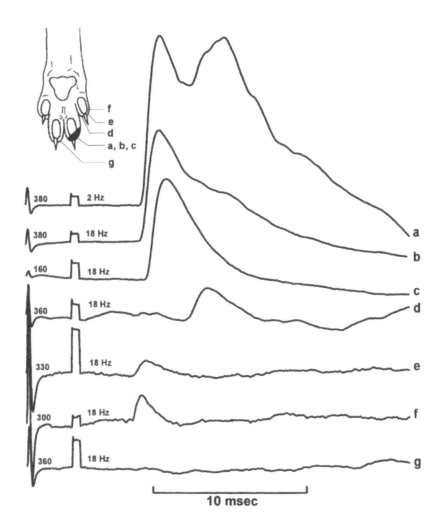

Figure 8.5 Synaptic potentials **(a–e)** recorded intracellularly from a spinocervical tract neuron in the cat dorsal horn following electrical stimulation (100 μsec), via a tungsten microelectrode, of different points on the skin, both within and outside the region from which impulses were evoked by brushing hairs (the receptive field). Each trace is the averaged response to 100 consecutive stimuli, applied at the frequency indicated at left. The stimulus occurs at the start of each trace (note stimulus artifact). The receptive field of the cell, at the tip of digit 3, and the stimulated points on the skin are indicated in the paw illustration. The number at the left of each trace represents the stimulus current in μA. **(a)** The EPSP obtained by stimulating at the centre of the receptive field at a slow stimulus rate of 2 Hz, contains two major components. **(b)** Increasing the stimulus frequency to 18 Hz causes the later, di- and poly-synaptically evoked component to fail, while the early, fast, mono-synaptic EPSP is relatively unaffected. **(d–f)** Stimulation at various points on digit 2 at a frequency of 18 Hz evokes small monosynaptic EPSPs with fast rise times and different, fixed latencies. Note that the EPSPs in **(e)** and **(f)** have peak amplitudes of approximately 100 μV, while that in **(b)** has a peak amplitude of about 3000 μV. No response could be evoked from the tip of digit 4 **(g)**. Calibration pulses: 200 μV (a, b, c, d, e, g) and 40 μV (f). (Wilson *et al.*, 1994).

(B) PLASTICITY IN THE SOMATOSENSORY (SI) CEREBRAL CORTEX

The primary somatosensory (SI) cortex is composed of cytoarchitectonic areas 3a, 3b, 1 and 2. Within each of these areas there is a complete topographic representation of the body, but neurons within each of them respond to different submodalities of somatic inputs. Thus area 3a is activated primarily by muscle spindle afferents while in area 3b cells respond to simple cutaneous stimuli. In area 1 neurons respond best to complex (e.g. moving) cutaneous stimuli and in area 2 neurons respond mainly to deep (joint) input (reviewed by Snow and Wilson, 1991).

As discussed above, our studies on the dorsal horn of the spinal cord suggested that plasticity in the somatosensory map at that level may result from changes in efficacy of pre-existing primary afferent projections to the dorsal horn. Within the SI cortex, studies in both animals (reviewed by Snow and Wilson, 1991) and humans (Borsook *et al.*, 1998) have indicated that the plasticity of the somatosensory representation is also the result of changes in efficacy of pre-existing widespread afferent projections.

The expression of a localized cutaneous stimulus in the responses of a localized set of SI neurons is therefore the result of a dynamic process — a process that determines the anatomical identity of the area stimulated and the modality of the stimulus (reviewed by Snow and Wilson, 1991; Ramachandran *et al.*, 1992). We might then ask what brain mechanisms can operate upon this dynamic equilibrium in normal individuals to modify the awareness of specific areas of the body.

Since the comparatively recent discovery (reviewed by Saper, 1987) of the diffuse extrathalamic cortical regulatory systems, a great deal of work has been done to elucidate the role of these systems in overall brain function. The diffuse extrathalamic cortical regulatory systems involve the cortical projections of the ventral tegmental system (dopamine), the raphe nuclei (serotonin), locus coeruleus (noradrenalin) and the pontine and basal forebrain systems (acetylcholine). The precise function of these systems is still controversial, but the noradrenergic locus coeruleus (LC) seems to be gaining favor as a system that controls vigilance, or attention (Aston-Jones *et al.*, 1991).

Activation of the LC releases noradrenalin at sites throughout all of the cerebral cortex and most of the rest of the brain. Although the presence of synaptic specializations in immunocytochemical studies of cortical noradrenergic axonal varicosities has been reported (Olschowka *et al.*, 1981), the relative incidence of these, compared to nonsynaptic varicosities, is controversial, and it is generally considered that the neurotransmitter released from LC axons in the cerebral cortex acts as a neurohumor, diffusing throughout the local neuropil (Saper, 1987). The effects of noradrenaline have been studied extensively in brain slices, where the neurotransmitter acting at α-adrenergic receptors has been shown to produce long-lasting enhancement of excitability and responsiveness to depolarization in neocortical neurons, by suppression of spike-rate adaptation (reviewed by McCormick *et al.*, 1993).

There have been few controlled experiments on the effect of noradrenalin on the responses of cortical neurons to sensory input, although some neurons in SI cortex have been reported to respond to noradrenalin by a reduction in the level of spontaneous discharge coupled with an increase in their response to cutaneous stimulation (Waterhouse *et al.*, 1981).

In an attempt to extend these studies we recorded the responses of single SI neurons in the urethane-anaesthetized rat to brief, tactile cutaneous stimuli generated by an electronically controlled air-jet (Snow *et al.*, 1993, 1994; see Fig. 8.6). Neurons showed a

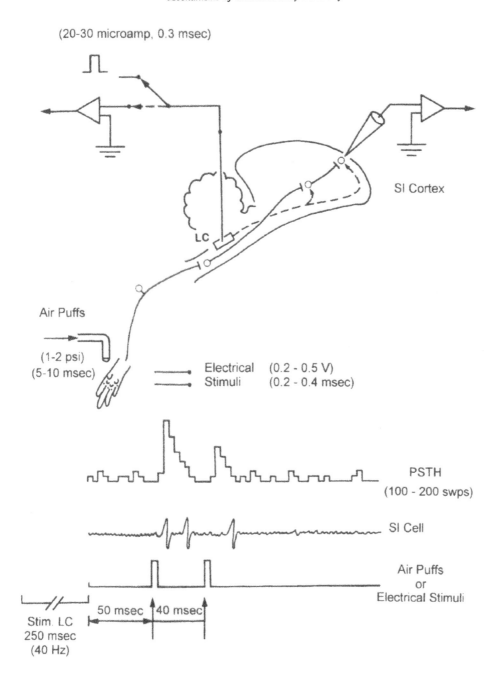

Figure 8.6 Effect of repetitive electrical stimulation of the noradrenergic locus coeruleus on the response of SI cortical neurons in the urethane-anaesthetized rat to natural stimulation of the contralateral forepaw. Diagram of the experimental design showing the experimental setup (top) and the stimulation and data gathering procedure (bottom). See text for further explanation. LC=locus coeruleus; PSTH=poststimulus-time-histogram. (Snow *et al.*, 1994).

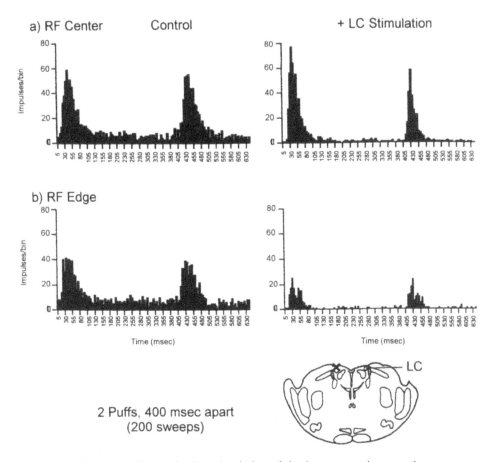

Figure 8.7 Different effects of microstimulation of the locus coeruleus on the responses of a SI neuron to air puffs applied at the centre (**a**) and the edge (**b**) of the receptive field. Data is presented as PSTHs of the responses of the SI unit to 200, 1/second presentations of 2 air puffs applied at an inter-puff interval of 400 msec. (**a**) Facilitation and temporal focusing of the responses evoked by puffs applied to the receptive field centre (left PSTH/ Control), when each presentation of the puffs was preceded by application of a train of 20mA pulses to the locus coeruleus (see inset and Fig. 8.5) (right PSTH/+LC Stimulation). (**b**) Inhibition of the response evoked by puffs applied to the edge of the receptive field (left PSTH/Control), when each presentation of the puffs was preceded by application of a train of 20mA pulses to the locus coeruleus (see inset and Fig. 8.5) (right PSTH/+LC Stimulation). Note the suppression of spontaneous activity following stimulation of the locus coeruleus (right PSTHs/+ LC Stimulation). (Snow *et al.*, 1994).

variety of responses to electrical stimulation of the LC, an observation borne out by the studies of Warren and Dykes (1996) on the responses of SI neurons in the cat to ionto-phoretic application of noradrenalin. In our studies we initially plotted the RF of the cortical neurons before studying the effects of LC stimulation on the application of tactile stimuli at various sites across the RF. It soon became clear that in some neurons electrical stimulation of the LC could differentially affect input from various sites across the RF.

This observation underscores the importance of subliminal synaptic inputs to cortical neurons from areas of skin that exceed their RF.

Of particular interest, however, were a group of neurons that decreased their level of spontaneous firing during LC conditioning. In some of these, LC stimulation seemed to focus the generation of action potentials in the cortical neuron to a very restricted period of time following the application of the tactile stimulus (Snow *et al.*, 1993, 1994) — a phenomenon that we have called *temporal focusing* (Fig. 8.7). In these neurons we have the potential for synchronizing the responses of a specific subset of neurons within a single cortical column. When this is approached, the potential for that module to successfully activate other modules with which it is associated, and thus to generate synchronous firing of neurons across wider regions of the cortex as may be necessary for perceptual binding (Singer, 1993), is greatly increased.

Thus in the effects of the LC we begin to see an example of how the brain itself can operate on regions of neuropil where the number of effective connections is considerably less than the total number. It is upon this substrate that the brain's own systems can operate. It is through this substrate that the brain can control its own state, manipulate selective attention to input and respond to injury. These are all epiphenomena of the neural circuitry of the brain revealed to us by unusual manipulations and operations. They enable us to see the brain and some aspects of consciousness as a self regulating, yet plastic system that works at any time within strict anatomical constraints.

REFERENCES

Aston-Jones, G., Chiang, C. and Alexinsky, T. (1991). Discharge of noradrenergic locus coeruleus neurons in behaving rats and monkeys suggests a role in vigilance. *Progress in Brain Research*, **88**, 501–520.

Borsook, D., Becerra, L., Fishman, S., Edwards, A., Jennings, C. L., Stojanovic, M., Papinicolas, L., Ramachandran, V. S., Gonzalez, R. G. and Breiter, H. (1998). Acute plasticity in the human somatosensory cortex following amputation. *NeuroReport*, **9**, 1013–1017.

Brown, A.G., Rose, P.K. and Snow, P.J. (1977). The morphology of hair follicle afferent fibre collaterals in the spinal cord of the cat. *Journal of Physiology (London)*, **272**, 779–797.

Brown, A.G., Brown, P.B., Fyffe, R.E.W. and Pubols, L.M. (1983). Effects of dorsal root section on spinocervical tract neurones in the cat. *Journal of Physiology (London)*, **337**, 589–608.

Brown, A.G., Koerber, H.R. and Noble, R. (1987). An intracellular study of spinocervical tract cell responses to natural stimuli and single hair afferent fibres in cats. *Journal of Physiology (London)*, **382**, 331–354.

Devor, M. and Wall, P.D. (1978). Reorganisation of spinal cord sensory map after peripheral nerve injury. *Nature*, **276**, 75–76.

Koerber, H.R. and Brown, P.B. (1982). Somatotopic organization of hindlimb cutaneous nerve projections to cat dorsal horn. *Journal of Neurophysiology*, **48**, 481–489.

Koerber, H.R. and Mirnics, K. (1995). Morphology of functional long-ranging primary afferent projections in the cat spinal cord. *Journal of Neurophysiology*, **74**, 2336–2348.

Lewin, G.R., Mckintosh, E. and McMahon, S.B. (1994). NMDA receptors and activity–dependent tuning of the receptive fields of spinal cord neurons. *Nature*, **369**, 482–485.

McCormick, D.A., Wang, Z. and Huguenard, J. (1993). Neurotransmitter control of neocortical neuronal activity and excitability. *Cerebral Cortex*, **3**, 387–398.

Meyers, D.E.R. and Snow, P.J. (1984). Somatotopically inappropriate projections of single hair follicle afferent fibres to the cat spinal cord. *Journal of Physiology (London)*, **347**, 59–73.

Olschowka, J.A., Molliver, M.E., Grzanna, R., Rice, F.L. and Coyle, J.T. (1981). Ultrastructural demonstration of noradrenergic synapses in the rat central nervous system by dopamine–hydroxylase immunocytochemistry. *Journal of Histochemistry and Cytochemistry*, **29**, 271–280.

Ramachandran, V.S., Stewart, M. and Rogers-Ramachandran, D.C. (1992). Perceptual correlates of massive cortical reorganization. *Neuroreport*, **3**, 583–586.

Saper, C.B. (1987). Diffuse cortical projection systems: anatomical organization and role in cortical function. In *Handbook of Physiology, Section 1, The Nervous System, Vol. V, Higher Functions of the Brain, Part 2,* edited by V.B. Mountcastle, F. Plum, and S.R. Geiger, pp. 169–209. Bethesda: American Physiological Society.

Singer, W. (1993). Synchronization of cortical activity and its putative role in information processing and learning. *Annual Reviews of Physiology,* **55,** 349–374.

Snow, P.J. and Wilson, P. (1991). Plasticity in the Somatosensory System of Mature and Developing Mammals — The Effects of Injury to the Central and Peripheral Nervous System. *Progress in Sensory Physiology, Volume 11.* Heidelberg: Springer-Verlag.

Snow, P.J., Andre, P. and Pompieano, O. (1993). Responses of SI cortical neurons to natural stimulation during microstimulation in the locus coeruleus region. *International Journal of Neuroscience,* **80,** 104.

Snow, P.J., Andre, P. and Pompeiano, O. (1994). Microstimulation of the locus coeruleus and the responses of SI neurons. *Proceedings of the Australian Neuroscience Society,* **5,** 67.

Warren, R.A and Dykes, R.W. (1996). Transient and long-lasting effects of iontophoretically administered norepinephrine on somatosensory cortical neurons in halothane-anesthetized cats. *Canadian Journal of Physiology and Pharmacology,* **74,** 38–57.

Waterhouse, B.D., Moises, H.C. and Woodward, D.J. (1981). Alpha-receptor-mediated facilitation of somatosensory cortical neuronal responses to excitatory synaptic inputs and iontophoretically applied acetylcholine. *Neuropharmacology,* **20,** 907–920.

Wilson, P. (1987). Absence of mediolateral reorganization of dorsal horn somatotopy after peripehral deafferentation in the cat. *Experimental Neurology,* **95,** 432–447.

Wilson, P. and Miller, M.J. (1994). Spinal arborizations of functionally identified group II cutaneous primary afferent fibres in the rat revealed by intra-axonal injection of Neurobiotin. *Proceedings of the Australian Neuroscience Society,* **5,** 194.

Wilson, P. and Snow, P.J. (1987). Reorganization of the receptive fields of spinocervical tract neurons following denervation of a single digit in the cat. *Journal of Neurophysiology,* **57,** 803–818.

Wilson, P. and Snow, P.J. (1991). The effects of neonatal deafferentation on the somatotopic organization of lumbar spinocervical tract cells in the cat. *Journal of Neurophysiology,* **66,** 762–776.

Wilson, P. and Snow, P.J. (1993). Morphology of Aβ hair folicle afferent collaterals in dorsal horn of cats with neonatal chronic denervation of digits. *Journal of Neurophysiology,* **70,** 2399–2410.

Wilson, P., Meyers, D.E.R. and Snow, P.J. (1986). The detailed somatotopic organization of the dorsal horn in the lumbosacral enlargement of the cat spinal cord. *Journal of Neurophysiology,* **55,** 604–617.

Wilson, P., Kitchener, P.D. and Snow, P.J. (1994). Long-range projections from group II hair follicle afferent fibres to spinocervical tract neurones in the cat lumbar dorsal horn. *Proceedings of the Australian Neuroscience Society,* **5,** 64.

Wilson, P., Kitchener, P.D. and Snow, P.J. (1996). Intraaxonal injection of Neurobiotin reveals the long-range projections of Aβ hair follicle afferent fibers to the cat dorsal horn. *Journal of Neurophysiology,* **76,** 242–254.

CHAPTER 9

ADAPTIVE PROPERTIES OF LOCAL CIRCUITS REVEALED BY PERIPHERAL DENERVATION

Harris D. Schwark

Department of Biological Sciences, University of North Texas, Denton, Texas, USA

The somatosensory system is an adaptive system: perturbations in signals from peripheral receptors can lead to adaptive changes in the response characteristics of central somatosensory neurons. A dramatic example of this is the reorganization of cortical body representations following amputation or denervation of digits in primates (e.g. Kaas, 1991; Buonomano and Merzenich, 1998). Similar map reorganizations occur in subcortical body representations (Garraghty and Kaas, 1991; Pettit and Schwark, 1993; Florence and Kaas, 1995; Rasmusson, 1996). At the level of single cells, these map reorganizations involve changes in receptive field size and/or location (hereafter referred to as receptive field reorganization). When primary afferents are temporarily inactivated by subcutaneous lidocaine injection, receptive fields of neurons in the central nervous system reorganize within minutes, suggesting that these changes are the result of changes in existing connections rather than from the formation of new ones (Calford and Tweedale, 1991a; Pettit and Schwark, 1993; Panetsos *et al.*, 1995). Such reorganization of receptive fields should result in changes in somatosensory acuity, and may reflect a mechanism by which the somatosensory system adapts to the requirements of a particular sensory environment. Experimental evidence consistent with such a role can be found in changes in body representations following tactile discrimination training in monkeys (e.g. Recanzone *et al.*, 1992) and nursing behavior in rats (Xerri *et al.*, 1994).

The neural mechanisms that underlie adaptive changes in receptive fields are unknown. Current evidence suggests that a reduction in afferent input results in a decrease in levels of tonic inhibition, which unmasks previously ineffective synapses. This chapter describes research directed towards understanding the mechanisms that underlie receptive field reorganization in dorsal column nuclei (DCN) neurons following temporary inactivation of primary afferents. Although receptive field reorganization has been described at all levels of the somatosensory system (Nakahama *et al.*, 1966; Calford and Tweedale, 1991a; Pettit and Schwark, 1993; Panetsos *et al.*, 1995), the DCN are an attractive site to study because they receive direct input from primary afferents, and contain the first synapses in the lemniscal mechanoreceptive pathway. Reorganization of DCN receptive fields may therefore reflect processes that take place within the nuclei, rather than reorganization that arises at lower levels and is imposed upon neurons upstream.

Figure 9.1 illustrates the effects of small subcutaneous injections of lidocaine into the receptive fields of cat DCN neurons. These injections resulted in an immediate cessation of stimulus-driven activity from the injection site. Within minutes, the receptive fields expanded so that areas of the skin that were normally ineffective in eliciting action

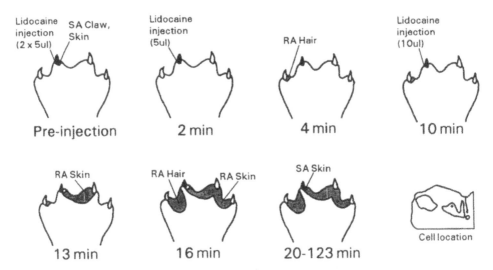

Figure 9.1 Receptive field reorganization following lidocaine injection. Four minutes after the initial lidocaine injections, a new receptive field appeared. The new field had response characteristics different from those of the original field. After an additional lidocaine injection, the new receptive fields expanded to cover parts of 4 digits. The new receptive fields remained throughout the recording session (123 minutes).

potentials now became effective (Pettit and Schwark, 1993). In many neurons the response characteristics (i.e. modality and adaptation characteristics) of the expanded regions were different than those of the original receptive field, suggesting that previously silent synapses were becoming effective. These tests were done only on neurons that responded to light tactile stimulation, and every neuron tested underwent receptive field reorganization. Similar results have since been described in the DCN of the rat (Panetsos *et al.*, 1995; Panetsos *et al.*, 1997). Because such receptive field reorganization can apparently arise independently in somatosensory cortex (Clarey *et al.*, 1996), and because the somatosensory cortex sends direct projections to the DCN, we tested the possibility that receptive field reorganization in the DCN depended upon cortical inputs. Even after ablation of somatosensory and motor cortex, lidocaine injections resulted in receptive field reorganization in every DCN neuron tested.

Although lidocaine injection apparently silences all primary afferents at the site of injection, complete blockade is not required to induce receptive field reorganization. Calford has produced receptive field reorganization in somatosensory cortex by bathing primary afferent nerves in capsaicin to block C fibers (Calford and Tweedale, 1991b), and we found similar results in the DCN (Pettit and Schwark, 1996). Small subcutaneous injections of capsaicin resulted in receptive field reorganization in every DCN neuron tested, even though these injections often had no obvious effect on stimulus-driven activity elicited from the original receptive field (Fig. 9.2). These capsaicin injections probably inactivated C fibers as well as some thinly myelinated afferents (Baumann *et al.*, 1991; LaMotte *et al.*, 1992). It is unlikely that the receptive field reorganization was due to activation of C fibers by capsaicin (as occurs in the dorsal horn, e.g. Cook *et al.*, 1987),

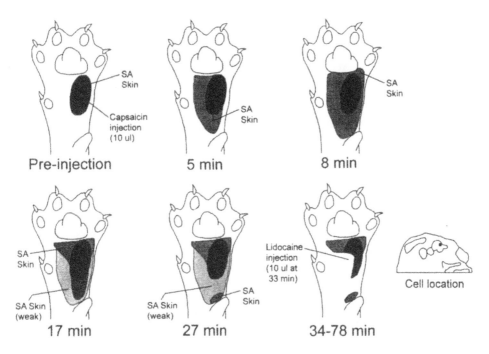

Figure 9.2 Receptive field reorganization following capsaicin injection. A single capsaicin injection in the center of the receptive field resulted in an expansion of the receptive field within 5 min. The response characteristics of the newly responsive areas were the same as the original field. The sensitivity of the neuron to stimulation in the expanded receptive field varied with time, which was also observed in some other neurons. Thirty-three minutes after the capsaicin injection, an injection of lidocaine was made in the same location. Responsiveness from a portion of the receptive field disappeared, but no further expansion was observed.

because such activation occurs immediately, yet the reorganized receptive field appeared only after a delay of several minutes, when capsaicin causes inactivation of some afferents. These results suggest that loss of tonic activity in a subset of primary afferents can result in receptive field reorganization in the DCN.

Because many C fibers release neuropeptides such as substance P, Calford has suggested that these neuropeptides might play a role in restricting receptive field size (Calford and Tweedale, 1991b). The slow inactivation of neuropeptide effects that has been reported in the DCN (Krnjevic and Morris, 1974) is compatible with the delay in the onset of receptive field reorganization. Although approximately half of the neurons in cat DCN respond to iontophoretically-applied substance P (Krnjevic and Morris, 1974), substance P-containing afferents terminating in the DCN are relatively sparse (Conti *et al.*, 1990). To identify the potential sites of action of substance P in the DCN, we studied the distribution of substance P binding (Schwark *et al.*, 1998). The highest density of substance P binding was located over the cell nests, which are in the middle portion of each nucleus. These cell nests contain neurons that respond to light tactile stimulation and project to the

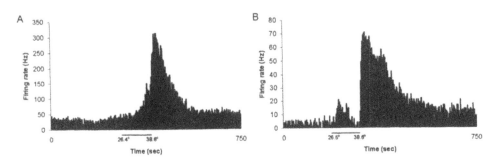

Figure 9.3   The response of two DCN neurons to skin warming. Both neurons had tactile receptive fields on the forepaw. **A:** Warming the receptive field of this neuron to 38.6°C resulted in a rapid, approximately 6-fold increase in spontaneous firing rate. **B:** The spontaneous firing rate of this neuron increased somewhat during warming. However, when the warming stimulus was removed, the neuron underwent a rapid, approximately 6-fold increase in spontaneous firing rate.

thalamus (Cheema *et al.*, 1983). Similar patterns of binding were observed in the DCNs of rat, cat, monkey and human. Thus, an anatomical substrate for substance P action exists in the DCN, but the specific primary afferents involved have not yet been identified.

The best candidates for tonically active fibers that are susceptible to capsaicin block are the slowly adapting type II mechanoreceptors (Burgess *et al.*, 1968; Amassian and Giblin, 1974) and warm thermoreceptors (Stolwijk and Wexler, 1971; Handwerker and Neher, 1976). To test the possibility that warm thermoreceptors project to the DCN, we measured the effects of skin warming on the activity of cat DCN neurons (Schwark *et al.*, 1997). Stimulation was done by radiant heating and cooling to eliminate confounds of tactile stimulation. In a sample of 29 DCN neurons with tactile receptive fields, 22 responded to warming or cooling. Examples of these responses are shown in Figure 9.3. Thus, temperature-sensitive primary afferents influence DCN neuron activity, and could play a role in setting up tonic inhibition within the DCN that restricts receptive field size. Dykes and Craig (Dykes and Craig, 1998) failed to find changes in DCN receptive fields with changes in skin temperature, although they induced temperature changes with contact stimuli, which may have interacted with mechanoreceptors.

Recently, it was shown that inactivation of the dorsal horn of the spinal cord also results in receptive field reorganization in cat DCN (Dykes and Craig, 1998). Primary afferent inputs to the DCN were presumably unaffected because some of the inactivations were produced by injection of cobalt chloride (Malpeli *et al.*, 1986; Dykes and Craig, 1998). Thus, it appears that DCN receptive field size can be affected by afferent input that synapses in the dorsal horn and enters the DCN, perhaps through the postsynaptic dorsal column pathway. These results raise the interesting possibility that tactile responses may be modified by activity in the dorsal horn, perhaps arising from noxious stimulation. The possibility that tactile information might be gated by dorsal horn activity has been proposed by Apkarian and colleagues (Apkarian *et al.*, 1994).

The experimental results described above suggest that loss of active inputs to the DCN results in receptive field reorganization. However, it is important to bear in mind that the experiments have revealed pathways that, when inactivated, are *sufficient,* but not

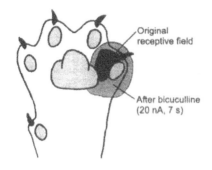

Figure 9.4 An example of receptive field enlargement in a DCN neuron during iontophoresis
of the GABA$_A$ receptor blocker bicuculline
methiodide. Response characteristics.

necessary to induce receptive field reorganization. Indeed, all the data presented so far are
compatible with the idea that a strong enough reduction in afferent activity in any pathway
can induce receptive field reorganization. It is probable, however, that there are limits to
the extent of inactivation that will produce these results. No receptive field reorganization
in cat DCN was found when the entire median nerve was inactivated by cooling (Zhang
and Rowe, 1997). This result underscores the complicated nature of the mechanisms that
underlie receptive field reorganization.

Although experimental data suggest that somatosensory receptive fields are restricted
by local, tonic inhibition, there is little direct evidence for an inhibitory influence on
receptive field structure in the DCN. In somatosensory cortex, blockade of GABA$_A$
receptors results in receptive field enlargement in approximately 85% of rapidly adapting
neurons and 35% of slowing adapting neurons (Dykes *et al.*, 1984; Alloway *et al.*, 1989).
In thalamus, however, blockade of GABA$_A$ receptors results in receptive field enlargement
in only 19% of rapidly adapting neurons and 11% of slowly adapting neurons (Hicks
et al., 1986). Thus, GABAergic inhibitory influences on receptive fields at subcortical
levels may be weaker than in the cortex. To test this idea, we studied receptive fields of
DCN neurons during blockade of GABA$_A$ or GABA$_B$ receptors. Such blockade resulted
in receptive field enlargement in most DCN neurons (Figs. 9.4, 5). During GABA$_A$ receptor
blockade 80% of the neurons underwent receptive field enlargement (79% of the rapidly
adapting neurons and 86% of the slowly adapting neurons). Blockade of GABA$_B$ receptors
resulted in receptive field enlargement in 58% of the neurons. These results are consistent
with the idea that tonic, GABAergic inhibition plays a role in restricting receptive field
size in the DCN.

Additional experimental evidence for a role of tonic inhibition in restricting receptive
field size has been obtained from peripheral inactivation studies in the rat DCN. Lidocaine
injections resulted in decreased activity in DCN neurons that could not be antidromically
activated from the thalamus (Panetsos *et al.*, 1997). If some of these neurons were inhibitory neurons, this may be direct evidence for a loss in tonic inhibition. Also, when cat dorsal
horn neurons are inactivated by cobalt chloride injections, which spare fibers of passage,
the responsiveness of DCN neurons increases (Dykes and Craig, 1998).

A second mechanism that could be involved in restricting receptive field size is shunting
of dendritic potentials by open channels at active synapses. Modeling experiments suggest
that the levels of spontaneous activity can have a significant effect on a neuron's cable
properties (Bernander *et al.*, 1991). High levels of spontaneous activity effectively shorten
the dendritic length constant due to the large number of open channels at active synapses.

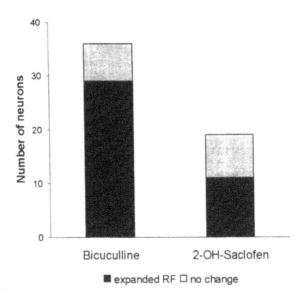

Figure 9.5 The frequency of receptive field enlargement in DCN neurons during blockade of GABA_A receptors with bicuculline methiodide or GABA_B receptors with 2-OH-saclofen.

As a result, synapses located on distal dendrites are ineffective. As spontaneous activity levels decline, fewer channels are open and the dendritic length constant increases, and distal synapses become more effective. To determine if this mechanism might be acting in the DCN during lidocaine injections, we analyzed the changes in spontaneous activity following lidocaine or capsiaicin injection. We found no consistent pattern of change that corresponded with the onset of receptive field reorganization. Similar results have been obtained for thalamic projecting neurons in the rat DCN (Panetsos *et al.*, 1997).

Which synaptic connections are masked by tonic inhibition in the DCN are not known. At least two types of connections may be involved: synapses between primary afferents and DCN neurons and synapses between DCN neurons. In the somatosensory thalamo-cortical system the anatomical extent of individual afferents exceeds the range predicted by receptive field boundaries (Snow *et al.*, 1988), and it is likely that this is true at other levels (e.g. Fyffe *et al.*, 1986; Hirai *et al.*, 1988). Thus receptive field enlargement might involve increased effectiveness of synaptic input from these afferents. Alternatively, receptive field enlargement might arise from changes in intrinsic, lateral connections (e.g. Dinse *et al.*, 1993). In any case, it seems likely that receptive field reorganization results from an unmasking of normally ineffective synapses, as has been proposed previously (Wall, 1977). To begin to test the involvement of these synaptic connections, we have examined patterns of correlated firing between pairs of neurons during receptive field reorganization. Some preliminary results are illustrated in Figure 9.6. Cross correlograms revealed a sharp peak at a lag time of approximately 1 ms, suggestive of a monosynaptic connection between the neurons. A second, broader peak at 4 ms was due to burst-like firing in one of the neurons. Lidocaine injection immediately reduced the correlated firing

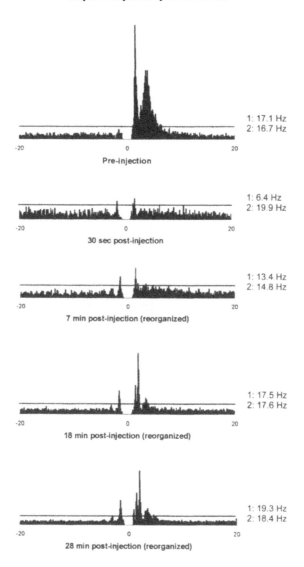

Figure 9.6 Cross-correlograms derived from two neurons recorded with a single electrode. Each neuron had a slowly adapting response to hair movement, and their receptive fields overlapped. The correlograms have been normalized so that they may be compared directly. The solid lines are 2 standard deviations above the mean correlation strength of the flanks. The spontaneous firing rates of each neuron are shown on the right side of the figure. Prior to lidocaine injection, the correlogram had a narrow peak at a lag of 1 ms, suggesting that there was a monosynaptic connection between the neurons. The broader peak at 4 ms was due to the burst-like firing pattern of one of the neurons. Immediately after lidocaine injection the strength of correlated firing (height of the peaks) declined dramatically. With the appearance of receptive field reorganization, the original peak in the correlogram returned. In addition, a new peak in the correlogram appeared at a lag of −1 ms, suggesting that a new, monosynaptic connection in the opposite direction became effective as a result of the lidocaine injection.

between the neurons. As the receptive field reorganized the original pattern of correlation reappeared, and a new peak appeared at a time lag of −1 ms, suggesting that a monosynaptic connection in the direction opposite to the original became effective. Such changes in correlated activity are compatible with the idea that previously ineffective synapses are strengthened as a result of lidocaine injection. However, it has been demonstrated recently that action potentials in DCN neurons can faithfully reflect action potentials in a single primary afferent (Gynther *et al.*, 1995). Thus, the appearance of a new peak in the correlogram could reflect unmasking of synapses between a primary afferent and the DCN neuron. Additional experiments are needed to examine these possibilities.

CONCLUSIONS

The experimental data reviewed above suggest that receptive fields in the DCN are quite sensitive to afferent input. Loss of afferent activity results in receptive field reorganization, usually expressed as enlargement of the receptive field. Which (if any) afferent input is critical for this effect is not known. Indeed, the data obtained so far are compatible with the proposal that decreased activity in any pathway can elicit the effect. The data on the general mechanisms involved are more clear. Loss of afferent activity reduces tonic inhibition in the DCN, and direct blockade of inhibitory synapses results in receptive field enlargement similar to that seen following afferent deprivation. An implicit assumption in our studies of receptive field reorganization is that, just as declines in afferent activity lead to receptive field enlargement, increases in afferent activity might lead to smaller receptive fields, and hence greater acuity. This might contribute to experience-dependent increases in tactile acuity.

ACKNOWLEDGMENTS

The author thanks Drs Oleg Ilyinksy and Michael Pettit, and Cullen Tennison for their important contributions to the experiments, and Dr Jannon Fuchs for valuable discussions. The research was supported by the University of North Texas and NSF grant IBN-9221956.

REFERENCES

Alloway, K.D., Rosenthal, P. and Burton, H. (1989). Quantitative measurements of receptive field changes during antagonism of GABAergic transmission in primary somatosensory cortex of cats. *Experimental Brain Research*, **78**, 514–532.

Amassian, V.E. and Giblin, D. (1974). Periodic components in steady-state activity of cuneate neurones and their possible role in sensory coding. *Journal of Physiology (London)*, **243**, 353–385.

Apkarian, A.V., Stea, R.A. and Bolanowski, S.J. (1994). Heat-induced pain diminishes vibrotactile perception: A touch gate. *Somatosensory and Motor Research*, **1**, 259–267.

Baumann, T.K., Simone, D.A., Shain, C.N. and LaMotte, R.H. (1991). Neurogenic hyperalgesia: the search for the primary cutaneous afferent fibers that contribute to capsaicin-induced pain and hyperalgesia. *Journal of Neurophysiology*, **66**, 212–227.

Bernander, Ö., Douglas, R.J., Martin, K.A.C. and Koch, C. (1991). Synaptic background activity influences spatiotemporal integration in single pyramidal cells. *Proceedings of the National Academy of Sciences, USA*, **88**, 11569–11573.

Buonomano, D.V. and Merzenich, M.M. (1998). Cortical plasticity: From synapses to maps. *Annual Review of Neuroscience*, **21**, 149–186.

Burgess, P.R., Petit, D. and Warren, R.M. (1968). Receptor types in cat hairy skin supplied by myelinated fibers. *Journal of Neurophysiology*, **31**, 833–848.

Calford, M.B. and Tweedale, R. (1991a). Acute changes in cutaneous receptive fields in primary somatosensory cortex after digit denervation in adult flying fox. *Journal of Neurophysiology*, **5**, 178–187.

Calford, M.B. and Tweedale, R. (1991b). C-fibres provide a source of masking inhibition to primary somatosensory cortex. *Proceedings of the Royal Society of London [Biol.]*, **243**, 269–275.

Cheema, S., Whitsel, B.L. and Rustioni, A. (1983). The corticocuneate pathway in the cat: relations among terminal distribution patterns, cytoarchitecture, and single neuron functional properties. *Somatosensory Research*, **1**, 169–205.

Clarey, J.C., Tweedale, R. and Calford, M.B. (1996). Interhemispheric modulation of somatosensory receptive fields: Evidence for plasticity in primary somatosensory cortex. *Cerebral Cortex*, **6**, 196–206.

Conti, F., DeBiasi, S., Giuffrida, R. and Rustioni, A. (1990). Substance P-containing projections in the dorsal columns of rats and cats. *Neuroscience*, **34**, 607–621.

Cook, A.J., Woolf, C.J., Wall, P.D. and McMahon, S.B. (1987). Dynamic receptive field plasticity in rat spinal cord dorsal horn following C-primary afferent input. *Nature*, **325**, 151–153.

Dinse, H.R., Recanzone, G.H. and Merzenich, M.M. (1993). Alterations in correlated activity parallel ICMS-induced representational plasticity. *NeuroReport*, **5**, 173–176.

Dykes, R.W. and Craig, A.D. (1998). Control of size and excitability of mechanosensory receptive fields in dorsal column nuclei by homolateral dorsal horn neurons. *Journal of Neurophysiology*, **80**, 120–129.

Dykes, R.W., Landry, P., Metherate, R. and Hicks, T.P. (1984). Functional role of GABA in cat primary somatosensory cortex: shaping receptive fields of cortical neurons. *Journal of Neurophysiology*, **52**, 1066–1093.

Florence, S.L. and Kaas, J.H. (1995). Large-scale reorganization at multiple levels of the somatosensory pathway follows therapeutic amputation of the hand in monkeys. *Journal of Neuroscience*, **15**, 8083–8095.

Fyffe, R.E.W., Cheema, S.S. and Rustioni, A. (1986). Intracellular staining study of the feline cuneate nucleus. I The terminal patterns of primary afferent fibers. *Journal of Neurophysiology*, 561268–1283.

Garraghty, P.E. and Kaas, J.H. (1991). Functional reorganization in adult monkey thalamus after peripheral nerve injury. *NeuroReport*, **2**, 747–750.

Gynther, B.D., Vickery, R.M. and Rowe, M.J. (1995) Transmission characteristics for the 1:1 linkage between slowly adapting type II fibers and their cuneate target neurons in cat. *Experimental Brain Research*, **105**, 67–75.

Handwerker, H.O. and Neher, K.D. (1976). Characterisitcs of C–fibre receptors in the cat's foot responding to stepwise increase of skin temperature to noxious levels. *Pflugers Archives*, **365**, 221–229.

Hicks, T.P., Metherate, R., Landry, P. and Dykes, R.W. (1986). Bicuculline-induced alterations of response properties in functionally identified ventroposterior thalamic neurons. *Experimental Brain Research*, **63**, 248–264.

Hirai, T., Schwark, H.D., Yen, C.-T., Honda, C.N. and Jones, E.G. (1988). Morphology of physiologically characterized medial lemniscal axons terminating in cat ventral posterior thalamic nucleus. *Journal of Neurophysiology*, **60**, 1439–1459.

Kaas, J.H. (1991). Plasticity of sensory and motor maps in adult mammals. *Annual Review of Neuroscience*, **14**, 137–167.

Krnjevic, K. and Morris, M.E. (1974). An excitatory action of substance P on cuneate neurons. *Canadian Journal of Physiology and Pharmacology*, **52**, 736–744.

LaMotte, R.H., Lundberg, L.E.R. and Torebjörk, H.E. (1992). Pain, hyperalgesia and activity in nociceptive C units in humans after intradermal injection of capsaicin. *Journal of Physiology (London)*, **448**, 749–764.

Malpeli, J.G., Lee, C., Schwark, H.D. and Weyand, T.G. (1986). Cat area 17. I. Patterns of thalamic control of cortical layers. *Journal of Neurophysiology*, **56**, 1062–1073.

Nakahama, H., Nishioka, S. and Otsuka, T. (1966). Excitation and inhibition in ventrobasal thalamic neurons before and after cutaneous input deprivation. In: *Progress in Brain Research. Vol 21A. Correlative Neurosciences*, edited by T. Tokizane. and J.P. Schade, pp. 180–196. Amsterdam: Elsevier.

Panetsos, F., Nuñez, A. and Avendaño, C. (1995). Local anaesthesia induces immediate receptive field changes in nucleus gracilis and cortex. *NeuroReport*, **7**, 150–152.

Panetsos, F., Nuñez, A. and Avendaño, C. (1997). Electrophysiological effects of temporary deafferentation on two characterized cell types in the nucleus gracilis of the rat. *European Journal of Neuroscience*, **9**, 563–572.

Pettit, M.J. and Schwark, H.D. (1993). Receptive field reorganization in dorsal column nuclei during temporary denervation. *Science*, **262**, 2054–2056.

Pettit, M.J. and Schwark, H.D. (1996). Capsaicin-induced rapid receptive field reorganization in cuneate neurons. *Journal of Neurophysiology*, **75**, 1117–1125.

Rasmusson, D.D. (1996) Changes in the response properties of neurons in the ventroposterior lateral thalamic nucleus of the racoon after peripheral deafferentation. *Journal of Neurophysiology*, **75**, 2441–2450.

Recanzone, G.H., Merzenich, M.M. and Jenkins, W.M. (1992). Frequency discrimination training engaging a restricted skin surface results in an emergence of a cutaneous response zone in cortical area 3a. *Journal of Neurophysiology*, **67**, 1057–1070.

Schwark, H.D., Pettit, M.J. and Fuchs, J.L. (1998). Distribution of substance P receptor binding in dorsal column nuclei of rat, cat, monkey, and human. *Brain Research*, **786**, 259–262.

Schwark, H.D., Tennison, C.F. and Ilyinsky, O.B. (1997). Influence of skin temperature on cuneate neuron activity. Society for Neuroscience, *Abstracts*, **23**, 2340.

Snow, P.J., Nudo, R.J., Rivers, W., Jenkins, W.M. and Merzenich, M.M. (1988). Somatotopically inappropriate projections from thalamocortical neurons to the SI cortex of the cat demonstrated by the use of intracortical microstimulation. *Somatosensory Research*, **5**, 349–372.

Stolwijk, J.A.J. and Wexler, I. (1971). Peripheral nerve activity in response to heating cat's skin. *Journal of Physiology (London)*, **214**, 377–392.

Wall, P.D. (1977). The presence of ineffective synapses and the circumstances which unmask them. *Philosophical Transactions of the Royal Society of London*, **278**, 361–372.

Xerri, C., Stern, J.M. and Merzenich, M.M. (1994). Alterations of the cortical representation of the rat ventrum induced by nursing behavior. *Journal of Neuroscience*, **14**, 1710–1721.

Zhang, S.P. and Rowe, M.J. (1997). Quantitative analysis of cuneate neurone responsiveness in the cat in association with reversible, partial deafferentation. *Journal of Physiology (London)*, **505**, 769–783.

CHAPTER 10
LIMITS ON SHORT-TERM PLASTICITY IN SOMATOSENSORY CORTEX

M.B. Calford

Psychobiology Laboratory, Division of Psychology, The Australian National University, Canberra, ACT Australia

It is now clear, from a large number of studies, that disruption of a subset of the inputs to the sensory representations in cortex of adult mammals induces a rapid unmasking of larger neuronal receptive fields and an increased neuronal responsiveness. Within the somatosensory cortex, such demonstrations indicate that a capacity for short-term plasticity results from the unmasking of normally unexpressed inputs to a cortical locus. Despite the clear demonstration of such unmasking, an explanation for how this occurs is not immediately apparent. Aside from the studies directly demonstrating short-term plasticity, there have been anatomical (Landry and Deschênes, 1981; Garraghty and Sur, 1990), physiological (Snow *et al.*, 1988) and pharmacological (e.g. Dykes *et al.*, 1984) demonstrations of cortical afferents that are not normally expressed as part of neural receptive fields. The rapid unmasking phenomena indicate that these inputs are inherently viable and thus must normally be actively suppressed. The only candidate source for an inhibition which can account for this suppression is that produced by the GABAergic interneurons located within a given cortical area. This short review will examine the implications that studies of short-term plasticity require for the nature of that inhibition, and the limitations of the capacity for plasticity that is dependent upon these normally unexpressed inputs.

THE NEED FOR TONIC INHIBITION

In addition to the role of local inhibition in limiting the extent of receptive fields which is implied from short-term unmasking studies, there are a number of direct physiological manifestations of the action of inhibitory interneurons in the somatosensory cortex. The most obvious of these are lateral inhibition and post-discharge inhibition. Lateral inhibition, which is apparent for about 10% of cat S1 neurons when plotted directly (Mountcastle and Powell, 1959) and for about 50% of neurons when a dual stimulation forward masking paradigm is used (Laskin and Spencer, 1979), produces an inhibitory field which extends beyond the boundaries of the excitatory receptive field. The forward masking work, however, showed that maximal inhibition occurs within the receptive field and on average peaked at the same point as maximal excitation (Fig. 10.1). It is inviting then to consider that disruption of stimulus driven lateral inhibition resulting from a peripheral nerve denervation, or by direct application of an inhibitory receptor antagonist to cortex, may be an explanation for receptive field expansion. However, there are two reasons for dismissing such an explanation, one empirical and one theoretical.

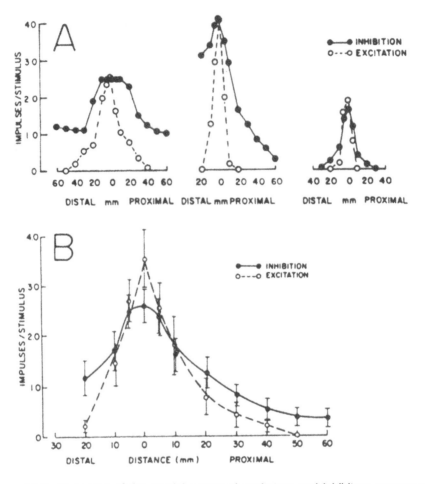

Figure 10.1 Examples of the spatial extent of excitatory and inhibitory components of neural receptive fields as seen at upper levels of the somatosensory pathway. **A:** shows neural response profiles demonstrating lateral inhibition as revealed by the use of two air-puff stimuli in a delayed forward masking paradigm. **B:** shows the averaged response profiles for a sample of neurons. Note that maximal excitation and maximal inhibition occur to stimulation of the same location. Neurons were recorded in cat vpl (thalamus) hairy-skin forepaw representation (from Jänig *et al.*, 1979).

- Empirically it has been shown that stimulus driven lateral inhibition is not apparent for all cortical cells (Laskin and Spencer, 1979), whereas almost all cortical neurons so studied show receptive field expansion with appropriate peripheral denervation of their receptive field area (e.g. Byrne and Calford, 1991). However, the major empirical objection to explanations that depend upon disruption of stimulus driven lateral inhibition exists because the geometrical extent of lateral inhibition is not sufficient to account for masking of the very large receptive fields which can be readily unmasked. In fact Laskin and Spencer (1979) and Jänig *et al.* (1979), working in thalamus, showed that when averaged over their samples the extent of inhibitory domains extended only marginally beyond that of excitatory domains.

- The basis of the theoretical problem is that whereas a peripheral denervation of a small body part leads to expansion of the receptive fields of some cortical neurons onto adjacent body areas, the denervation has no direct effect on these body areas. As discussed above, the areas around a normal receptive field can contribute inhibitory inputs and these have the potential to mask weaker excitatory inputs. The outcome of stimulation of a given body part should then reflect the balance of these competing influences on a given cortical cell. But, given that there are no peripheral interconnections between body areas, why should denervation of a nearby body part affect this balance? In addition, primary mechanoreceptor afferents are silent when not stimulated and therefore their denervation would, in itself, have no influence on the balance of excitation and inhibition resulting at a cortical neuron for stimulation of a nearby area.

The problems that these considerations raise for understanding how immediate topographical plasticity occurs have not been widely recognized, and only two solutions have been suggested:

1. that the denervation produces a discharge in the directly affected axons which acts to trigger a disinhibition (Kolarik *et al.*, 1994);
2. that there are some inputs to a cortical locus which provide a tonic signal and that these contribute a topographically-appropriate tonic level of inhibition (via local interneurons) which is disrupted by a denervation (Calford and Tweedale, 1991b; Pettit and Schwark, 1993, 1996; Dykes and Craig, 1998).

The first of these proposals links short-term changes in cortex to short-term changes that have been shown in second-order polymodal neurons of the dorsal horn, where intense stimulation of peripheral C-fibers leads to expansion of the mechanoreceptor component of the receptive field of these cells (e.g. Cook *et al.*, 1987). It is conceivable (but not shown) that some injury discharge may occur when there is a denervation that is capable of triggering such an event. However, such nerve cuts do not produce short-term receptive field plasticity in dorsal horn mechanoreceptor recipient neurons (Snow and Wilson, 1991); thus if there is a site at which nerve injury produces some release of peptides or other cofactors which may modify inhibition, it must be higher in the pathway. In my laboratory we initially considered explanations of this type, but when recording from cutaneous receptive neurons in cortex (Calford and Tweedale, unpublished observations after Calford and Tweedale, 1991a,b) or cuneate nucleus (Martin and Calford, unpublished observations after Martin, 1993) no massive increases in spontaneous rate or any other manifestation of an injury discharge have been noted. In addition, attempts to produce receptive field changes of single neurons in flying-fox SI, with electrical-stimulation tetani (at amplitudes above C-fibre threshold) to the median nerve were unsuccessful (Calford and Tweedale, 1991b; Calford and Tweedale, unpublished observations). The solution to the problem of unmasking presented by Kolarik *et al.* (1994) was supported by a review of studies in which denervations were effected by nerve cuts or amputation. Irrespective of the above considerations, the explanation does not appear useful for instances in which functional denervation has been obtained with local anaesthesia (e.g. Calford and Tweedale, 1991a,c;

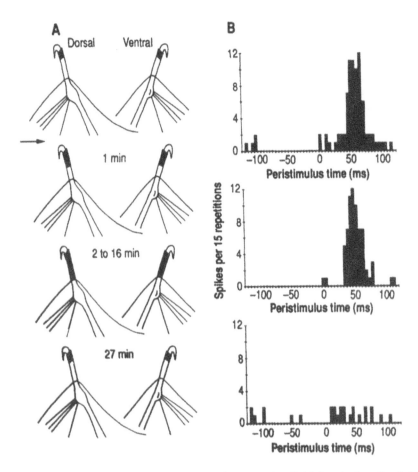

Figure 10.2 Panel A: Transient changes to the receptive field (located on the left thumb) of a neuron in right primary somatosensory cortex of a flying fox following amputation of the right thumb (not shown). **Panel B:** Summed responses to 15 repetitions of a brush stimulus to the left thumb (upper), both thumbs (centre) and right thumb (lower), indicating that the neuron was excited by stimulation of the left thumb but neither excited nor inhibited by stimulation of the right thumb (from Calford and Tweedale, 1990).

Panestos *et al.*, 1995, see also chapter by Schwark this book) where there is no 'denervation' signal.

Regarding the latter explanation of topographically appropriate tonic inhibition, three possible sources of drive for such interneuron mediated inhibition were investigated: ipsilateral body surface, interhemispheric connections, and peripheral C-fibres.

In an investigation with flying foxes and macaque monkeys (Calford and Tweedale, 1990, 1991c) the receptive fields of single neurons, or of a small group of neurons, in primary somatosensory cortex were monitored after denervation of the body area contralateral to their respective receptive fields (ipsilateral to the recording site). This procedure, conducted with anaesthetized animals, induced a rapid expansion of the receptive field

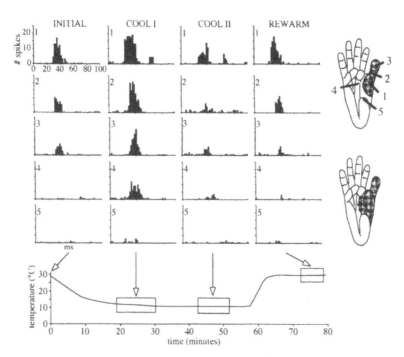

Figure 10.3 Transient expansion of the receptive field of a neuron in primary somatosensory cortex (3b) in macaque monkey (foot representation) induced by cooling inactivation of the corresponding location in the opposite cerebral hemisphere. Each panel presents the response of the neuron to stimulation at the 5 positions indicated on the drawing of the foot. Arrows at the base of the panels point to the period and opposite cortex temperature (focal) during which recordings were collected. At right are shown the initial and the maximal receptive fields. An expanded receptive field and increased responsiveness were found in the 'COOL I' period; an unmasking effect, indicating that interhemispheric connections contribute to local inhibition in area 3b. However, continuing the experiment in a steady state (COOL II) produced a reversal of this unmasking which is interpreted as due to an induced synaptic plasticity which increased the effectiveness of local inhibition (from Clarey *et al.*, 1996).

(Fig. 10.2). It was established prior to the denervation that the neurons under study had no stimulus-driven excitatory or inhibitory influences from stimulation of the ipsilateral body surface — as apparent to extracellular recording. Nevertheless, the fact that these receptive fields expanded indicates that there must be a net tonic inhibitory input sourced from the ipsilateral body surface. The pathway for the transmission of this input is unclear.

Clarey *et al.* (1996) showed that interhemispheric callosal connections also provide an input which acts via local interneurons as a net tonic inhibition. This interpretation was required when it was found that neurons in primary somatosensory cortex (area 3b) of flying fox and macaque monkey show a rapid expansion of receptive field size and an increase in responsiveness with focal cooling of the mirror image position in the contralateral hemisphere (Fig. 10.3). These initial unmasking effects indicate a capacity for a rapid functional plasticity within cortex itself, brought about by the removal of a subset of the inputs to a cortical locus. The effect was readily reversible with warming of the site of deactivation. The lack of a callosal interconnectivity between distal representations (Jones and Powell, 1969; Jones *et al.*, 1978) within area 3b created a difficulty in explaining these effects. However similar effects are found with cooling of the caudal somatosensory field (area 1). This suggests that changes in activity in the callosal pathway connecting the area 1 representations are relayed to area 3b via an ipsilateral connection (Clarey *et al.*, 1992). Whatever the pathway, an explanation for these effects must involve removal of a source of tonic input which provides a stimulus to local inhibitory circuits. It is known that callosal axons have tonic activity (Swadlow, 1991).

In consideration of a possible source of tonic peripheral activity that may provide a source of input to central inhibitory neurons, C-fibre activity has been blocked using the selective neurotoxin, capsaicin. The receptive fields of cutaneous-stimulation responsive single neurons in SI of cat (Calford and Tweedale, 1991b), flying-fox (Calford and Tweedale, 1991b) and mouse (Nussbaumer and Wall, 1985), vpl of raccoon (Rasmusson *et al.*, 1993) and cuneate nucleus of cat (Pettit and Schwark, 1996) have been shown to expand with peripheral capsaicin application. Applied subcutaneously, or directly to a peripheral nerve, capsaicin rapidly blocks C-fiber conduction with a concomitant expansion of cortical receptive fields (Fig. 10.4). Since mechanoreceptor A-fibers are unaffected, responses in the area of the original receptive field are maintained. Both the peripheral conduction block and the receptive field expansion exceed the period of an acute experiment. The clear implication of these experiments is that some sub-group of peripheral C-fibers provides a source of activity that ultimately drives inhibition which contributes to shaping the receptive fields of central pathway neurons. It is unclear which group of C-fibers are involved, but they must have some level of tonic activity (for discussion of this point, see Dykes and Craig, 1998). Dykes and Craig (1998) have shown that, within matched body part representations, receptive fields of neurons in the dorsal column nuclei expand after blockade of ipsilateral dorsal horn cellular responsiveness. This suggests that the pathway for propagation of the peripheral tonic activity is via the ipsilateral funiculus. Dykes and Craig (1998) point out that termination is probably in border zones of the DCN which contain GABAergic inhibitory neurons projecting into the lemniscal relay regions (Fyffe *et al.*, 1986).

In the investigation of the ipsilateral effects of peripheral denervation, where the denervation was induced with local anaesthetic injections the indirect effect of receptive field expansion in ipsilateral cortex had a similar timecourse to the direct effect of receptive field expansion (or expression of a transient new receptive field) for affected neurons on contralateral cortex (Calford and Tweedale, 1990). However where the denervation was permanent the expression of new or expanded receptive fields in contralateral cortex lasted for the period of the experiment while the ipsilateral effect was a transient phenomenon lasting around 20 minutes (Fig. 10.3A). As we know that there are multiple sources of input to local inhibitory interneuron circuits which can provide tonic stimulation, it is plausible that some form of gain control acts to stabilize the local activity and restores the normal receptive field dimensions. It is clear, however, that such a mechanism does not always operate — for the very large new receptive fields induced in contralateral cortex by a nerve cut or amputation shrink very slowly in the formation of a new topographic map (around 1 week; Calford and Tweedale, 1988). Similarly the large receptive fields

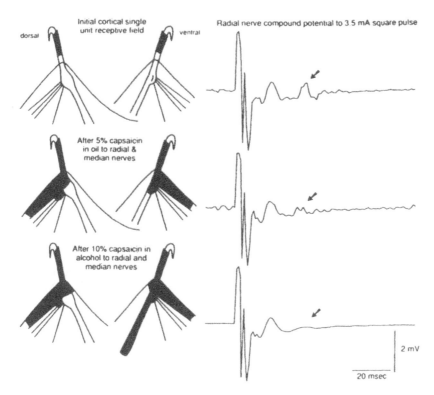

Figure 10.4 Expansion of the mechanoreceptor receptive field of a cortical neuron with blockage of C-fiber activity in a peripheral nerve. The left panel shows dorsal and ventral views of part of one wing of a flying fox, with the receptive field indicated in black. The right panel shows the compound action potential (recorded with hook electrodes) of the radial nerve to a square wave stimulus above C-fiber threshold. The C-fiber wave (arrowed) was diminished after application of capsaicin directly to the exposed radial and median nerves. This produced an immediate expansion of the receptive field of the cortical neuron (from Calford and Tweedale, 1991b).

induced rapidly by C-fibre denervation with capsaicin do not shrink over the period of acute-recording (Calford and Tweedale, 1991b) and have long-term effects (Nussbaumer and Wall, 1985).

A similar transient effect was found with topical cooling of the representation of the mirror image body place in somatosensory cortex in the Clarey *et al.* (1996) study. It was found that after around 20 minutes of cooling of the contralateral cortex the initial effects of an increased responsiveness and an enlargement of the receptive field of a neuron reversed (Fig. 10.3). Since this occurred when the experimental manipulation was in a steady state it was interpreted as representing an induced neuronal plasticity — most probably an increase in the efficacy of local interneuron-based inhibition. While this change may conceivably occur in a number of ways it is intriguing to consider the parsimonious explanation that there is a mechanism for maintaining some optimal or normal level of responsivity through a gain control which modulates inhibitory synaptic efficacy.

LIMITS ON SHORT-TERM PLASTICITY

Ultimately the limit on short-term unmasking-based plasticity resulting from discrete peripheral denervation must be set by the underlying anatomy. However it is not easy to predict, in a given paradigm, what these limits may be and whether the limits on the short-term effect are exceeded with longer periods of recovery. Jain *et al.* (1995) reported neither short- nor long-term recovery of responsiveness in adult rat primary somatosensory cortex (S1) denervated by section of the contralateral dorsal column at midthoracic levels. Clearly such a denervation would affect a large area of the central representations. However, in both short- and long-term studies the limits do not appear predictable by either a simple consideration of the involved representation in cortical area or of the peripheral extent of the denervation; and there has been no suggestion that consideration of peripheral dermatomal boundaries or relativities can be predictive of plasticity limits. Thus in some studies where comparison with other reports would predict significant reorganization, limited representational plasticity has been reported following peripheral denervation (e.g. Li *et al.*, 1994; Waite, 1984). In the squirrel monkey the cortical area affected by median nerve section is slightly larger than the representation of two digits — yet the former undergoes considerable short-term and near complete long-term representational plasticity (Merzenich *et al.*, 1983a,b) while the latter retains a central unresponsive region (Merzenich *et al.*, 1984). Such differences have been attributed to the precise nature of peripheral innervation and central representational contiguities rather than to the absolute size of the representation or to the method of denervation (Garraghty *et al.*, 1994; Garraghty and Kaas, 1991).

This issue has also been addressed by studies of the visual cortex. Calford and Rosa and colleagues have demonstrated extensive topographical plasticity in primary visual cortex of the cat (up to 9 mm across the representation) with both short-term (Schmid *et al.*, 1995; Calford *et al.*, 1999) and long-term retinal lesions (Schmid *et al.*, 1996). However, in the same laboratory, equivalent reorganization across the boundary of the monocular representation was not apparent (Rosa *et al.*, 1995; immediate effect to 16 months recovery) even though such reorganization would be less than 1 mm across the cortical representation. Thus, in both somatosensory and visual cortex, it is clear that some physiological representational boundaries are also boundaries that limit unmasking-based plasticity.

Comparing the limits on short-term and long-term plasticity the conclusion to be reached from studies prior to 1991 was that both manifestations probably depended upon the same underlying basis, this being the degree of divergence in topography provided by the underlying afferent projection. While the mechanism for the short-term effect appeared to be a mechanistic unmasking provided by a direct disruption of inhibition, the more extensive effects seen in some long-term studies (Pons *et al.*, 1991; Garraghty *et al.*, 1991) and the refinement of the short-term unmasking seen in others (Merzenich *et al.*, 1983a,b; Calford and Tweedale, 1988) provided evidence of a role for synaptic plasticity (Kano *et al.*, 1991). This conclusion was further strengthened by work showing that training can induce receptive field and map changes in monkey S1 (Recanzone *et al.*, 1992).

A major influence on the search for mechanisms of representational plasticity in the somatosensory cortex was provided by the 1991 report from Pons and colleagues that with long-term limb deafferentation in macaques (C2–C5 dorsal rhizotomies) the extensive hand representation of area 3b (10 to 14 mm in linear extent) was seen to represent the face (mainly chin). In this study the deafferentation was provided by dorsal root rhizotomies for a period of greater than 10 years. Clearly this result was incompatible with any of the potential mechanisms for plasticity that depended upon divergence in the afferent projection to cortex. The scale of the changes suggested some source of axonal sprouting as a likely explanation. Alternative explanations, within the dogmatic framework that significant neuronal growth does not occur in the adult central nervous system, which were suggested included:

1. a telescoping effect of a small degree of plasticity at each level being amplified by divergence in the projection along the lemniscal pathway (Pons *et al.*, 1991);
2. an amplification of the effect of corticocortical projections from other cortical areas which have a less precise topography (Calford, 1991 after work in raccoon; Doetsch *et al.*, 1988; Smits *et al.*, 1991)
3. that the representation of the back of the head (occiput) medial to the hand representation includes a secondary representation of the face, thus allowing the interpretation that reorganization of 'hand representation' into 'face representation' requires a maximum extent of 3–5 mm across cortex (Lund *et al.*, 1994).

The subsequent finding of induced axonal sprouting in the visual cortex (Darian-Smith and Gilbert, 1994) did not provide a solution to the large-scale reorganization problem, since the described sprouting (of within-field corticocortical projections) was confined to the geometric limits of the original projections. Irrespective, the reported sprouting is a remarkable finding. Darian-Smith and Gilbert used binocular laser lesions (matched) to remove the dominant input to a region of the visual cortex in adult cats. Examining this region around 8 months later they showed a dramatically increased density of projections from the adjacent cortex into the lesion projection zone (the region of cortex normally representing the area of the retina that was lesioned) relative to similar projections in a control animal. The increased density, seen with anterograde transport of biocytin, was interpreted as resulting from *axonal sprouting of intrinsic excitatory neurons* and the formation of new terminal branches. It appeared that the higher density projections did not extend further across cortex than the longest projections in normal cortex. Thus the findings of extensive short-term representational plasticity after restricted retina detachments (Schmid *et al.*, 1995) but of smaller receptive fields and increased responsiveness with long-term recovery from retinal lesions (Schmid *et al.*, 1996) are consistent with this growth being within the geometrical limits of the original projection. A main influence of the Darian-Smith and Gilbert work has been to highlight intrinsic corticocortical projections as a source of the normally unexpressed afferents to a cortical locus.

Psychophysical and imaging studies with human arm amputees found effects that appeared to parallel the plasticity seen in the monkey arm deafferentation studies. Ramachandran *et al.* (1992) and Halligan *et al.* (1993) reported that many arm amputees, when appropriately asked, describe secondary sensations referred to their phantom limb when stimulated on the cheeks, lips or chin. Some of these secondary sensations are topographically mapped onto the face. Parallels were drawn between this finding and the extensive topographical plasticity seen in the arm-deafferented monkeys. Thus one explanation for the secondary sensations was that with expansion of the face representation, stimulation of the face produced activation within an area of primary somatosensory cortex (at least) that once represented the arm, hand and digits. Necessary for this explanation

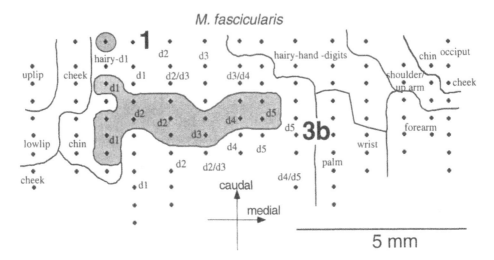

Figure 10.5 Map of body representation in somatosensory area 3b in an anesthetized macaque monkey. Multi- and single-unit recordings at the indicated positions located within the caudal bank of the central sulcus. The shading covers a previously unreported area in which neurons were found to have a primary receptive field on the digits and a secondary receptive field on the lips or chin (Calford, Brinkman, Krubitzer, Elston, Rosa, Tweedale and Wright, unpublished).

was the implication that while the face representation may have expanded, the identity of the remapped cortex as 'hand' was not entirely lost. This may fit with the neuromatrix hypothesis of Melzack (1990) concerning how stimulation of specific brain regions is linked to perceptions of specific body parts by a genetically deterministic structure. Alternatively it may indicate that, while reorganization takes place in primary somatosensory cortex, there is no reorganization of some other cortical area more responsible for the perception of body image. This latter region would thus be required to receive inappropriate activity in the inputs from a reorganized primary somatosensory cortex. A further possible explanation is suggested by work from this laboratory, which has described a dual hand/face representation in primary somatosensory cortex of normal monkeys.

   With extracellular recording from primary somatosensory cortex (area 3b) of anaesthetized macaque monkeys it was found that many neurons within the 'hand' representation had secondary receptive fields on the lower part of the face (Fig. 10.5). Stimulation thresholds within the secondary receptive field were higher than those on the digits, nevertheless neurons gave vigorous responses to cutaneous stimuli or movement of hairs. More than half of these secondary receptive fields were located on, or extended to, the ipsilateral chin or lower lip.

   One of the considerations prompted by the finding of dual digit/face receptive fields in normal monkey cortex is to re-evaluate whether it is necessary to invoke cortical plasticity as an explanation for the psychophysical effects after amputation. Rather than resulting from a reorganization of the digit representation, the secondary sensations referred to the phantom hand with stimulation of the face (Ramachandran *et al.*, 1992) may result from activation of a normal population of neurons with dual sensitivity. However such a possibility does not deny the physiological manifestation of large-scale reorganization such as seen in monkeys with arm deafferentation (Pons *et al.*, 1991) or human amputees (Flor *et al.*, 1995). Indeed, comparison of short-term and long-term arm amputation in macaque monkeys revealed that there is a complete filling in of responsiveness in the long-term and only limited short-term unmasking (Calford *et al.*, 1995). Nevertheless, the finding of this secondary face area within the hand representation further reduces the extent of reorganization required to explain the long-term deafferentation results. Coupled with the consideration of the medially located occiput representation (Lund *et al.*, 1994), required reorganization is then within expected limits of intrinsic corticocortical connections. Considering the relative geometries of the medial occiput/face area, the area with overlapping hand and face representation and the area which would be 'deafferented' by a C2–C5 dorsal rhizotomy of the map in Figure 10.5 reveals that no part of the cortex which would lose its hand/arm input is more than 2.5 mm from an original face representation. Gandevia and Phegan (1999) have demonstrated a perceptual effect that appears to be explicable by this overlapping representation. When a body area is locally anaesthetized it is perceived as being larger than normal (think of the effects of dental anesthesia on the perception of the gum and lips). Gandevia and Phegan measured this effect with size estimation methods for local anesthesia of the thumb. Their subjects also reported a secondary changed perception such that their lips also appeared expanded. Both effects were easily demonstrated and were sufficiently clear that even drawings of the affected thumb and lips revealed the expanded percept (Fig. 10.6).

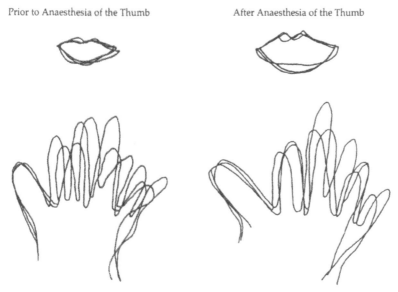

Figure 10.6 Drawings made, with the non-preferred hand, by subjects before and during local anesthesia of the thumb of the preferred hand. The perceived size of a body part is reflected in the drawings. In addition to the direct effect on the perceived size of the thumb, the lips are also perceived as larger during the period of local anesthesia (from Gandevia and Phegan, 1999).

ACKNOWLEDGMENTS

Thanks to Anita Disney for comments on the manuscript. The collection of original data presented in this chapter was supported by grants from the National Health and Medical Research Council of Australia and the Australian Research Council.

REFERENCES

Byrne, J.A. and Calford, M.B. (1991). Short-term expansion of receptive fields in rat primary somatosensory cortex after hindpaw digit denervation. *Brain Research*, **565**, 218–224.

Calford, M.B. (1991). Curious cortical change (News and Views). *Nature*, **352**, 759–760.

Calford, M.B., Krubitzer, L.A., Rosa, M.G.P., Clarey, J.C., Tweedale, R. and Brinkman, J.C. (1995). Unmasking of ipsilateral face representation in primary somatosensory cortex after long- and short-term denervation of the forelimb in adult monkey. *Society for Neuroscience Abstracts*, **21**, 52.4.

Calford, M.B., Schmid, L.M. and Rosa, M.G.P. (1999). Monocular focal retinal lesions induce short-term topographic plasticity in adult cat visual cortex. *Proceedings of the Royal Society of London B: Biological Series*, **266**, 1–9.

Calford, M.B. and Tweedale, R. (1988). Immediate and chronic changes in responses of somatosensory cortex in adult flying-fox after digit amputation. *Nature*, **332**, 446–448.

Calford, M.B. and Tweedale, R. (1990). Interhemispheric transfer of plasticity in the cerebral cortex. *Science*, **249**, 805–807.

Calford, M.B. and Tweedale, R. (1991a). Acute changes in cutaneous receptive fields in primary somatosensory cortex after digit denervation in adult flying fox. *Journal of Neurophysiology*, **65**, 178–187.

Calford, M.B. and Tweedale, R. (1991b). C-fibres provide a source of masking inhibition to primary somatosensory cortex. *Proceedings of the Royal Society of London [B]*, **243**, 269–275.

Calford, M.B. and Tweedale, R. (1991c). Immediate expansion of receptive fields of neurons in area 3b of macaque monkeys after digit denervation. *Somatosensory and Motor Research*, **8**, 249–260.

Clarey, J.C., Tweedale, R. and Calford, M.B. (1992). Pathways of interhemispheric transfer revealed by focal cooling of somatosensory cortex. *Society for Neuroscience Abstracts*, **18**, 1544.

Clarey, J.C., Tweedale, R. and Calford, M.B. (1996). Interhemispheric modulation of somatosensory receptive fields: evidence for plasticity in primary somatosensory cortex. *Cerebral Cortex*, **6**, 196–206.

Cook, A.J., Woolf, C.J., Wall, P.D. and McMahon, S.B. (1987). Dynamic receptive field plasticity in rat spinal cord dorsal horn following C-primary afferent input. *Nature*, **325**, 151–153.

Darian-Smith, C. and Gilbert, C.D. (1994). Axonal sprouting accompanies functional reorganization in adult cat striate cortex. *Nature*, **368**, 737–740.

Doetsch, G.S., Standage, G.P., Johnston, K.W. and Lin, C.S. (1988). Intracortical connections of two functional subdivisions of the somatosensory forepaw cerebral cortex of the raccoon. *Journal of Neuroscience*, **8**, 1887–900.

Dykes, R.W. and Craig, A.D. (1998). Control of size and excitability of mechanosensory receptive fields in dorsal column nuclei by homolateral dorsal horn neurons. *Journal of Neurophysiology*, **80**, 120–129.

Dykes, R.W., Landry, P., Metherate, R. and Hicks, T.P. (1984). Functional role of GABA in cat primary somatosensory cortex: shaping receptive fields of cortical neurons. *Journal of Neurophysiology*, **52**, 1066–1093.

Flor, H., Elbert, T., Knecht, S., Wienbruch, C., Pantev, C., Birbaumer, N., Larbig, W. and Taub, E. (1995). Phantom-Limb Pain as a Perceptual Correlate of Cortical Reorganization Following Arm Amputation. *Nature*, **375**, 482–484.

Fyffe, R.E., Cheema, S.S. and Rustioni, A. (1986). Intracellular staining study of the feline cuneate nucleus. I. Terminal patterns of primary afferent fibers. *Journal of Neurophysiology*, **56**, 1268–83.

Gandevia, S.C. and Phegan, C.M.L. (1999). Perceptual distortions of the human body image produced by local anaesthesia, pain and cutaneous stimulation. *Journal of Physiology*, **514**, 609–614.

Garraghty, P.E., Hanes, D.P., Florence, S.L. and Kaas, J.H. (1994). Pattern of Peripheral Deafferentation Predicts Reorganizational Limits in Adult Primate Somatosensory Cortex. *Somatosensory and Motor Research*, **11**, 109–117.

Garraghty, P.E. and Kaas, J.H. (1991). Large-scale functional reorganization in adult monkey cortex after peripheral nerve injury. *Proceedings of the National Academy of Science USA*, **88**, 6976–6980.

Garraghty, P.E. and Sur, M. (1990). Morphology of single intracellularly stained axons terminating in area 3b of macaque monkeys. *Journal of Comparative Neurology*, **294**, 583–593.

Guillemot, J.P., Richer, L., Ptito, M., Guilbert, M. and Lepore, F. (1992). Somatosensory receptive field properties of corpus callosum fibres in the raccoon. *Journal of Comparative Neurology*, **321**, 124–132.

Halligan, P.W., Marshall, J.C., Wade, D.T., Davey, J. and Morrision, D. (1993). Thumb in cheek? Sensory reorganization and perceptual plasticity after limb amputation. *NeuroReport*, **4**, 233–236.

Jänig, W., Spencer, W.A. and Younkin, S.G. (1979). Spatial and temporal features of afferent inhibition of thalamocortical relay cells. *Journal of Neurophysiology*, **42**, 1450–1460.

Jain, N., Florence, S.L. and Kaasm J.H. (1995). Limits on plasticity in somatosensory cortex of adults rats: hindlimb cortex is not reactivated after dorsal column section. *Journal of Neurophysiology*, **73**, 1537–1546.

Jones, E.G., Coulter, J.D. and Hendry, S.H.C. (1978). Intracortical connectivity of architectonic fields in the somatic sensory, motor and parietal cortex of monkeys. *Journal of Comparative Neurology*, **181**, 291–348.

Jones, E.G. and Powell, T.P.S. (1969). Connexions of the somatic sensory cortex of the rhesus monkey II contralateral cortical connexions. *Brain*, **92**, 717–730.

Kano, M., Lino, K. and Kano, M. (1991). Functional reorganization of adult cat somatosensory cortex is dependent upon NMDA receptors. *NeuroReport*, **2**, 77–80.

Kolarik, R.C., Rasey, S.K. and Wall, J.T. (1994). The consistency, extent, and locations of early-onset changes in cortical nerve dominance aggregates following injury of nerves to primate hands. *Journal of Neuroscience*, **14**, 4269–4288.

Landry, P. and Deschênes, M. (1981). Intracortical arborizations and receptive fields of identified ventrobasal thalamocortical afferents to the primary somatic sensory cortex in the cat. *Journal of Comparative Neurology*, **199**, 345–371.

Laskin, S.E. and Spencer, W.A. (1979). Cutaneous masking. II. Geometry of excitatory and inhibitory receptive fields of single units in somatosensory cortex of the cat. *Journal of Neurophysiology*, **42**, 1061–1082.

Li, C.X., Waters, R.S., Oladehin, A., Johnson, E.F., Mccandlish, C.A. and Dykes, R.W. (1994). Large Unresponsive Zones Appear in Cat Somatosensory Cortex Immediately After Ulnar Nerve Cut. *Canadian Journal of Neurological Sciences*, **21**, 233–247.

Lund, J.P., Sun, G.D. and Lamarre, Y. (1994). Cortical reorganization and deafferentation in adult macaques. *Science*, **265**, 546–548.

Martin, R.L. (1993). The representation of the body surface in the gracile, cuneate and spinal trigeminal nuclei of the little red flying fox (*Pteropus scapulatus*). *Journal of Comparative Neurology*, **335**, 335–442.

Melzack, R. (1990). Phantom limbs and the concept of a neuromatirx. *Trends in Neuroscience*, **13**, 88–92.

Merzenich, M.M., Kaas, J.H., Wall, J., Nelson, R.J., Sur, M. and Felleman, D. (1983a). Topographic reorganization of somatosensory cortical areas 3b and 1 in adult monkeys following restricted deafferentation. *Neuroscience*, **8**, 33–55.

Merzenich, M.M., Kaas, J.H., Wall, J.T., Sur, M., Nelson, R.J. and Felleman, D.J. (1983b). Progression of change following median nerve section in the cortical representation of the hand in areas 3b and 1 in adult owl and squirrel monkeys. *Neuroscience*, **10**, 639–665.

Merzenich, M.M., Nelson, R.J., Stryker, M.P., Cynader, M.S., Schoppmann, A. and Zook, J.M. (1984). Somatosensory cortical map changes following digit amputation in adult monkeys. *Journal of Comparative Neurology*, **224**, 591–605.

Mountcastle, V.B. and Powell, T.P. (1959). Neural mechanisms subserving cutaneous sensibility, with special reference to the role of afferent inhibition in sensory perception and discrimination. *Bulletin of the Johns Hopkins Hospital*, **105**, 210–232.

Nussbaumer, J.C. and Wall, P.D. (1985). Expansion of receptive fields in the mouse cortical barrelfield after administration of capsaicin to neonates or local application on the infraorbital nerve in adults. *Brain Research*, **360**, 1–9.

Panestos, F., Nuñez, A. and Avendaño, C. (1995). Local anaesthesia induces immediate receptive field changes in nuclues gracilis and cortex. *NeuroReport*, **7**, 150–152.

Pettit, M.J. and Schwark, H.D. (1993). Receptive field reorganization in dorsal column nuclei during temporary denervation. *Science*, **262**, 2054–6.

Pettit, M.J. and Schwark, H.D. (1996). Capsaicin-induced rapid receptive field reorganization in cuneate neurons. *Journal of Neurophysiology*, **75**, 1117–25.

Pons, T.P., Garraghty, P.E., Ommaya, K., Kaas, J.H., Taub, E. and Mishkin, M. (1991). Massive cortical reorganization after sensory deafferentation in adult macaques. *Science*, **252**, 1857–1860.

Ramachandran, V.S., Stewart, M. and Rogers-Ramachandran D.C. (1992). Perceptual correlates of massive cortical reorganization. *NeuroReport*, **3**, 583–586.

Rasmusson, D.D., Louw, D.F. and Northgrave, S.A. (1993). The immediate effects of peripheral denervation on inhibitory mechanisms in the somatosensory thalamus. *Somatosensory and Motor Research*, **10**, 69–80.

Recanzone, G.H., Merzenich, M.M., Jenkins, W.M., Grajski, K.A. and Dinse, H.R. (1992). Topographic reorganization of the hand representation in cortical area 3b of owl monkeys trained in a frequency-discrimination task. *Journal of Neurophysiology*, **67**, 1031–1056.

Rosa, M.G.P., Schmid, L.M. and Calford, M.B. (1995). Responsiveness of cat area-17 after monocularinactivation — limitation of topographic plasticity in adult cortex. *Journal of Physiology (London)*, **482**, 589–608.

Schmid, L.M., Rosa, M. and Calford, M.B. (1995). Retinal-Detachment Induces Massive Immediate Reorganization in Visual-Cortex. *NeuroReport*, **6**, 1349–1353.

Schmid, L., Rosa, M., Calford, M. and Ambler, J. (1996). Visuotopic reorganization in the primary visual cortex of adult cats following monocular and binocular retinal lesions. *Cerebral Cortex*, **6**, 388–405.

Smits, E., Gordon, D.C., Witte, S., Rasmusson, D.D. and Zarzecki, P. (1991). Synaptic potentials evoked by convergent somatosensory inputs in raccoon somatosensory cortex: substrates for plasticity. *Journal of Neurophysiology*, **66**, 688–695.

Snow, P.J., Nudo, R.J., Rivers, W., Jenkins, W.M. and Merzenich, M.M. (1988). Somatotopically inappropriate projections from thalamocortical neurons to the SI cortex of the cat demonstrated by the use of intracortical microstimulation. *Somatosensory Research*, **5**, 349–372.

Snow, P.J. and Wilson, P. (1991). Plasticity in the somatosensory system of developing and mature mammals — The effects of injury to the central and peripheral nervous system. In *Progress in Sensory Physiology Vol. 11*. Heidelberg: Springer-Verlag.

Swadlow, H.A. (1991). Efferent neurons and suspected interneurons in second somatosensory cortex of the awake rabbit: receptive fields and axonal properties. *Journal of Neurophysiology*, **66**, 1392–409.

Waite, P.M.E. (1984). Rearrangement of neuronal responses in the trigeminal system of the rat following peripheral nerve section. *Journal of Physiology*, **352**, 425–445.

CHAPTER 11

CORTICAL PLASTICITY: GROWTH OF NEW CONNECTIONS CAN CONTRIBUTE TO REORGANIZATION

Sherre L. Florence and Jon H. Kaas

Department of Psychology, Vanderbilt University, Nashville, TN, USA

"It appears that most of the earlier reports of the high functional plasticity of the nervous system will go down in the record as unfortunate examples of how an erroneous medical scientific opinion, once implanted, snowball(s) until it biases experimental observations and crushes dissenting interpretations." (Sperry, 1959).

As evidenced in the quote above by Sperry there was a longstanding belief that the organization of the adult brain is relatively fixed. However, it is clear that the processing machinery of the brain must have the ability to change, even in the mature brain, to reflect behavioral recoveries after damage, the learning of new skills, and perceptual reorganizations after sensory distortions. Indeed, we now know that the adult brain is highly adaptable.

Some of the most persuasive early evidence of the potential for the adult brain to change came from studies of the effects of injury on the organization of sensory representations in the primary somatosensory pathway. In somatosensory cortex, where there are large two-dimensional representations of the receptor surfaces, even small changes in neuronal response properties can be detected. This made it easy to demonstrate that populations of neurons, when deprived of their normal sources of activation, become immediately responsive to remaining sources of sensory input (Fig. 11.1). The earliest evidence of this immediate cortical reorganization was shown by Merzenich and colleagues (Merzenich *et al.*, 1983a, 1983b); since then, similar changes have been documented under a number of circumstances and in a variety of species (for review see Kaas, 1991). Similar rapidly occurring changes have been reported, on a smaller scale, in other parts of the somatosensory pathway, including the spinal cord (e.g. Basbaum and Wall, 1976; Wall, 1977; Mendell *et al.*, 1978; Devor and Wall, 1981; Pubols and Brenowitz, 1981; Wilson and Snow, 1987), the dorsal column nuclei (e.g. Wall and Egger, 1971; Dostrovsky *et al.*, 1976; Millar *et al.*, 1976; McMahon and Wall, 1983; Pettit and Schwark, 1993; Northgrave and Rasmusson, 1996) and thalamus (e.g. Wall and Egger, 1971; Pollin and Albe-Fessard, 1979; Garraghty and Kaas, 1991; Nicolelis, 1993; Rasmusson, 1993, 1996a, 1996b; Shin *et al.*, 1995). The magnitudes of these rapidly occurring changes were small (on the order of half

---

Address for correspondence: Dr. Sherre L. Florence, 301 Wilson Hall, Department of Psychology, Vanderbilt University, Nashville, TN 37240, USA

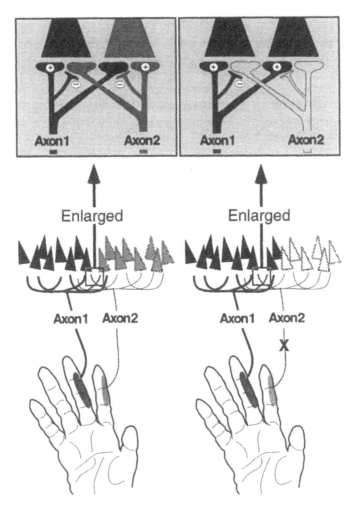

Figure 11.1 Schematic drawing of overlapping afferent inputs to sensory cortex, and the mechanism of afferent driven inhibition. Normally, the response properties of cortical neurons reflect only a portion of their total excitatory inputs (shown by + symbol) as a result of inhibition (shown by − symbol) from competing afferents. If the activivation of one of the afferents is reduced, as occurs after nerve injury, the inhibitory influence on non-deprived afferents is released and the full extent of their excitatory influence becomes apparent (shown to the right).

a millimeter or less), and were attributed to the rapid unmasking, or the potentiation, of previously existing connections (for review see Wall, 1977; Devor, 1983). The unmasking is thought to result from the down-regulation of activity-driven inhibition on previously sub-threshold inputs (Fig. 11.1).

Further modifications of the sensory representations can develop hours to weeks after the sensory denervation. Thus, unmasking of suppressed inputs is not the only mechanism of reorganization. Instead, a number of additional mechanisms for the strengthening of weak synapses have been defined. These include Hebbian mechanisms for use-dependent modulations of synaptic strengths (Hebb, 1949; Stent, 1973), and activity-based regulations of neurotransmitters and receptors (e.g. Lipton and Kater, 1989). These processes may begin soon after the deprivation occurs, but they take much longer than the changes revealed by unmasking to fully emerge. For example, early use-dependent changes in neuronal responses are apparent within 24 hours after a sensory manipulation (Diamond *et al.*, 1994), but continue to develop for 3 to 30 days (Armstrong-James *et al.*, 1994). Moreover, the modifications in protein transcription that precede changes such as the down-regulation of the inhibitory neurotransmitter GABA (Garraghty *et al.*, 1991) or the reduction of choline acetyltransferase (Dykes *et al.*, 1995) may take considerably longer and do not fully emerge for weeks or months.

Since a strictly point-to-point hierarchy of connections from the peripheral receptors to cortex would not allow for even limited adjustments in receptive field locations, these processes of reorganization depend on some divergence and convergence of connections. Indeed, anatomical evidence shows that there is considerable overlap among connections in the somatosensory pathway. For example, the extent of the primary afferent arbors from the ventroposterior nucleus in the thalamus to cortical area 3b are much larger than the grain of the somatotopic map (Garraghty *et al.*, 1989; Garraghty and Sur, 1990; see also Snow *et al.*, 1988; Rausell and Jones, 1995). However, convergence and divergence of connections are not sufficiently extensive to support large-scale changes in topographic organization. Most injuries that produce denervations of more than a millimeter or so in cortical sensory representations produce zones of inactivation in cortex that remain for months. Amputation of multiple digits on the hand (Merzenich *et al.*, 1984), and ligation of combinations of nerves to the hand (Garraghty *et al.*, 1994), are examples of injuries that have led to long-lasting denervations in sensory cortex. Evidence such as these indicate that there is a spatial limit to the amount of topographic change that can be produced by the plasticity mechanisms described above. The 'distance limit' has been estimated to be about 1.5 mm (Merzenich *et al.*, 1984), and most of the evidence for topographic shifts in sensory representations are well within this anatomical constraint. However, recent studies have shown that, under certain circumstances, reorganizations occur that are so extensive (up to 10 millimeters and more) they cannot be explained on the basis of any of the previously described plasticity mechanisms. Instead, the changes can only be attributed to the growth of new connections. The possibility of new growth is the focus of this review.

LOCAL AXON SPROUTING IN THE CENTRAL NERVOUS SYSTEM

The first clear evidence for new growth in the mature brain involved modifications in the local distributions of synapses. One of the earliest demonstrations of this type of sprouting was by Raisman (1969) who found new synapse formation after partial deafferentation of the lateral septal nuclei of adult rats. Raisman cleverly took advantage of the distinctive termination patterns of two types of afferents to the septal nuclei to demonstrate the formation of new synapses (for review see Raisman, 1985). One type of afferent input to the lateral septal nuclei courses through the medial forebrain bundle, and forms synapses directly on the soma of the neurons in the septal nuclei. The other afferent type arises in the hippocampus, courses through the fimbria and makes synaptic connections with the dendrites of the septal neurons. Raisman found that if the medial forebrain bundle was lesioned, thereby depriving septal neurons of one source of inputs, the fimbrial inputs sprouted and formed synapses on the cell soma, the domain normally occupied by the lesioned afferents (for review see Raisman, 1985).

A similar strategy was used by Tsukahara and colleagues (1974) to show that new synapses can be functionally active. In the red nucleus in cats, they found that afferents made new synapses in an atypical location on the neuron (i.e. closer to the cell body) if the inputs that normally occupied that location were lesioned. The time course of the excitatory postsynaptic potentials following stimulation of the reorganized afferents possessed a fast-rising component that resembled the time course of the lesioned inputs (for review see Tsukahara, 1985). Thus, not only did the reorganized afferents come to occupy the location typically held by another input type, but the activation pattern of the afferents took on functional characteristics of the lesioned inputs.

Although these demonstrations showed relatively irrefutable evidence for sprouting in the adult central nervous system, the new growth only occurred locally. Thus, the impact of the changes was small. As a result, it continued to be argued that the adult brain was, for the most part, structurally stable and unamenable to change. Considerably more evidence would be necessary before the deeply entrenched disbelief in the possibility of new growth in the adult brain would be reconsidered.

EVIDENCE FOR EXTENSIVE AXON GROWTH

The most substantial growth of new processes was demonstrated by the laboratory of Aguayo who induced sprouting experimentally (Aguayo, 1985; Aguayo *et al.*, 1990). They transplanted a peripheral nerve segment into the central nervous system, and found that neurons near the site of the transplant sprouted axon processes several millimeters into the grafted nerve. The outcome was attributed in part to growth-promoting elements in the peripheral nerve sheath, since at that time there was a general consensus that the environment in the central nervous system was unfavorable for axon growth. Much earlier studies by Cajal (1928) helped establish this notion by showing that although axons near the site of a spinal cord lesion in young cats and dogs sprouted new processes, the new growth atrophied and degenerated within weeks after the injury. The environment in the peripheral nervous system was thought to be much more permissive to new growth, and this advantage was conferred onto CNS neurons by the peripheral nerve transplants that were used by Aguayo and colleagues.

During the work in Aguayo's laboratory it became apparent that there also were internal factors within neurons that dictate the capacity for sprouting. They demonstrated the importance of the internal constraints by making peripheral nerve transfers to many different regions of the brain and cataloguing, for each area, the types of neurons that innervated the nerve transfers and the distance over which the sprouts traveled (for review see Aguayo, 1985). The simple assumption was that if the only requirement for new growth was a favorable environment, then the environment provided by the peripheral nerve transplant should confer the same advantages to all neurons near the transplant. If this were the case, all neuron types would be equally likely to sprout into the transplants. However, they found considerable differences in the capacity to sprout across regions of the brain and among cell types within a region (for review see Aguayo, 1985). Additionally, the likelihood of new growth appeared to be greatest if the cell was directly injured by the peripheral nerve transfer (Aguayo, 1985). This demonstrated that the environment was not the only factor involved in sprouting, but that additionally there are regulating factors within the neurons that, in situations like injury, can prompt some adult neurons to grow.

Recent evidence confirms that injury can trigger new growth in the adult brain, even without altering the environment within the central nervous sysytem. Instead, injury to peripheral branches of sensory nerves seems to facilitate new growth of the central projections of the injured neurons. In macaque monkeys, for example, we have found that crushing a sensory nerve to the hand can prompt sprouting of the central terminals of injured afferent axons beyond their normal territory in the dorsal column of the spinal cord (Florence *et al.*, 1993). Afferent axons from sensory receptors in the skin of the hand of monkeys normally terminate in a punctate topographic manner in the dorsal horn of the spinal cord and in the cuneate nucleus of the brain stem. This can be shown by making small injections of neuroanatomical tracers subcutaneously into the skin of the hand, and observing the distribution of the centrally transported label (Florence *et al.*, 1989, 1991). In monkeys that had a crush to one of the nerves of the hand, however, an injection into a digit on the injured hand produced more widespread label than normal; the label extended into spinal cord territories normally occupied by inputs from adjacent digits (Florence *et al.*, 1993). This enlargement of the input pattern likely resulted from sprouting of the terminal arbors of injured sensory afferents.

Similar conclusions have come from studies in rats, although there continues to be considerable controversy about what variables lead to sprouting and whether it even occurs (for reviews of early work see Pubols and Brenowitz, 1981; Wall, 1981; Devor, 1983; Micevych *et al.*, 1986; Goldberger and Murray, 1988; for later discussion, see McNeill and Hulsebosch, 1987; Molander *et al.*, 1988; Pubols and Bowen, 1988; LaMotte *et al.*, 1989; Polistina *et al.*, 1990; LaMotte and Kapadia, 1993; Mannion *et al.*, 1996; Doubell *et al.*, 1997). Compelling evidence for new growth in the rat spinal cord resulted from a multiple injury paradigm used by McMahon and Kett-White (1991). Their approach was to produce a zone of chronic denervation in the dorsal horn, and determine whether remaining inputs to the dorsal horn could sprout into the denervated area. The critical component of their experiment was that in some animals the chronic denervation was combined with peripheral nerve injury to 'prime' the remaining sensory afferents for growth into the denervated area (McMahon and Kett-White, 1991). The chronic injury was the 'spared root' preparation in which dorsal root ganglia were lesioned on either side of one intact dorsal root. The distribution of the inputs through this remaining dorsal root was labeled with a tracer that revealed afferent termination patterns. After the chronic injury alone, there was no evidence that the terminal distribution of the afferents in the remaining dorsal root had changed; however, if the nerve carrying these remaining sensory fibers was crushed coincident with the spared root preparation, the terminal distribution of the afferents became much larger than normal (McMahon and Kett-White, 1991). This indicated that the afferents had sprouted relatively long distances into the denervated portion of the dorsal horn.

Although the sprouting demonstrated by McMahon and Kett-White (1991) involved new growth within individual layers of the dorsal horn in rats, injury also can trigger sprouting of sensory afferents into different layers of the spinal cord. In normal adult rats, the large myelinated and the small unmyelinated sensory afferents have different laminar termination patterns that can be detected by tracers that preferentially label only one of the afferent types (Robertson and Grant, 1985). For example, cholera toxin subunit B bound to HRP (BHRP) labels the large myelinated sensory afferents in rats, and normally does not label inputs to layer II, the target of the small unmyelinated afferents. However after peripheral nerve injury, injection of BHRP into the injured nerve produced dense reaction product in layer II (Woolf *et al.*, 1992; 1995; Shortland and Woolf, 1993; Mannion *et al.*, 1996; Doubell *et al.*, 1997). The change in the pattern of labeling indicated that large myelinated afferents had sprouted into layer II. This was confirmed in a study by Koerber *et al.* (1994) where individual axons were injected and reconstructed for morphological analysis. Additionally, markers for the small unmyelinated afferents (which normally target layer II of the spinal cord) appear in deeper spinal cord layers (Cameron *et al.*, 1992), so the small afferents also appear to sprout as a result of a peripheral injury.

Although the evidence for new growth described thus far suggests that injury is an effective trigger for the growth of new connections, sprouting can also occur under many conditions where neurons are not directly involved in injury. Uninjured neurons sprout into territory that has been deprived of at least some of its normal excitatory input, much as was described by the laboratories of Raisman (1985) and Tsukahara (1985). This type of growth has been called collateral sprouting, and it occurs under a variety of conditions and in a number of regions of the brain, including the spinal cord (e.g. LaMotte, *et al.*, 1989; Polistina *et al.*, 1990; Woolf *et al.*, 1992; LaMotte and Kapadia, 1993; Koerber *et al.*, 1994; Mannion *et al.*, 1996; Doubell *et al.*, 1997), hippocampus (Crutcher *et al.*, 1981; Benedetti *et al.*, 1983; Crutcher, 1987; Rossi *et al.*, 1991, 1994) and cortex (Darian-Smith and Gilbert, 1994; Florence *et al.*, 1997) as well as in sympathetic neurons (McLachlan *et al.*, 1993; Chung *et al.*, 1996). Furthermore, increases in activation patterns (as opposed to the decreases in activity that result from injury) also have been shown to result in the growth of more extensive neuronal processes. The best known evidence for the importance of increased activity on cell structure is from the Greenough laboratory, which showed proliferation of dendritic branches on cortical neurons in adult rats after training in a reaching task (Greenough *et al.*, 1985; see also Schallert and Jones, 1993; Jones and Schallert, 1994; Jones *et al.*, 1996).

Despite overwhelming evidence that sprouting of new connections was possible in the adult central nervous system, the spatial extent of the new growth was restricted so that the potential for any significant functional contribution through this mechanism was doubted. Moreover, since most of the reorganization of sensory representations in cortex could be easily explained by local changes in synaptic strength, there was no impetus to invoke sprouting as an important factor in the functional reorganization of pathways in the adult central nervous system. This situation dramatically changed in the early 1990s when it was demonstrated that remarkable cortical reorganization occurred in monkeys, which was so massive it could not be explained by any of the plasticity mechanisms based on local adjustments in synaptic efficacy.

NEW GROWTH AS A MECHANISM OF FUNCTIONAL REORGANIZATION

In monkeys that had total deafferentation of an upper extremity many years earlier, Pons and coworkers (1991) used microelectrode mapping techniques to show that the large zone of cortex where the forearm was normally represented was not silent, as might have been expected, since all sensory inputs had been removed. Instead, a markedly expanded representation of the face occupied the deprived cortical territory (Fig. 11.2; Pons *et al.*, 1991). Because the changes in cortical organization were spatially so extensive (on the order of 10 mm), they could not be explained by modifications of existing connections. Thus, it was speculated that new growth must have occurred (for alternative interpretation,

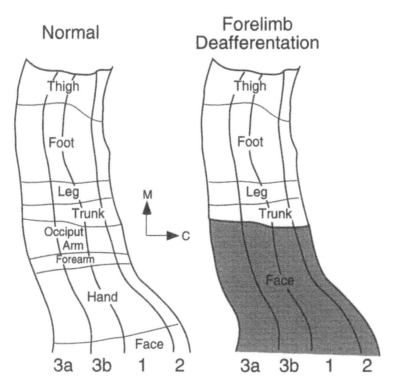

Figure 11.2 Schematics of the topographic representations of the body surface in soma-tosensory cortex (areas 3a, 3b, 1 and 2) in normal macaque monkeys (to the left) and after long-standing forelimb deafferentation (to the right). In normal macaques, the forelimb representation occupies a large extent of somatosensory cortex between the representa-tions of the face, laterally, and the trunk, medially. After cervical dorsal rhizotomy, which eliminates all sensory information from the forelimb and adjoining neck region, the deafferented part of cortex contains an expanded representation of the face. The extent of this cortical reorganization is too massive to be explained by local changes in synaptic efficacy. Body part borders are marked by horizontal lines, and areal borders are indicated by thicker vertical lines. C, caudal; M, medial. (Taken from Pons *et al.*, 1991).

see Lund *et al.*, 1994). Unfortunately, since the mapping method reveals only the functional makeup of neurons, Pons *et al.* (1991) were unable to investigate whether sprouting of connections was responsible for the change in cortical organization.

Not long after the finding by Pons and colleagues was published, Ramachandran and coworkers found that in humans with forelimb amputation, touch to the face could prompt distinct sensations of touch to the phantom forelimb (Ramachandran, 1992; Ramachandran *et al.*, 1993). Later, non-invasive imaging studies showed that the region of cortex where the hand is normally represented is reactivated in individuals with limb amputations by inputs from the face or the arm (Elbert *et al.*, 1994; Flor *et al.*, 1995; Yang *et al.*, 1994), much as was found in the monkeys studied by Pons *et al.* (1991). Moreover, the abnormal sensory perceptions were attributed to the reorganized topography of the forelimb repre-sentations in cortex. Thus, the data sparked a renewed interest in the search for mechanisms

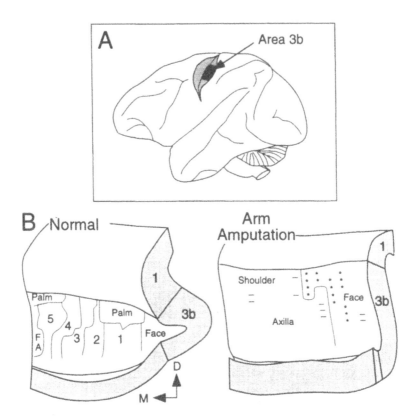

Figure 11.3 **A:** Lateral brain view showing the location of primary somatosensory cortex, area 3b, on the posterior bank of the central sulcus is macaque monkeys. The approximate position of the hand representation in indicated by gray shading. **B:** Expanded views of area 3b that are rotated to a nearly frontal plane so that the topography of the sensory representation can be shown in an *en face* view. Both drawings illustrate the portion of area 3b that is normally occupied by the hand representation. In normal monkeys (left), digit 1 is represented laterally, followed in medial sequence by representations of digits 2–5, the wrist, forearm and arm. After arm amputation (right), this region becomes reactivated by an expanded cutaneous representation of the face laterally, and the area bordering the amputated stump medially. At some recording sites, neurons responded to sensory stimulation to both the face and shoulder (indicated by asterisks), and at some recording sites, neurons responded only to non-cutaneous sensory stimulation (indicated by dashes). A, anterior; D, dorsal; M, medial. (Taken from Florence and Kaas, 1995).

such as new growth that could produce such massive functional changes in the adult central nervous system.

We had the opportunity to directly test whether new growth might be involved in large-scale cortical reorganization by combining both anatomical and electrophysiological analyses in monkeys with previous amputation of the hand or arm for veterinary treatment of accidental injuries (Florence and Kaas, 1995). Consistent with the findings of Pons *et al.* (1991), we found that the derived region of cortex was not deactivated (Fig. 11.3). Instead, most of the neurons had acquired new receptive fields on either the remaining stump of the arm or on the face (Fig. 11.3). At some of the recording sites, neurons even had receptive fields on both the arm and face, a combination not found in normal monkeys (Fig. 3). Moreover, the total extent of the cortical reorganization was very large (on the order of about 6 mm), and the pattern of the change could not be explained simply by adjustments in the strength of any known previously existing connections. Thus, some sprouting of connections was the only feasible explanation.

If the growth of new connections into the deprived hand representation was responsible for the cortical change, we predicted that it would occur in one of the subcortical relay nuclei, where the scale of the topographic representations is considerably smaller. For example, in either the somatosensory thalamus or brain stem, changes in the distribution of connections on the order of a few millimeters could produce changes of a much larger *apparent* size when relayed to cortex. Secondly, we knew from our earlier work that peripheral injuries can lead to up-regulation of growth-associated proteins (e.g. GAP-43) in the brain stem terminations of peripheral nerve afferents (Jain *et al.*, 1995). Thus, we reasoned that the most likely site of sprouting, if it occurred in monkeys with amputation, would be in the somatosensory brain stem, specifically the cuneate nucleus which relays sensory information from the forelimb (e.g. Florence *et al.*, 1989, 1991).

To detect changes in the distribution of inputs to the cuneate nucleus in the amputee monkeys, we made small injections of the neuroanatomical marker, cholera toxin subunit B conjugated to horseradish peroxidase (BHRP), into the skin of the remaining arm just proximal to the stump. Additionally, BHRP injections were made into the matched locations of the arm on the other side of the body, so that for each monkey there was a within-animal comparison. A marked difference was readily apparent in the distribution of label on the side of the amputation compared to the controls (Fig. 11.4; see also Florence and Kaas, 1995). Label from the wrist, forearm and arm on the control side formed an arch dorsally in the nucleus; whereas on the side of the amputation the label was widely distributed, and occupied ventral regions of the cuneate nucleus that normally receive inputs from the digits and palm of the hand (Fig. 11.4). This indicates that inputs from the remaining part of the forelimb had sprouted into the zone where the deafferented hand inputs terminate. This reorganization could lead to new patterns of activation in the cuneate nucleus, such that the cells normally activated by stimulation of the hand come to be activated by stimuli to the arm. If this change were expressed at the cortical level, a large expanse of reorganization would be likely. Thus, originally we attributed the changes we had seen in the functional organization of somatosensory cortex to changes that took place more peripherally in the distribution of primary afferent terminations.

Because the density of the expanded inputs observed in the cuneate nucleus after amputation was sparse, we reasoned that the impact of the reorganization at this level would have the most effect on higher-order processing stations if potentiated by additional changes that occur at thalamic and cortical levels. These higher-level changes could be limited to the sorts of local alterations that appear to mediate the use-dependent changes in sensory representations. However, we did not rule out the possibility that sprouting also could occur at other levels of the pathway after such a profound injury. Thus, in a second series of monkeys, we have studied the patterns of corticocortical and thalamocortical connections that had either amputation or injury to the hand. To test whether the changes in cortical topography after forelimb injury were accompanied by structural changes in the deprived sensory representation, we injected minute volumes of neuroanatomical tracers into a part of the reorganized cortical map (i.e. into the zone that would likely have

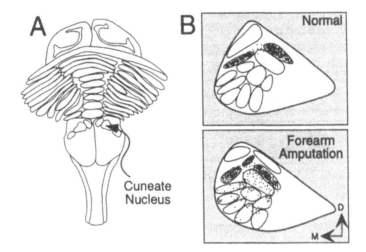

Figure 11.4 A: Caudal view of the brain showing the approximate location of the cuneate nucleus in the brain stem. **B:** Drawings of frontal sections through the cuneate nucleus showing the distribution of label following matched, bilateral injections of horseradish peroxidase conjugated cholera toxin subunit B into the forearms of a macaque monkey which had had a prevoius forearm amputation. The distribution of label after injections of the normal arm is shown to the top (right). Note that the label is confined to the dorsal half of the nucleus. In contrast, injections of the amputated arm in the same location as on the normal side label a much more extensive region of the cuneate nucleus, extending ventrally where inputs from the digits and hand are concentrated (bottom right). This change in the distribution of label has been interpreted as sprouting of the forearm afferents into the deprived hand representation. Both nuclei have been oriented so that medial is to the left and dorsal is to the top for purposes of comparison. D, dorsal; M, medial. (Taken from Florence and Kaas, 1995).

contained the hand representation). In normal monkeys, such an injection labels a zone in cortex that is roughly topographically matched to the region injected; for example, if the representation of D1 in area 1 is injected, the projection is concentrated in the D1 representation in area 3b and extends only a millimeter or so into neighboring representations (Fig. 11.5; Florence *et al.*, 1998). The total mediolateral extent of label produced by a mm diameter injection in normal monkeys ranges from 3–4 mm. However, in the three monkeys with longstanding injuries to the hand, the distribution of cortical label was much more widely distributed (Fig. 11.5). For example, in a monkey that had amputation of the fingers on one hand, an injection that was matched in size to those in the normal monkeys produced a labeled zone that spanned 6.2 mm in total mediolateral extent. In a monkey that had a paralyzed hand, the span of label was even larger (more than 7 mm), and in a monkey with a complete arm amputation, the labeled zone was larger still, extending 8.9 mm mediolaterally (Fig. 11.7). This indicates that, as a result of the injury, cortical neurons had sprouted new axon processes over relatively long distances into and out of the deprived region of cortex.

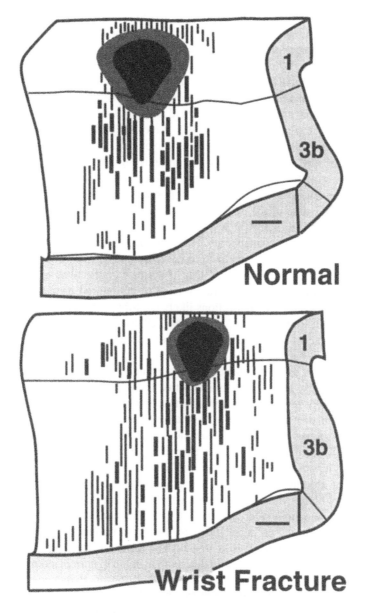

Figure 11.5 Drawings of the distribution of label in primary somatosensory cortex, area 3b, in a normal macaque monkey and in a macaque that had a long-standing wrist fracture after an injection of wheat germ agglutinnin conjugated to horseradish peroxidase (WGA-HRP) into area 1. Plots of the distribution of label were made in series of parasagittal sections through somatosensory cortex, then the drawings were aligned and rotated using the Bioquant Image Analysis System (R&M Biometrics, Nashville, TN) to give an *en face* view of the hand representation in area 3b. The locations of label in each section processed for WGA-HRP reaction produce has been indicated by a vertical black line; thicker lines indicate dense label. Scale bars correspond to 1 mm. (Taken from Florence *et al.*, 1998).

Similar evidence for the extension of cortical connections in regions of deactivation come from experiments where bilateral retinal lesions deprive a region of primary visual cortex of its normal source of activation. Injections of tracers in non-deprived cortex, near the region of the deactivation, produced more dense intrinsic connections than in normal animals (Darian-Smith and Gilbert, 1994). This change was attributed to axon sprouting, and it was thought to subserve the reorganization of the deprived region in cortex that leads to an expanded representation of the visual field adjacent to the lesion (Kaas *et al.*, 1990; Chino *et al.*, 1992; Gilbert and Wiesel, 1992). Importantly, the new growth observed in visual cortex was limited to local proliferation of axon processes; the new connections did not extend beyond the normal radius of the pre-existing network of connections. Thus, although the sprouting likely had an important influence on the reorganization of primary visual cortex, the extent of the growth was topographically confined.

By contrast, the extent of the new cortical connections in monkeys with longstanding forelimb injuries was remarkably widespread, and extended well beyond the normal limit of somatosensory cortical connections. These connections may provide an additional source of activation for the deprived hand representation; however, the specific functional impact of these connections remains unclear. If the new connections are excitatory, they could provide activation of cells in cortex related to the injured or missing hand. If this were the case, the reorganized connections likely contribute to 'phantom-limb' sensations (e.g. Melzack, 1990; Ramachandran *et al.*, 1992, 1993) or perhaps to phantom pain (e.g. Flor *et al.*, 1995; Knecht *et al.*, 1996). Alternatively, if the new connections are inhibitory, they may help reduce the abnormal activity that produces phantom sensations or pain. The symptoms of phantom limb and phantom pain are transitory, and over time may diminish or disappear (Melzack, 1990). This indicates that there may be processes that counteract the unwanted activity responsible for phantom sensations, and cortical connections such as these are candidates for this sort of regulatory role.

Remarkably, extensive new growth does not appear to occur at all levels of the somatosensory system. As mentioned above, after longstanding injuries to the hand or arm, primary afferent inputs sprout into the deprived representation in the cuneate nucleus (and in the spinal cord, although these data were not discussed); the injury also leads to sprouting in somatosensory cortex. Yet there does not appear to be any detectable expansion of the thalamocortical connections (Florence *et al.*, 1998). The cortical injections in the monkeys that had forelimb injuries produced label in the ventroposterior lateral nucleus of the thalamus, the primary relay nucleus to area 3b. The thalamic label in the injured monkeys was comparable in width, density and mediolateral location to normal monkeys. Thus, there appears to be little or no effect of the forelimb injury on the thalamocortical projection pattern. Similarly, no changes have been found in the distribution of connections in the visual relay nucleus of the thalamus, the lateral geniculate nucleus, after retinal lesions (Stelzner and Keating, 1977; Eysel, 1982; see also Darian-Smith and Gilbert, 1995). Thus, thalamic neurons appear to have little potential for new growth. This possibility is supported by the earlier work of Aguayo (for review see Aguayo, 1985) which showed that, in the thalamus, the majority of neurons that sprouted into peripheral nerve transplants were from the reticular nucleus.

The remarkable difference in the ability of neurons at some levels of the brain to sprout, but not in others, raises the question of how the signal to sprout is relayed. This seems to be an especially perplexing challenge given that after amputation or retinal lesion the signal must be transmitted through the thalamus, which shows little if any response to the sensory deprivation. The best known mechanism for a transynaptic relay of experience-dependent changes is the Hebbian/Stent model for synaptic modifications based on correlations and discorrelations in activity patterns (Hebb, 1949; Stent, 1973). Perhaps activity-dependent modifications can act as a trigger for new growth in regions of the brain that have the potential to sprout. Alternatively, or perhaps in conjunction with the activity-dependent effects, there may be injury-related molecular signals that can be transported transynaptically to initiate growth at cortical levels. As reviewed briefly below, molecular signals play an important role in synaptic plasticity, particularly with regard to new growth.

MOLECULAR CONSTRAINTS ON NEW GROWTH

An important next step in plasticity research is to determine what internal and external events trigger the growth. Toward this end, there has been a virtual explosion of work on neurotrophic factors and growth-associated proteins, both of which appear to play impor-tant roles in adult neuroplasticity. Both are normally present in high concentrations during development and are markedly down-regulated after maturity, but can be re-expressed after injury. Neurotrophic factors include a family of proteins known to be associated with growth, including nerve growth factor (NGF), that exert their effects on neurons through cell surface receptors (for review see Levi-Montalcini, 1987). Important sources of neurotrophins are the target organs, such as muscle, which are thought to synthesize and secrete the proteins for regulation of the nervous innervation (Levi-Montalcini, 1987). In addition, neurotrophins are synthesized by neuronal elements and they presumably regulate the survival and differentiation of developing neurons and maintain some of the populations of adult neurons (for review see Thoenen, 1995).

Of the growth factors studied, the best known is the membrane-bound protein, growth associated protein 43 (GAP-43), whose synthesis was found to be dramatically upregulated in injured neurons. During nerve regeneration in rabbits (Skene and Willard, 1981a), toads (Skene and Willard, 1981b) and fish (Benowitz *et al.*, 1981; Benowitz and Lewis, 1983) the upregulation of GAP-43 followed a time course that mirrored that of axon elongation, so it was clearly indicated as an important component of the growth response. Consistent with the role of GAP-43 in growth was the evidence that in preparations where nerve regeneration did not occur after injury, such as in the mammalian optic nerve, the concentration of GAP-43 was not elevated (for review see Skene, 1989). Thus, not only was there a positive correlation between the presence of GAP-43 and regeneration, but also the converse negative correlation. More recent studies have shown that GAP-43 is also elevated in the central axon processes of dorsal root neurons whose peripheral processes have been injured, for example in the dorsal horn of adult rats (Woolf *et al.*, 1990) and in the cuneate nucleus of infant monkeys (Jain *et al.*, 1995). It is assumed that these central processes are capable of new growth during the period of elevated GAP-43 levels (for discussion see Jain *et al.*, 1995).

The mechanism by which injury triggers new growth appears to involve a retrograde signal that is transported from the site of injury to the cell body; in the soma an upregulation of the expression of one or more proteins acts to 'switch' neurons from a mode in which only local axon proliferation is produced into an axon elongation mode of growth (Smith and Skene, 1997). Given differences among neuronal types in the capacity for growth, this mechanism may be cell specific. For example, the potential of some cells for growth may lie in the ability to be 'switched' into a growth mode. Alternatively, perhaps the effective-ness of the different growth-promoting proteins is highly specific for different types of cells. Studies to test this possibility are well underway, and involve the delivery of high concentrations of neurotrophic factors into the peripheral (Fitzgerald *et al.*, 1985; Lewin *et al.*, 1992; Cohen-Cory and Fraser, 1995; Sawai *et al.*, 1996) or central nervous systems (e.g. Isaacson *et al.*, 1995; Mamounas *et al.*, 1995; Oyesiku and Wigston, 1996; Tuszynski *et al.*, 1996), or most recently by producing transgenic mice bred to produce excess quantities of neurotrophic factors (e.g. Holtmaat *et al.*, 1995). The findings confirm that although the presence of the growth-promoting factors can prompt some cells to proliferate new growth, there are vast differences in the effectiveness of neurotrophic factors across cell types. Apparently, cells must possess receptors specific for a neurotrophic factor in order for it to be incorporated into the cell body, where its action is mediated, and different cell types possess different receptors (for review see Ebadi *et al.*, 1997). Therefore, before neurotrophic factors can be effectively used to promote growth for therapeutic purposes, for example in stroke or neurological disease, the action of each of the neurotrophic factors on specific cell types must be identified.

Finally, there also appear to be growth inhibitors that may counteract the facilitory action of neurotrophic factors. A class of molecules have been identified that repulse growing axon tips (for review see Schwab, 1996). The role of these inhibitors during development appears to be axon guidance, like 'Do Not Enter' signs for growing axons. Elimination of myelin, where some of the inhibitory factors exist, led to highly aberrant growth of corticospinal axons in young rats (for review see Schwab, 1996). In the adult brain the role of growth inhibitors seems to the stabilization of existing connections. For example, Schwab and colleagues (see Schwab, 1996) showed that elimination of myelin resulted in considerable collateral sprouting.

To maximize axon growth in the adult central nervous system, the ideal experimental condition would involve simultaneous administration of growth-promoting agents and blockade of growth inhibitors. In the case of injury, which for some cells is akin to administration of growth factors, blockade of growth inhibitors might allow for much more substantial growth than has been shown to date. Schwab's group tested this theory by eliminating the myelin-forming substrate (since growth inhibitors are located in myelin) or by administration of antibodies against growth inhibitors, in rats that had hippocampal, spinal cord or optic tract lesions (for review see Schwab, 1996). They found considerably more extensive growth than if inhibitors had not been blocked. More recently, Bregman *et al.* (1995) used this same experimental paradigm to show dramatically improved recovery from spinal cord injury using antibodies to growth inhibitors. Thus it appears that the ideal circumstances for new growth in the adult nervous system, and ultimately perhaps maximal recovery from injury, are the availability of growth promoters and blockade of growth inhibitors.

CONCLUSIONS

Although the most common mechanisms for reorganization in the adult brain involve modifications of existing connections, the growth of new connections (sprouting) also appears to be a factor in functional neuroplasticity. Most recently, new growth has been implicated in the reactivation of large populations of neurons in the somatosensory pathway. For example, growth occurs in the spinal cord and lower brain stem, and can involve the central sprouting of both injured and uninjured peripheral nerve axons when nearby synaptic territories are vacated by the removal of other peripheral nerve inputs. Similarly in cortex, local-circuit neurons have the potential for considerable new growth. Extensions of intracortical axons following peripheral denervations can achieve quite dramatic proportions. In contrast, neurons in the thalamus display no potential for new growth, even in circumstances such as sensory denervation, where vacant synaptic space is likely to be abundant. An important future direction for research will be to understand what initiates growth of new connections in cortex after peripheral disturbances, and why little or no effect of the peripheral manipulation is detected in thalamus or the thalamic relay to cortex.

We do not know yet if the potential for new growth is present in other cortical areas, but the possibility seems likely. In cases of cortical lesions, such as those of motor cortex (Nudo and Milliken, 1996), the processes of undamaged neurons may grow to compensate for the roles of the lost neurons. Perhaps also, neurons whose processes are damaged by the lesions sprout new connections. That such changes are likely is suggested by the considerable behavioral recoveries that follow restricted cortical damage (Cohen *et al.*, 1993; Castro-Alamancos and Borrell, 1995; Nudo *et al.*, 1996). Clearly, the potential for such changes in cortical circuitry needs to be evaluated further.

The functional consequences of new growth in the central nervous system also need to be defined. The effects of growth are likely to be varied, producing the desired therapeutic outcomes in some instances, but possibly leading to misperceptions and unwanted outcomes, such as pain, in others. Thus, it is important to determine further the conditions and extents of new growth, and obtain an understanding of how new growth can be promoted or prevented to maximize therapeutic potential.

REFERENCES

Aguayo, A.J. (1985). Axonal regeneration from injured neurons in the adult mammalian central nervous system. In *Synaptic Plasticity*, edited by C. W. Cotman, pp. 457–484. New York: Guilford Press.

Aguayo, A.J., Bray, G.M., Rasminsky, M. Zwimpfer, T. Carter, D. and Vidal-Sanz, M. (1990). Synaptic connections made by axons regenerating in the central nervous system of adult mammals. *Journal of Experimental Biology*, **153**, 199–224.

Armstrong-James, M., Diamond, M.E. and Ebner, F.F. (1994). An innocuous bias in whisker use in adult rats modifies receptive fields of barrel cortex neurons. *Journal of Neuroscience*, **14**, 6978–6991.

Basbaum, A.I. and Wall, P.D. (1976). Chronic changes in the response of cells in adult cat dorsal horn following partial deafferentation: The appearance of responding cells in a previously non-responsive region. *Brain Research*, **116**, 181–204.

Benedetti, F., Montarolo, P.G., Strata, P. and Tosi, L. (1983). Collateral reinnervation in the olivocerebellar pathway in the rat. *Birth Defects: Original Article Series*, **19**, 461–464.

Benowitz, L.I. and Lewis, E.R. (1983). Increased transport of 44,000- to 49,000-dalton acidic proteins during regeneration of the goldfish optic nerve: A two-dimensional gel analysis. *Journal of Neuroscience*, **3**, 2153–2163.

Benowitz, L.I., Shashoua, V.E. and Yoon, M.G. (1981). Specific changes in rapidly transported proteins during regeneration of the goldfish optic nerve. *Journal of Neuroscience*, **1**, 300–307.

Bregman, B.S., Kunkel-Bagden, E., Schnell, L., Dai, H.N., Gao, D. and Schwab, M.E. (1995). Recovery from spinal cord injury mediated by antibodies to neurite growth inhibitors. *Nature*, **378**, 498–501.

Cajal, S. and Ramon, Y. (1928). *Degeneration and Regeneration of the Nervous System*. London: Oxford University Press.

Castro-Alamancos, M.A. and Borrell, J. (1995). Functional recovery of forelimb response capacity after forelimb primary motor cortex damage in the rat is due to the reorganization of adjacent areas of cortex. *Neuroscience*, **68**, 793–805.

Chino, Y.M., Kaas, J.H., Smith, E.L. III., Langston, A.L. and Cheng, H. (1992). Rapid reorganization of cortical maps in adult cats following restricted deafferentation in retina. *Vision Research*, **32**, 789–796.

Chung, K., Lee, B.H., Yoon, Y.W. and Chung, J.M. (1996). Sympathetic sprouting in the dorsal root ganglia of the injured peripheral nerve in a rat neuropathic pain model. *Journal of Comparative Neurology*, **376**, 241–252.

Cohen, L., Brasil-Neto, J.P., Pascual-Leone, A. and Hallett, M. (1993). Plasticity of cortical motor output organization following deafferentation, cerebral lesions, and skill acquisition. In *Electrical and magnetic stimulation of the brain and spinal cord*, edited by O. Devinsky, A. Beric and M. Dogali, pp.187–200. New York: Raven Press, Ltd.

Cohen-Cory, S. and Fraser, S.E. (1995). Effects of brain-derived neurotrophic factor on optic axon branching and remodeling *in vivo*. *Nature*, **378**, 192–194.

Crutcher, K.A. (1987). Sympathetic sprouting in the central nervous system: a model for studies of axonal growth in the mature mammalian brain. *Brain Research*, **434**, 203–233.

Crutcher, K.A., Brothers, L. and Davis, J.N. (1981). Sympathetic noradrenergic sprouting in response to central cholinergic denervation; a histochemical study of neuronal sprouting in the rat hippocampal formation. *Brain Research*, **210**, 115–128.

Darian-Smith, C. and Gilbert, C.D. (1994). Axonal sprouting accompanies functional reorganization in adult cat striate cortex. *Nature*, **368**, 737–740.

Darian-Smith, C. and Gilbert, C.D. (1995). Topographic reorganization in the striate cortex of the adult cat and monkey is cortically mediated. *Journal of Neuroscience*, **15**, 1631–1647.

Devor, M. (1983). Plasticity of spinal cord somatotopy in adult mammals: Involvement of relatively ineffective synapses. *Birth Defects: Original article series*, **19**, 287–314.

Devor, M. and Wall, P.D. (1981). Plasticity in the spinal cord sensory map following peripheral nerve injury in rats. *Journal of Neuroscience*, **1**, 679–684.

Diamond, M.E., Huang, W. and Ebner, F.F. (1994). Laminar comparison of somatosensory cortical plasticity. *Science*, **265**, 1885–1888.

Dostrovsky, J.O., Millar, J. and Wall, P.D. (1976). The immediate shift of afferent drive of dorsal column nucleus cells following deafferentation: a comparison of acute and chronic deafferentation in gracile nucleus and spinal cord. *Experimental Neurology*, **52**, 480–495.

Doubell, T.P., Mannion, R.J. and Woolf, C.J. (1997). Intact sciatic myelinated primary afferent terminals collaterally sprout in the adult rat dorsal horn following section of a neighboring peripheral nerve. *Journal of Comparative Neurology*, **380**, 95–104.

Dykes, R.W., Avendano, C. and Leclerc, S.S. (1995). Evolution of cortical responsiveness subsequent to multiple forelimb nerve transections: An electrophysiological study in adult cat somatosensory cortex. *Journal of Comparative Neurology*, **354**, 333–344.

Ebadi, M., Bashir, R.M., Heidrick, M.L., Hamada, F.M., El Refaey, H., Hamed, A. Helal, G., Baxi, M.D., Cerutis, D.R. and Lassi, N.K. (1997). Neurotrophins and their receptors in nerve injury and repair. *Neurochemistry International*, **30**, 347–374.

Elbert, T., Flor, H., Birbaumer, N., Knecht, S., Hampson, S., Larbig, W. and Taub, E. (1994). Extensive reorganization of the somatosensory cortex in adult humans after nervous system injury. *NeuroReport*, **5**, 2593–2597.

Eysel, U.T. (1982). Functional reconnections without new axonal growth in a partially denervated visual relay nucleus. *Nature*, **299**, 442–444.

Fitzgerald, M., Wall, P.D., Goedert, M. and Emson, P.C. (1985). Nerve growth factor counteracts the neurophysiological and neurochemical effects of chronic sciatic nerve section. *Brain Research*, **332**, 131–141.

Flor, H., Elbert, T. Knecht, S. Wienbruch, C. Pantev, C. Birbaumer, N. Larbig, W. and Taub, E. (1995). Phantom-limb pain as a perceptual correlate of cortical reorganization following arm amputation. *Nature*, **375**, 482–484.

Florence, S.L., Garraghty, P.E., Carlson, M. and Kaas, J.H. (1993). Sprouting of peripheral nerve axons in the spinal cord of monkeys. *Brain Research*, **601**, 343–348.

Florence, S.L. and Kaas, J.H. (1995). Large-scale reorganization at multiple levels of the somatosensory pathway follows therapeutic amputation of the hand in monkeys. *Journal of Neuroscience*, **15**, 8083–8095.

Florence, S.L., Taub, H.B. and Kaas, J.H. (1998). Large-scale sprouting of cortical connections after peripheral injury in adult macaque monkeys. *Science*, **282**, 1117–1121.

Florence, S.L., Wall, J.T. and Kaas, J.H. (1989). Somatotopic organization of inputs from the hand to the spinal grey and cuneate nucleus of monkeys with observations on the cuneate nucleus of humans. *Journal of Comparative Neurology*, **286**, 48–70.

Florence, S.S., Wall, J.T. and Kaas, J.H. (1991). Central projections from the skin of the hand in squirrel monkeys. *Journal of Comparative Neurology,* **311,** 563–578.

Garraghty, P.E., Hanes, D.P., Florence, S.L. and Kaas, J.H. (1994). Pattern of peripheral deafferentation predicts reorganizational limits in adult primate somatosensory cortex. *Somatosensory and Motor Research,* **11,** 109–117.

Garraghty, P.E. and Kaas, J.H. (1991). Functional reorganization in adult monkey thalamus after peripheral nerve injury. *NeuroReport,* **2,** 747–750.

Garraghty, P.E., LaChica, E.A. and Kaas, J.H. (1991). Injury-induced reorganization of somatosensory cortex is accompanied by reductions in GABA staining. *Somatosensory and Motor Research,* **8,** 347–354.

Garraghty, P.E., Pons, T.P., Sur, M. and Kaas, J.H. (1989). The arbors of axons terminating in the middle cortical layers of somatosensory area 3b in owl monkeys. *Somatosensory and Motor Research,* **6,** 401–411.

Garraghty, P.E. and Sur, M. (1990). Morphology of single intracellularly stained axons terminating in area 3b of macaque monkeys. *Journal of Comparative Neurology,* **294,** 583–593.

Gilbert, C.D. and Wiesel, T.N. (1992). Receptive field dynamics in adult primary visual cortex. *Nature,* **356,** 150–152.

Goldberger, M.E. and Murray, M. (1988). Patterns of sprouting and implications for recovery of function. *Advances of Neurology,* **47,** 361–383.

Greenough, W.T., Larson, J.R. and Withers, G.S. (1985). Effects of unilateral and bilateral training in a reaching task on dendritic branching of neurons in the rat motor-sensory forelimb cortex. *Behavioral and Neural Biology,* **44,** 361–314.

Hebb, D.O. (1949). *The Organization of Behavior: A Neuropsychological Theory.* New York: Wiley.

Holtmaat, A.J.G.D., Dijhuizen, P.A., Oestreicher, A.B., Romijn, H.J., Van der Lugt, N.M.T., Berns, A., Maragolis, F.L., Gispen, W.H. and Verhaagen, J. (1995). Directed expression of the growth-associated protein B-50/GAP-43 to olfactory neurons in transgenic mice results in changes in axon morphology and extraglomerular fiber growth. *Journal of Neuroscience,* **15,** 7953–7965.

Isaacson, L.G., Ondris, D. and Crutcher, K.A. (1995). Plasticity of mature sensory cerebrovascular axons following intracranial infusion of nerve growth factor. *Journal of Comparative Neurology,* **361,** 451–460.

Jain, N., Florence, S.L. and Kaas, J.H. (1995). GAP–43 expression in the medulla of macaque monkeys: changes during postnatal development and the effects of early median nerve repair. *Developmental Brain Research,* **90,** 24–34.

Jones, T., Kleim, J. and Greenough, W. (1996). Synaptogenesis and dendritic growth in the cortex opposite unilateral sensorimotor cortex damage in adult rats: a quantitative electron microscopic examination. *Brain Research,* **733,** 142–148.

Jones, T.A. and Schallert, T. (1994). Use-dependent growth of pyramidal neurons after neocortical damage. *Journal of Neuroscience,* **14,** 2140–2154.

Kaas, J.H. (1991). Plasticity of sensory and motor maps in adult mammals. (1991). *Annual Review of Neuroscience,* **14,** 137–167.

Kaas, J.H., Krubitzer, L.A., Chino, Y.M., Langston, A.L., Polley, E.H.A. and Blair, N. (1990). Reorganization of retinotopic cortical maps in adult mammals after lesions of the retina. *Science,* **248,** 229–231.

Knecht, S., Henningsen, H., Elbert, T., Flor, H., Hohling, C., Pantev, C. and Taub, E. (1996). Reorganizational and perceptual changes after amputation. *Brain,* **119,** 1213–1219.

Koerber, H.R., Mirnics, B.P.B. and Mendell, L.M. (1994). Central sprouting and functional plasticity of regenerated primary afferents. *Journal of Neuroscience,* **14,** 3655–3677.

LaMotte, C.C. and Kapadia, S.E. (1993). Deafferentation induced terminal field expansion of myelinated saphenous afferents in the adult rat dorsal horn and the nucleus gracilis following pronase injection of the sciatic nerve. *Journal of Comparative Neurology,* **330,** 83–94.

LaMotte, C.C., Kapadia S.E. and Kocol, C.E. (1989). Deafferentation induced expansion of sciatic terminal field labeling in the adult rat dorsal horn following pronase injection of the sciatic nerve. *Journal of Comparative Neurology,* **288,** 311–325.

Levi-Montalcini, R. (1987). The nerve growth factor 35 years later. *Science,* **237,** 1154–1162.

Lewin, G.R., Winter, J. and McMahon, S.B. (1992). Regulation of afferent connectivity in the adult spinal cord by nerve growth factor. *European Journal of Neuroscience,* **4,** 700–707.

Lipton, S.A. and Kater, S.B. (1989). Neurotransmitter regulation of neuronal outgrowth, plasticity and survival. *Trends in Neurosciences,* **12,** 265–270.

Lund, J.P., Sun, G.-D. and Lamarre, Y. (1994). Cortical reorganization and deafferentation in adult macaques. *Science*, **265**, 544–548.

Mamounas, L.A., Blue, M.E., Siuciak, J.A. and Altar, C.A. (1995). Brain-derived neurotrophic factor promotes the survival and sprouting of serotonergic axons in rat brain. *Journal of Neuroscience*, **15**, 7929–7939.

Mannion, R.J., Doubell, T.P., Coggeshall, R.E. and Woolf, C.J. (1996). Collateral sprouting of injured primary afferent A-fibers into the superficial dorsal horn of the adult rat spinal cord after topical capsaicin treatment to the sciatic nerve. *Journal of Neuroscience*, **16**, 5189–5195.

McLachlan, E.M., Janig, W., Devor, M. and Michaelis, M. (1993). Peripheral-nerve injury triggers noradrenergic sprouting within dorsal-root ganglia. *Nature,* **363**, 543–546.

McMahon, S.B. and Kett-White, R. (1991). Sprouting of peripheral regenerating primary sensory neurones in the adult central nervous system. *Journal of Comparative Neurology*, **304**, 307–315.

McMahon, S.B. and Wall, P.D. (1983). Plasticity in the nucleus gracilis of the rat. *Experimental Neurology*, **80**, 195–207.

McNeill, D.L. and Hulsebosch, C.E. (1987). Intraspinal sprouting of rat primary afferents after deafferentation. *Neuroscience Letters*, **81**, 57–62.

Melzack, R. (1990). Phantom limbs and the concept of a neuromatrix. *Trends in Neuroscience*, **13**, 88–92.

Mendell, L.M., Sassoon, E.M. and Wall, P.D. (1978). Properties of synaptic linkage from 'distant' afferents onto dorsal horn neurons in normal and chronically deafferented cats. *Journal of Physiology*, **285**, 299–310.

Merzenich, M.M., Kaas, J.H., Wall, J.T., Nelson, J.R., Sur, M. and Felleman, D.J. (1983a). Topographic reorganization of somatosensory cortical areas 3b and 1 in adult monkeys following restricted deafferentation. *Neuroscience*, **8**, 33–55.

Merzenich, M.M., Kaas, J.H. Wall, J.T. Sur, M. Nelson, R.J. and Felleman, D.J. (1983b). Progression of change following median nerve section in the cortical representation of the hand in areas 3b and 1 in adult owl and squirrel monkeys. *Neuroscience*, **10**, 639–665.

Merzenich, M.M., Nelson, R.J., Stryker, M.P., Cynader, M.S., Schoppmann, A. and Zook, J.M. (1984). Somatosensory cortical map changes following digit amputation in adult monkeys. *Journal of Comparative Neurology*, **224**, 591–605.

Micevych, P.E., Rodin, B.E. and Kruger, L. (1986). The controversial nature of the evidence for neuroplasticity of afferent axons in the spinal cord. In *Spinal afferent processing*, edited by T.L. Yaksh, pp. 417–443. New York: Plenum Press.

Millar, J., Basbaum, A.F. and Wall, P.D. (1976). Restructuring of the somatotopic map and appearance of abnormal neuronal activity in the gracile nucleus after partial deafferentation. *Experimental Neurology*, **50**, 658–672.

Molander, C., Kinnman, E. and Aldskogius, H. (1988). Expansion of spinal cord primary sensory afferent projection following combined sciatic nerve resection and saphenous nerve crush: A horseradish peroxidase study in the adult rat. *Journal of Comparative Neurology*, **276**, 436–441.

Nicolelis, M.A.L., Lin, R.C.S. Woodward, D.J. and Chapin, J.K. (1993). Induction of immediate spatiotemporal changes in thalamic networks by peripheral block of ascending cutaneous information. *Nature*, **361**, 533–536.

Northgrave, S.A. and Rasmusson, D.D. (1996). The immediate effects of peripheral deafferentation on neurons of the cuneate nucleus in raccoons. *Somatosensory and Motor Research*, **13**, 103–113.

Nudo, R.J., Wise, B.M., SiFuentes, F. and Milliken, G.W. (1996). Neural substrates for the effects of rehabilitative training on motor recovery after ischemic infarct. *Science*, **272**, 1791–1794.

Nudo, R.L. and Milliken, G.W. (1996). Reorganization of movement representations in primary motor cortex following focal ischemic infarcts in adult squirrel monkeys. *Journal of Neurophysiology*, **75**, 2144–2149.

Oyesiku, N.M. and Wigston, D.J. (1996). Ciliary neurotrophic factor stimulates neurite outgrowth from spinal cord neurons. *Journal of Comparative Neurology*, **364**, 68–77.

Pettit, M.J. and Schwark, H.D. (1993). Receptive field reorganization in dorsal column nuclei during temporary denervation. *Science*, **292**, 2054–2056.

Polistina, D.C., Murray, M. and Goldberger, M.E. (1990). Plasticity of dorsal root and descending serotoninergic projections after partial deafferentation of the adult rat spinal cord. *Journal of Comparative Neurology*, **299**, 349–363.

Pollin, B. and Albe-Fessard, D. (1979). Organization of somatic thalamus in monkeys with and without section of dorsal spinal tracts. *Brain Research*, **173**, 431–449.

Pons, T.P., Garraghty, P.E., Ommaya, A.K., Kaas, J.H., Taub, E. and Mishkin, M. (1991). Massive cortical reorganization after sensory deafferentation in adult macaques. *Science*, **252**, 1857–1860.

Pubols, L.M. and Bowen, D.C. (1988). Lack of central sprouting of primary afferent fibers after ricin deafferentation. *Journal of Comparative Neurology*, **275**, 282–287.

Pubols, L.M. and. Brenowitz, G.L (1981). Alteration of dorsal horn function by acute and chronic deafferentation. In *Spinal Cord Sensation*, edited by A.G. Brown and M. Rethelyi, pp.19–328. Edinburgh: Scottish Academic Press.

Raisman, G. (1969). Neuronal plasticity in the septal nuclei of the adult rat. *Brain Research*, **14**, 25–48.

Raisman, G. (1985). Synapse formation in the septal nuclei of adult rats. In *Synaptic Plasticity*, edited by C.W. Cotman, pp.13–38. New York: Guilford Press.

Ramachandran, V.S. (1993). Behavioral and magnetoencephalographic correlates of plasticity in the adult human brain. *Proceedings of National Academy of Science, USA* **90**, 10413–10420.

Ramachandran, V.S., Rogers-Ramachandran, D. and Stewart, M. (1992). Perceptual correlates of massive cortical reorganization. *Science*, **258**, 1159–1160.

Rasmusson, D.D. (1996a). Changes in the response properties of neurons in the ventroposterior lateral thalamic nucleus of the raccoon after peropheral deafferentation. *Journal of Neurophysiology*, **75**, 2441–2450.

Robertson, G. and Grant, G. (1985). A comparison between wheat germ agglutinin and choleragenoid-horseradish peroxidase as anterogradely transported markers in central branches of primary sensory neurones in the rat with some observations in the cat. *Neuroscience*, **14**, 895–905.

Rossi, F., Borsello, T. and Strata, P. (1994). Embryonic Purkinje cells grafted on the surface of the adult uninjured rat cerebellum migrate in the host parenchyma and induce sprouting of intact climbing fibres. *European Journal of Neuroscience*, **6**, 121–136.

Rossi, F., Wiklund, L., van der Want, J.J. and Strata, P. (1991). Reinnervation of cerebellar Purkinje cells by climbing fibres surviving a subtotal lesion of the inferior olive in the adult rat. *Journal of Comparative Neurology*, **308**, 513–535.

Sawai, J., Clarke, D.B., Kittlerova, P. Bray, G.M. and Aguayo, A.J. (1996). Brain-derived neurotrophic factor and neurotrophin-4/5 stimulate growth of axonal branches from regenerating retinal ganglion cells. *Journal of Neuroscience*, **16**, 3887–3894.

Schallert, T. and Jones, T.A. (1993). Exuberant neuronal growth after brain damage in adult rats: The essential role of behavioral experience. *Journal of Neural Transplantation and Plasticity*, **4**, 193–198.

Schwab, M.E. (1996). Bridging the gap in spinal cord regeneration. *Nature Medicine*, **2**, 976.

Shin, H.-C., Park, S., Son, J. and Sohn, J.-H. (1995). Responses from new receptive fields of VPL neurones following deafferentation. *NeuroReport*, **7**, 33–36.

Shortland, P. and Woolf, C.J. (1993). Chronic peripheral nerve section results in a rearrangement of the central axonal arborizations of axotomized A beta primary afferent neurons in the rat spinal cord. *Journal of Comparative Neurology*, **330**, 65–82.

Skene, J.H.P. (1989). Axonal growth-associated proteins. *Annual Reviews of Neuroscience*, **12**, 127–156.

Skene, J.H.P. and Willard, M. (1981a). Axonally transported proteins associated with axon growth in rabbit central and peripheral nervous system. *Journal of Cell Biology*, **89**, 96–103.

Skene, J.H.P. and Willard, M. (1981b). Changes in axonally transported proteins during axon regeneration in toad retinal ganglion cells. *Journal of Cell Biology*, **89**, 86–95.

Smith, D.S. and Skene, J.H.P (1997). A transcription-dependent switch controls competence of adult neurons for distinct modes of axon growth. *Journal of Neuroscience*, **17**, 646–658.

Snow, P.R., Nudo, R.J., Rivera, W., Jenkins, W.M. and Merzenich, M.M. (1988). Somatotopically inappropriate projections from thalamocortical neurons to the SI cortex of the cat demonstrated by the use of intracortical microstimulation. *Somatosensory Research*, **5**, 349–372.

Sperry, R.W. (1959). The growth of nerve circuits. *Scientific American*, **201**, 68–75.

Stelzner, D.J. and Keating, E.G. (1977). Lack of intralaminar sprouting of retinal axons in monkey LGN. *Brain Research*, **126**, 201–221.

Stent, G.S. (1973). A physiological mechanism for Hebb's postulate of learning. *Proceedings of the National Academy of Science USA*, **70**, 997–1001.

Thoenen, H. (1995). Neurotrophins and neuronal plasticity. *Science*, **270**, 593–598.

Tsukahara, N. (1985). Synaptic plasticity in the red nucleus and its possible behavioral correlates. In *Synaptic Plasticity*, edited by C. W. Cotman, pp. 201–230. New York: Guilford Press.

Tsukahara, N., Hiltborn, H. and Murakami, F. (1974). Sprouting of corticorubral synapses in red nucleus neurones after destruction of the nucleus interpositus of the cerebellum. *Experientia*, **30**, 57–58.

Tuszynski, M.H., Gabriel, K., Gage, F.H., Suhr, S., Meyer, S. and Rosetti, A. (1996). Nerve growth factor delivery by gene transfer induces differential outgrowth of sensory, motor, and noradrenergic neurites after adult spinal cord injury. *Experimental Neurology*, **137**, 157–173.

Wall, P.D. (1977). The presence of ineffective synapses and the circumstances which unmask them. *Philosophical Transactions of the Royal Society London B*, **278**, 361–372.

Wall, P.D. (1981). The nature and origins of plasticity in adult spinal cord. In *Spinal Cord Sensation*, edited by A.G. Brown and M. Rethelyi, pp. 297–308. Edinburgh: Scottish Academic Press.

Wall, P.D. and Egger, M.D. (1971). Formation of new connections in adult rat brains after partial deafferentation. *Nature*, **232**, 542–545.

Wilson, P. and Snow, P.J. (1987). Reorganization of the receptive fields of spinocervical tract neurons following denervation of a single digit in the cat. *Journal of Neurophysiology*, **57**, 803–818.

Woolf, C.J., Reynolds, M.L., Molander, C., O'Brien, C., Lindsay, R.M. and Benowitz, L.I. (1990). The growth-associated protein GAP-43 appears in dorsal root ganglion cells and in the dorsal horn of the rat spinal cord following peripheral nerve injury. *Neuroscience*, **34**, 465–478.

Woolf, C.J., Shortland, P. and Coggeshall, R.E. (1992). Peripheral nerve injury triggers central sprouting of myelinated afferents. *Nature*, **355**, 75–78.

Woolf, C.J., Shortland, P., Reynolds, M.L., Ridings, J., Doubell, T.P. and Coggeshall, R.E. (1995). Central regenerative sprouting: the reorganization of the central terminals of myelinated primary afferents in the rat dorsal horn following peripheral nerve section or crush. *Journal of Comparative Neurology*, **360**, 121–134.

Yang, T.T., Gallen, C., Schwartz, B., Bloom, F.E., Ramachandran, V.S. and Cobb, S. (1994). Sensory maps in the human brain. *Nature*, **368**, 592–593.

CHAPTER 12

LATERAL INTERACTIONS IN CORTICAL NETWORKS

O.V. Favorov[1], J.T. Hester[1], D.G. Kelly[2], M. Tommerdahl[1] and B.L. Whitsel[1,3]

[1]Department of Biomedical Engineering; [2]Department of Statistics and [3]Department of Physiology, University of North Carolina, Chapel Hill, NC, USA

Massive systems of connections synaptically link neurons over widely separated regions of somatosensory cortex, just as they do in other cortical areas. This paper suggests a particular theoretical framework for understanding these connections and the dynamics they engender. We propose that a fundamental function of lateral connections in the cortical network is to reflect the statistical structure of the sensory environment, thus endowing the network with predictive powers which enable it to draw inferences about the present situation from the input patterns. This function is made possible by a combination of three conditions:

1. Lateral connections are plastic, malleable by the sensory environment. Consequently, as a result of sensory experience, lateral connections come to reflect the orderly structure of the sensory environment, becoming an internal model of statistical predictive relations present in the environment.
2. The cortical network has complex nonlinear dynamics. Driven by its sensory experiences, such a near-chaotic network undergoes self-organization that selectively tunes cells to *combinations* of sensory events that are predictive of other such events. This enables lateral connections to capture more complex, higher-order statistical relationships in the environment.
3. Associative interactions among cortical representations of sensory events change from excitatory to inhibitory and vice versa, depending on whether the events are actually occurring or are recalled by association. This enables lateral connections to evoke representations of expected but not directly observed sensory events, to choose the most probable representation of ambiguous weak sensory events, and to emphasize outstanding events, which are anomalous with respect to the spatiotemporal context of events in which they are embedded.

INTRODUCTION

A cortical network is a complex system that contains multiple interacting circuits, generating multiple types of lateral interactions among cortical columns. The extent and importance of these lateral interactions is only now beginning to be fully appreciated,

Address for correspondence: Dr. O.V. Favorov, Department of Biomedical Engineering, CB# 7575, The University of North Carolina, Chapel Hill, NC 27599-7575, USA

spurred by recent major advances in anatomical and physiological experimental techniques. While this area of research is still in its early stages, and a general theory of lateral interactions is yet to emerge, we distinguish — based on our experimental and modeling work — three systems of lateral interactions that take place within a cortical area. They are:

(1) the most local, *minicolumnar* interactions, taking place among cells located within ca. 0.1mm of each other in the plane of cortical surface;

(2) intermediate, *macrocolumn-range* interactions, within ca. 0.5mm cortical columns; and

(3) *long-range* interactions, among more widely separated cortical columns. We will give a brief summary of each of these three systems and then introduce what we believe is a fundamental sensory information-processing task that is carried out by these lateral interactions.

SYSTEMS OF LATERAL INTERACTIONS

Minicolumnar Interactions

Minicolumns are 0.05 mm diameter cords of cells extending radially across all cortical layers, whose receptive field and functional properties distinguish each such minicolumn from its immediate neighbors. A common misconception is that cells located so closely have very similar receptive fields. In fact, most of the experimental literature in somatosensory, visual, auditory, motor, and associative cortical areas is in agreement that while neighboring cortical cells can show a remarkable uniformity in some of their receptive field properties (e.g. stimulus orientation in visual cortex), they can differ prominently in other of those properties (for review see Favorov and Kelly, 1996a). When receptive fields are considered *in toto*, in all their dimensions, neighboring neurons typically have little in common — a stimulus that is effective in driving one cell will frequently be much less effective in driving its neighbor.

This prominent local receptive field diversity is constrained, however, in the radial cortical dimension. Cells that make up individual radially oriented minicolumns have very similar receptive field properties (Abeles and Goldstein, 1970; Hubel and Wiesel, 1974; Albus, 1975; Merzenich *et al.*, 1981; Favorov and Whitsel, 1988; Favorov and Diamond, 1990; the evidence is summarized in Favorov and Kelly, 1996a). The differences in receptive field properties observed among neighboring neurons are mostly confined to neurons that belong to different, albeit neighboring, minicolumns. Minicolumns probably correspond to the 0.05 mm diameter radially oriented cords of neuronal cell bodies that are very noticeable in Nissl-stained cortical sections. They are also likely to be related to the ontogenetic columns of Rakic (1988), and also to the pyramidal cell modules defined by bundling of apical dendrites of pyramidal cells (Peters and Yilmaz, 1993).

Several elements of cortical microarchitecture might be responsible for the existence of minicolumns — *spiny stellate* cells (Jones, 1975, 1981; Lund, 1984) are likely to be one of them. Spiny stellates are excitatory intrinsic cells located in layer 4, and are major recipients of thalamocortical connections. In turn, they (especially the star pyramid subclass of spiny stellates) distribute afferent input radially via narrow, 0.05 mm bundles of axon collaterals to all other cells in the same minicolumn, thus imposing a uniform set of functional properties within each minicolumn.

Another class of intrinsic cortical cells that is also likely to contribute to the minicolumnar subdivisioning of the cortex is *double bouquet* cells. These are GABAergic cells with bodies and local dendritic trees in layer 2 and the upper part of layer 3. Their most distinctive characteristic is that their axons descend in tight 0.05 mm bundles of collaterals down through layers 3 and 4 and into layer 5, making synapses all along the way on the distal dendrites of pyramidal and spiny stellate cells, but avoiding the main shaft of apical dendrites (Jones, 1975, 1981; DeFelipe *et al.*, 1989; DeFelipe and Farinas, 1992). Because these cells are more likely to inhibit cells in adjacent minicolumns rather than in their own, they offer a mechanism by which adjacent minicolumns can inhibit each other.

Why do adjacent minicolumns have different receptive fields? We (Favorov and Kelly, 1994a,b; 1996a) have proposed that during perinatal development minicolumns actively drive their neighbors to establish afferent connections with different, only partially over-lapping sets of thalamic neurons. This is accomplished via lateral inhibitory interactions among adjacent minicolumns, subserved by the double bouquet cells. This mechanism offers an explanation for the existence of prominent receptive field differences among neighboring minicolumns, and for relative receptive field uniformity within minicolumns. The local receptive field differences, in turn, result in emergence of new functional properties that make neurons sensitive to the shape and temporal features of peripheral stimuli.

Because of the prominent differences in receptive field properties among neighboring minicolumns, even the simplest stimuli can be expected to evoke spatially complex minicolumnar patterns of activity in the engaged cortical region, with a mixture of active and inactive minicolumns (Favorov and Kelly, 1994b). This expectation has been experi-mentally confirmed in studies of stimulus-evoked activity in somatosensory cortex using either 2-deoxyglucose (2-DG) metabolic labeling (McCasland and Woolsey, 1988; Tommerdahl *et al.*, 1993) or near-infrared optical imaging of the intrinsic signal (Figs. 12.1 and 2). Our modeling studies (Favorov *et al.*, 1994b) predict that these minicolumnar activity patterns should be highly stimulus specific and carry detailed information about stimulus features.

Macrocolumn-range Interactions

These interactions are defined by the lateral extent of inhibitory connections. Axons of inhibitory cells, such as basket or chandelier cells, typically ramify and make connections within a few hundreds of micrometers from the soma, although axons of some basket cells can travel for 1 mm or more. Overall, most inhibitory connections are confined to lateral distances of ca. 0.5 mm, which is an approximate size of a cortical macrocolumn. Also, at such distances cells are linked by excitatory axon collaterals of pyramidal cells.

Traditionally, macrocolumn-range interactions are recognized as providing the classical lateral, or surround, inhibition (Mountcastle, 1957). We believe they also make another important contribution to cortical activity. Based on our modeling studies (Favorov and Kelly, 1996a, b; see also van Vreeswijk and Sompolinsky, 1996), we believe that the macrocolumn-range system of lateral interactions should have a strong propensity for generating complex, quasi-periodic or even chaotic, nonlinear dynamics.

While there are great technical difficulties in studying nonlinear dynamics *in vivo*, a tentative understanding can be gained from computer simulations of cortical models. Our

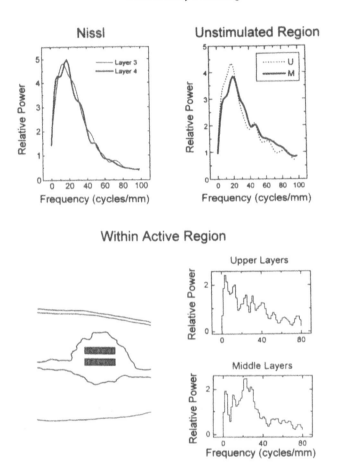

Figure 12.1 Minicolumnar pattern of activation of cat SI cortex: spatial frequency analysis of stimulus-evoked 2-DG labeling (modified from Tommerdahl *et al.*, 1993). Discrete Fourier transforms were computed across linear, tangentially oriented sectors of middle and upper cortical layers and plotted as periodograms. **Top left:** Average periodograms for optical density data sampled from Nissl-stained sections of SI cortex. Note that the distribution of spatial frequencies is the same in the upper and middle layers, and that the radial striation of Nissl-stained sections is clearly reflected by the prominent peak near the minicolumnar frequency of 20 cycles/mm. **Top right:** Average periodograms of 2-DG labeling in an unstimulated region of SI. Note a close similarity of these periodograms to those of Nissl-stained sections, demonstrating prominent minicolumnar periodicity of 20 cycles/mm due to radial striation of neuropil by cell bodies. **Bottom:** An outline of a patch of above-background 2-DG labeling in a section through an SI region activated by stimulus. Fourier transforms were computed across the two rectangular regions and plotted as periodograms on the right. Note that the periodogram of the middle layers is similar to those of the unstimulated region and of the Nissl-stained sections, in that it also has the peak near minicolumnar frequency. In contrast, in the upper-layer periodogram the peak is prominently shifted to lower spatial frequencies, indicating that at the level of upper layers the activated SI column consists of interdigitated radial strands of enhanced and reduced 2-DG label, suggesting that neighboring minicolumns are driven to markedly different degrees by stimuli used.

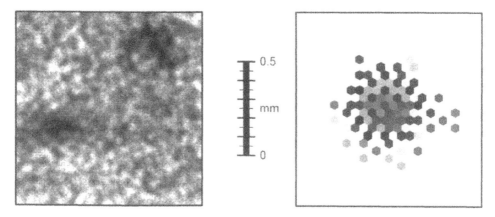

Figure 12.2 Minicolumnar pattern of activation of cat SI cortex: optical imaging of the stimulus-evoked intrinsic signal (**left**) and modeling prediction (**right**). The optical image was obtained in the near-infrared range (830 nm; for details about the method, see Tommerdahl *et al.*, 1996). It shows two activated cortical regions, the bottom one in area 3b, the top one in area 3a. Upon closer inspection, the active regions appear as a patchwork of minicolumn-sized spots. Note that these spots tend to be organized in short parallel strings, and that orientation of these strings varies across the activated region. The image on the right was generated by our minicolumnar model of SI network (Favorov and Kelly, 1994a, b). It shows a spatial pattern of active minicolumns, driven by a punctate stimulus. This pattern is similar to the one on the left in that in it active minicolumns are also organized in short parallel strings that run in different directions in different parts of the activated region.

modeling studies (Favorov *et al.*, 1995; Favorov and Kelly, 1996a) suggest that the complexity of a cortical network's dynamics is highly variable. It is very sensitive to the network's conditions and can be affected by changes in, for example, the stimulus used or its strength, or by block of AMPA, NMDA, GABA$_A$ receptors, etc. (a number of such examples are offered in Fig. 12.3). As Fig. 12.3 shows, different components of the network can make very different contributions to the complexity of dynamics. For example, blockade of glutaminergic AMPA receptors increases the complexity of dynamics, while the blockade of NMDA receptors, which are also glutaminergic, has an opposite effect. While such dependencies have not been evaluated systematically in experimental studies of the living cortex, some — e.g. stimulus strength and NMDA receptor block — have been confirmed in *in vivo* and *in vitro* experiments (Tolhurst *et al.*, 1981; Prince *et al.*, 1994; Whitsel *et al.*, 1997).

The most characteristic property of complex nonlinear dynamics — its sensitive dependence on the initial conditions — causes the network to be sensitive to the temporal order in which a set of stimuli are presented (Favorov and Kelly, 1996a). For example, different sequences of stimuli can generate, at the time of their completion, different spatial patterns of activity in the network, *even when the last stimuli in these sequences were identical*. This makes it possible, for example, to recognize, by spatial pattern alone, a temporal sequence of letters as a particular word, as opposed to its anagrams. Thus, a network with complex nonlinear dynamics can convert temporal information into a spatial code; it can represent spatially and temporally encoded information about the stimulus

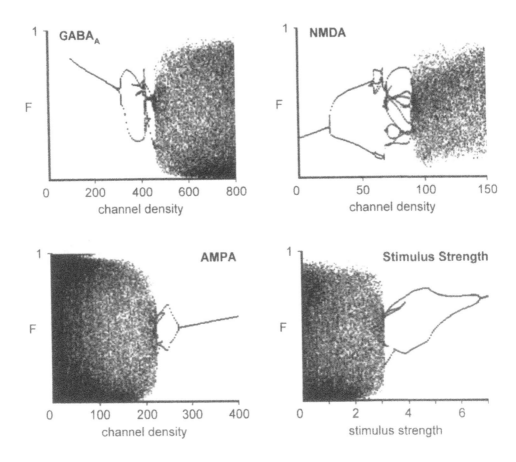

Figure 12.3 Bifurcation plots showing how the complexity of stimulus-evoked dynamics is affected by varying the stimulus strength or the relative strengths of GABA$_A$, NMDA, AMPA receptor-mediated conductances in our model of a cortical network (for details see Favorov and Kelly, 1996a). The plots describe activity of the same representative cell during continuous presence of a punctate stimulus in its receptive field, while the strength of one of the parameters studied was slowly increased in very small steps (strengths are expressed in nominal units). The vertical axis plots the instantaneous firing rates of the cell (scaled between 0 and 1) at which the first derivative of the firing rate is zero. The top-left plot, for example, shows that when GABA$_A$ receptors were almost completely blocked, the cell's firing rate in the stimulus presence was stable (a single point on the plot). At larger GABA$_A$ conductances, the cell remained stable (it only lowered its rate of firing). However, at some further increment in GABA$_A$ conductance, the cell became periodic (indicated by the appearance of two points on the plot). At still further increases in GABA conductance, the cell's activity underwent several transitions between periodic and quasi-periodic and then became chaotic (most of the vertical axis covered by dots). Note that for GABA$_A$ and NMDA, the complexity of dynamics increased with conductance increase, while for AMPA and stimulus strength it decreased.

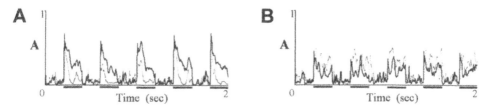

Figure 12.4 Stimulus-dependent coherence in activity fluctuations of a pair of cells in a cortical model under chaotic dynamics in the presence of random noise. **A:** Activities of the two cells are plotted superimposed, during intermittent presentation of a bar stimulus. Note that the cells had a clear tendency to be active simultaneously (correlation coefficient $\rho = 0.83$). **B:** Activities of the same two cells during intermittent presentation of the same bar stimulus, only it was rotated by 30 degrees. In the presence of this stimulus the two cells showed much less tendency to be active simultaneously ($\rho = -0.1$)

sequence by a purely spatial pattern that is specific for the entire time sequence, not just for the current stimulus.

Complex nonlinear dynamics in the network also manifests itself in complexity and diversity of the cells' temporal behaviors in response to even the most simple stimuli. Such temporal behaviors of cells can exhibit an orderly sensitivity to stimulus features (Favorov and Kelly, 1996b). This orderly sensitivity reveals itself in two ways. The first is that while, as a rule, cells differ in their temporal behaviors, in response to a given stimulus there will be some cells whose activities fluctuate more or less coherently (Fig. 12.4). Different stimuli will cause different groups of cells to have coherent activity fluctuations, i.e. the distribution of temporal coherence across the network is stimulus-specific. Such stimulus-specific coherence has been demonstrated in living cortex and is currently the subject of considerable experimental and theoretical interest (Aertsen *et al.*, 1989; Singer, 1995).

A second interesting property of the cells' temporal behaviors under complex nonlinear dynamics is that the stimulus-evoked temporal activity patterns of cells are also very sensitive to stimulus features. The temporal pattern of response of an active cell varies from one presentation of a stimulus to the next, but it varies around a waveform specific to that stimulus (Fig. 12.5). The time-course of response is sufficiently reproducible that it is possible to distinguish among stimuli based on a single presentation. In real cortex, the temporal patterns of evoked activity of cells are also stimulus specific. For example, in somatosensory cortex they reflect direction of stimulus motion (Ruiz *et al.*, 1995), in auditory cortex they reflect sound location (Middlebrooks *et al.*, 1994), and in visual cortex they reflect multidimensional stimulus properties (Richmond *et al.*, 1987; Richmond and Optican, 1990; Kruger and Becker, 1990; Shaw *et al.*, 1993).

To summarize, by generating complex nonlinear dynamics, macrocolumn-range lateral interactions can have profound effects on the functional properties of cells. The very limited overview of these effects attempted here gives only an initial appreciation of the vast richness of functional contributions that these lateral interactions can make — via nonlinear dynamics — to cortical behaviors and information processing.

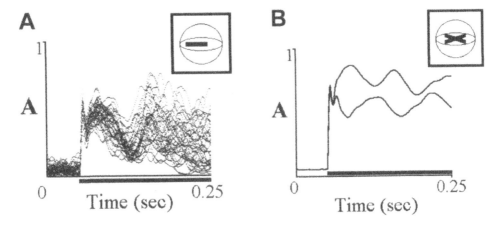

Figure 12.5 Stimulus specificity of temporal patterns of evoked activity in a cortical model under chaotic dynamics in the presence of random noise. **Left:** Superimposed time-courses of the response of an exemplary cell to 40 presentations of the same stimulus (shown in the inset over the outline of that cell's receptive field). When plotted together, these 40 time-courses reveal a particular complex waveform underlying them. Note that the dispersion among the individual time courses becomes gradually more pronounced with time after stimulus onset. **Right:** Average time-courses of the response of an exemplary cell to 200 presentations of two different stimuli. The stimuli are shown in the inset over the outline of that cell's receptive field. Note that these two stimulus bars were positioned similarly in relation to the receptive field borders and orientation, but evoked very different temporal response patterns.

LONG-RANGE INTERACTIONS

At cortical distances greater than approximately 0.5 mm, lateral connections are almost exclusively provided by pyramidal cells, whose axon collaterals can travel horizontally for very long distances (several millimeters) across a cortical area, forming clustered terminal arborizations along the way (Gilbert and Wiesel, 1983; DeFelipe *et al.*, 1986; Lund *et al.*, 1993; Burton and Fabri, 1995). The extreme length, large diameter, and focused terminations of these collaterals suggest that they form the major route for communication among cortical columns.

As studies in visual cortex show (Ts'o *et al.*, 1986; Gilbert and Wiesel, 1989; Malach *et al.*, 1993; Yoshioka *et al.*, 1996; Bosking *et al.*, 1997), long-range horizontal connections link cortical columns whose receptive fields, although non-overlapping, are similar in some other ways. For example, in visual cortex the linked cortical columns prefer stimuli with similar orientations. Within the targeted cortical columns, long-range connections terminate on both excitatory and inhibitory neurons, with no preference for either neuron class (Kisvarday *et al.*, 1986; McGuire *et al.*, 1991). Since excitatory and inhibitory cells in a column are heavily interconnected, the net physiological impact of long-range connections on the pyramidal cells can be quite complex. For example, when long-range connections are activated at low intensities, they evoke EPSPs in the target pyramidal cells, but at greater intensities they can evoke dominant IPSPs (Hirsch and Gilbert, 1991). This means that at high intensities long-range connections activate inhibitory cells and inhibit — via these active inhibitory cells — the pyramidal cells that occupy the targeted cortical column.

Long-range interactions are believed to be responsible for integration of sensory information coming to the cortical network from different parts of spatially extensive stimulus patterns (Gilbert, 1992; Singer, 1995; Grossberg *et al.*, 1997). This horizontal integration may underlie many aspects of perception, such as context-sensitive perceptual grouping and segregation of sensory inputs, separation of figures from their background, binding together different components of an image into a single percept, and perceptual constancies.

ASSOCIATIVE FUNCTIONS OF LATERAL INTERACTIONS

An important property of excitatory lateral connections is that they are plastic, malleable by the sensory environment not only in newborns but even in adults (reviewed by Singer, 1995). Their plasticity enables lateral connections to reflect statistical relations among naturally occurring sensory events. That is, because the strength of lateral connections depends on the correlation in the behaviors of the connected cells (the Hebbian rule; Brown *et al.*, 1990), the stronger the statistical association between two sensory events, the stronger the lateral connections that link the groups of neurons representing those events in the network. Thus, while sensory events are represented in the cortical network by activities of groups of neurons, the lateral connections between such groups can be viewed as embodying statistical relations between the sensory events represented by the groups. This enables neuronal groups to have predictive interactions with other groups with which they are significantly correlated.

The natural sensory environment is rich in statistical associations of various degrees of complexity. Sensory events tend to occur in some spatiotemporal combinations and not in others. This order reflects the fact that the natural environment is orderly; it has a strong and rich causal and, more generally, statistical structure. The statistical structure of the environment can be expected — based on the associative abilities of lateral connections — to be captured at some level in the matrix of lateral connections within the cortical network. In other words, the matrix of lateral connections in the cortical network might act as an internal model of the statistical interactions present in the sensory environment. That model's ability to predict might be exploited by a cortical sensory network in its task of extracting information from spatiotemporal patterns of receptor sheet activation. For example, when faced with a local stimulus, the network might take statistical expectancy guidance from that stimulus' context in determining how to represent it. This guidance would be especially useful if the stimulus by itself is ambiguous.

Recognition of the importance of associations for mental function dates back to Aristotle (Sorabji, 1968) and has been a subject of intense scrutiny among philosophers (e.g. Locke, Hume, Mill, Russell, Quine) and psychologists (e.g. James, Pavlov and Skinner). In recent times associations have been studied in artificial neural networks, leading to an appreciation of the ability of associative networks to generalize and to do similarity-based computations (e.g. Anderson, 1995). However, the idea of associative interactions has yet to be applied to the tasks of low- and intermediate-level analyses of sensory information carried out by the sensory cortical areas. The proposal made above — that lateral connections reflect the associative structure of the sensory environment, and use its predictive powers in sensory information processing — is conceptually appealing, and we want to determine how effective and useful such contributions might be.

Modeling the cortical network as a statistical predictive device, we find that, to be as effective as intended, associative lateral connections have to be combined with some additional mechanisms. In the next two sections we explain the need for these mechanisms and suggest how they are implemented in the cortical network.

COMPLEXITY OF STATISTICAL RELATIONS

The stimulus tuning properties of the cells in a cortical network set limits on which statistical relations present in the environment can be captured by the lateral connections. Since lateral connections are established on a cell-to-cell basis, between individual pairs of neurons, only those statistical relations that obtain among pairs of cells in the network will be reflected in the lateral connection matrix. Suppose, for the sake of argument, that cells in the network had very simple stimulus tuning properties, each cell being a detector of a line of a particular length, location and orientation on the skin. In such a network, therefore, only the most simple statistical relations — those between individual lines — can be reflected in its lateral connections. Even if the sensory environment is very structured, but this structure is based on higher-order statistical interactions, such a network will not be able to recognize this statistical structure and will not be able to use it in its processing of sensory information. Because the neurons of such a network have simple stimulus sensitivities, it also will have very limited associative capabilities. In order for a cortical network to be able to capture higher-order statistical interactions present in the environment, its cells must be selectively sensitive to various *combinations* of simple sensory events, not just individual events (such as bars, edges, etc.). In fact, the more complex the combinations of sensory events that individual cells are sensitive to, the more complex the statistical relations captured by the network's associative connections. Indeed, even in primary sensory cortical areas cells have rich diversity of functional properties and, of course, in higher-order cortical areas functional properties of cells become progressively more and more complex and diverse, leading to the prediction that progressively higher-level areas should be able to capture progressively higher-order statistical relations in the environment.

A number of mechanisms contribute to functional properties of cortical neurons. One well known mechanism is selective convergence of afferent inputs onto a cortical cell, such as convergence onto layer 4 cells in VI of sets of LGN cells with linearly arranged receptive fields, giving layer 4 cells elongated receptive fields and preference for lines or edges in a narrow range of orientations (Hubel and Wiesel, 1962; Chapman *et al.*, 1991; Reid and Alonso, 1995). Earlier in this paper we described another possible mechanism, based on inhibitory interactions among adjacent minicolumns (Favorov and Kelly, 1994a,b). However, an effective mechanism for capturing higher-order statistical relations in the environment might turn out to be one based on near-chaotic dynamics (Fig. 12.6). As described above, macrocolumn-range lateral interactions in the cortical network are a very likely generator of quasi-periodic or chaotic dynamics. By their nature, such complex dynamics are exquisitely sensitive to even small spatiotemporal details of afferent input patterns and in the stimulus presence they can greatly affect temporal behaviors and mean levels of activity of cells. The high sensitivity of dynamics to stimulus details greatly increases the complexity and variety of stimulus-reflecting behaviors in the network, creating among the network's cells a rich repertoire of sensitivities to combinations of sensory events (Fig. 12.7). Of course, near-chaotic dynamics do not tune a neuron exclusively to only one combination of sensory

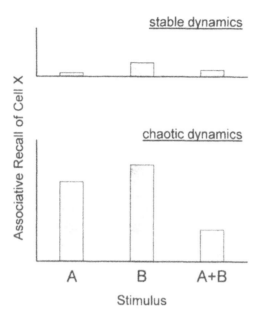

Figure 12.6 Demonstration of a superior ability of a network under chaotic dynamics to form higher-order associations. A network of 64 excitatory and 64 inhibitory cells with randomly assigned lateral connections was set up. Each cell received its own, independently controlled stimulus drive. In addition, another cell (cell X) was set up outside the network, with its own stimulus drive. This cell received Hebbian associative connections from all the excitatory cells in the network. Next, the system was presented with three stimuli. The first one, stimulus A, directly engaged a randomly chosen set of 16 cells in the network. The second one, stimulus B, directly engaged a completely different random set of 16 cells. The third one, stimulus A+B, was a combination of A and B, directly driving both sets of cells. When the network was presented with either A or B, cell X was also made active. However, when A+B was presented, cell X was kept inactive. Thus, the sensory environment of this system had a clear associative relationship of an *exclusive or* type between sensory events engaging the network and cell X. Connections from the network to cell X were adjusted in response to this sensory environment. Then the network — but not cell X — was stimulated, and the activity of cell X was measured. The top plot shows results obtained when the network had simple dynamics — each stimulus evoked a static pattern of activity in the network. Under such conditions, associative links failed to capture the orderly relationship between sensory event X and events A, B, and A+B. The bottom plot shows results obtained when the network had chaotic dynamics. The network now was able to establish associative connections to cell X that successfully captured the relationship between sensory event X and sensory events A, B, and A+B.

events; rather, single cells respond preferentially to multiple such combinations, which are different for different neurons. Thus, while no single cell can represent unequivocally any particular sensory event combination, this can be done collectively by a group of cells that share in common a preference for the same combination (a distributed representation).

While near-chaotic dynamics give cells greater sensitivity to various patterns of sensory events, these sensitivities emerge haphazardly. Although they are not well developed, they form a primary repertoire out of which a subset of complex sensitivities can be selectively

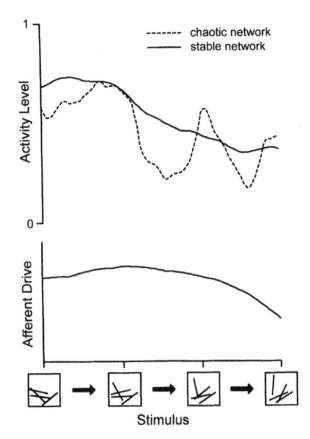

Figure 12.7 Chaotic dynamics enables cells in a cortical network model (Favorov and Kelly, 1996b) to discriminate among very similar *combinations* of sensory events. The network was presented with stimuli composed of 4 bars (bar=sensory event), and the average level of activity during stimulus presentation was computed for a representative cell. Locations and orientations of the 4 bars were changed consistently in 100 very small steps. The bottom plot shows how afferent drive to the cell varied across the stimulus sequence, the top plot shows the same for cell's response when the network was set to operate in chaotic or static regime. Note that under chaotic dynamics the cell exhibits far greater sensitivity to stimulus details.

enhanced by the network's sensory experiences. This is made possible by the fact that lateral connections in the network are plastic. In response to sensory experiences they will self-organize and, by doing so, will change the attractor structure of the dynamics and will consequently modify the cells' sensitivities to complex stimulus patterns. Our modeling studies (Fig. 12.8) suggest that this process is very beneficial: it allows the network to capture in its lateral connections higher-order statistical relations that obtain in the sensory environment. The reason for this is that self-organization of lateral connections and dynamics selectively enhances cells' sensitivities to those combinations of sensory events that have significant statistical relations with some other sensory events or their combinations, at the expense of other combinations that lack such relations. That is, out of the primary

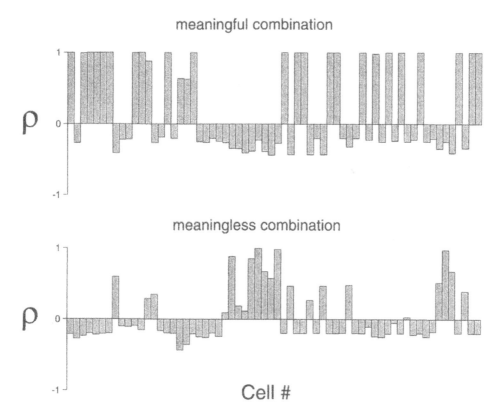

Figure 12.8   Self-organization of a chaotic network selectively enhances sensitivity of cells to those *combinations* of sensory events that have significant associations with some other sensory events. A model network with chaotic dynamics, similar to that in Figure 6, was repeatedly exposed to a number of pairs of stimuli (pair = sensory event combination) while its lateral connections were allowed to self-organize. One of these stimulus pairs (the meaningful combination) was consistently associated with another stimulus that activated a cell outside the network (representing a different cortical region). This cell had Hebbian associative connections with cells in the network. Following self-organization of connections, cells in the network significantly enhanced their tuning to the meaningful combination. The plots show, for 64 excitatory cells, how well their responses correlated (ρ – correlation coefficient) with the presence/absence of a particular stimulus pair. **Top plot:** Correlations with meaningful combination. **Bottom plot:** Correlations with a meaningless combination (i.e. not associated with the stimulus to the outside cell). Note that cells became much more correlated with the meaningful than with the meaningless combinations.

repertoire of cells' sensitivities, sensory experiences selectively enhance sensitivities to those sensory event combinations that have predictive powers (i.e. their presence implies something else about the situation) and therefore would be most useful in higher-order associative interactions in the network.

The idea that cortical cells acquire their complex functional properties by a two-step mechanism of (i) spontaneous *a priori* variations in neural connectivity among cells within local neuronal populations accidentally creating an over-abundant primary repertoire of diverse functional properties, and (ii) sensory experiences forming a secondary repertoire, closely tuned to the animal's sensory environment, by selecting and enhancing some of the properties available in the primary repertoire, has been originally proposed by Edelman, under the name of Neural Darwinism (1978, 1987). Here we suggest a neural implementation of Edelman's proposal. First, a very effective way of generating a rich primary repertoire of complex functional properties is for the cortical network to operate in a near-chaotic dynamic range. Second, in forming the secondary repertoire, cells' sensitivities are selected to those complex sensory events that have predictive powers. As a result:

1. groups of neurons in the network become able to represent complex sensory events;
2. the limited representational space available to the network is used most efficiently, devoting it to the most informative sensory events;
3. groups of neurons representing associated complex sensory events establish lateral connections; and
4. these lateral connections enable the network to capture higher-order statistical interactions in the environment.

SIGN OF ASSOCIATIVE INTERACTIONS

The traditional view of associative interactions is that if two sensory events tend to occur together, their neural representations should have excitatory links. The reason for having excitatory links is so that if only one event is experienced, it will evoke, by association, the other as well. However, a problem arises when both sensory events co-occur, as they frequently do. In this situation, their neuronal groups will respond to their respective events, but in addition they will also mutually excite each other via their lateral associative connections. As a result, the two groups will respond excessively. This creates a major problem. To explain, suppose the network is presented with a stimulus pattern that contains a number of positively associated sensory events and, embedded among those, is a sensory event that is unusual to find in such a context (i.e. it has negative associative relations with other events present). Perceptually, this unusual event stands out and catches our attention (a 'pop-out' phenomenon). That is, out-of-place sensory events are more salient than the events that fit in. However, a network with excitatory associative links will act in the opposite way — all the positively associated sensory events will mutually excite each other's neural representations and, as a result, the network's representation of this stimulus pattern will overemphasize these related events while minimizing the out-of-place event. An animal whose cortex is made up of such networks will be doomed to have its perception captive to the trivial while overlooking the unusual or outstanding. Thus, excitatory interactions between positively associated sensory events are very undesirable when these events are experienced together. In fact, it would be more consistent with the perceptual pop-out phenomena if neural representations of such events had *inhibitory* associative interactions, because then these events would suppress each other's representations in the network relative to the representation of an out-of-place sensory event, making that event more salient.

These considerations suggest that it would be most desirable if interactions between two associated sensory events could, depending on circumstances, change their sign: i.e.

be inhibitory when — as is more frequent — both events occur together, or be excitatory when — more rarely — only one of the two is being experienced. This would make associative interactions more complex than is traditionally assumed. Is there any experimental support for this proposal? We know from visual cortical studies that lateral interactions between cortical neuronal populations can, indeed, change their sign: when weakly activated the long-range horizontal connections evoke EPSPs in the target pyramidal cells, but when strongly activated, they can evoke dominant IPSPs (Hirsch and Gilbert, 1991). Switching the sign of lateral interactions has also been demonstrated on a more functional level in studies in which cells in visual cortex were presented with pairs of stimuli, one of which was applied to the cell's receptive field (central stimulus) while the other was applied outside the so-called classical receptive field (contextual stimulus). The effect of the contextual stimuli was found to vary depending on the strength of the central stimuli. Contextual stimuli typically enhanced responses of cells to weak central stimuli, but suppressed responses to strong central stimuli (Knierim and Van Essen, 1992; Kapadia *et al.*, 1995; Levitt and Lund, 1997). Thus, supporting our proposal, these studies show that contextual associative influences on a cell tend to be inhibitory in the presence of the cell's own strong stimulus, but they switch to excitatory when the cell's own stimulus is weakened.

What might be the mechanism of the proposed sign reversal of associative interactions? Stemmler *et al.* (1995) proposed that the reversal of contextual influences on weak and strong stimuli observed in the experimental studies above is due to excitatory cells being more responsive than inhibitory cells when inputs are weak, but less responsive than inhibitory cells when inputs are strong. While a potentially important contributor, this mechanism is limited in its capabilities. The most serious limitation is that it cannot generate strong associative recall: if a combination of sensory events strongly predicts the presence of another event that is not experienced directly, we would like to recall this event by activating its neuronal population. However, according to the mechanism of Stemmler *et al.*, strong associative inputs to this neuronal population will have greater impact on local inhibitory cells than on excitatory cells, which will prevent excitatory cells from responding.

We believe that Stemmler's mechanism, while correct in most of its details, misses an important component — double bouquet cells. These cells are very numerous inhibitory cells located superficially in the upper cortical layers and having axon collaterals that descend radially in a tight, 0.05 mm diameter, bundle through layers 2, 3, 4 and into layer 5, making numerous synaptic contacts in all these layers (Jones, 1975, 1981). An interesting feature of these cells is that while they make synapses on basal dendrites and also on the oblique side branches of apical dendrites of pyramidal cells, they completely avoid the main shaft of the apical dendrites (DeFelipe *et al.*, 1989; DeFelipe and Farinas, 1992). This means that, when they are active, double bouquet cells can greatly attenuate the contribution of basal and oblique dendrites to spike discharge activity of pyramidal cells by preventing excitatory inputs on those dendrites from reaching the soma. For a given pyramidal cell, double bouquet cells located in adjacent minicolumns will be especially effective, as they will synapse on the most proximal portions of that pyramid's basal and oblique dendrites and therefore will be able to functionally truncate entire dendritic branches.

Double bouquet cells might be the cells that control the sign of associative interactions in the cortical network. To explain how they might do it, let us first consider a situation in which two statistically associated stimuli are presented together. As we discussed above, we want their interactions to be inhibitory. Each stimulus will activate — via the thalamus — its own local group of minicolumns, including double bouquet cells located there. Double bouquet cells, in turn, will act on pyramidal cells in the same group of minicolumns and functionally truncate their basal and oblique dendrites. This will greatly reduce how much of the associative lateral input (coming from the other stimulus) will be able to reach somata of the pyramidal cells, since a great fraction of these inputs is located on the truncated dendrites. Assuming that inhibitory cells are much less affected by double bouquet cells (Somogyi and Cowey, 1981; DeFelipe *et al.*, 1989; DeFelipe and Farinas, 1992; Tamas *et al.*, 1997a, b), we find that associative inputs will be able to make much stronger excitatory contribution to firing of local inhibitory cells, such as basket and chandelier cells, than to pyramidal cells. As a result, their overall effect on pyramidal cells will be inhibitory.

Now let us turn to a situation in which only one of two statistically associated stimuli is present. In this situation we want associative interactions to be excitatory, so that the present stimulus will recall the missing one. The local group of minicolumns whose activity represents the missing stimulus will not be driven directly from the thalamus, but they will receive excitatory input from the present stimulus via associative lateral connections. For our mechanism we assume that associative lateral connections do not engage double bouquet cells. This is supported by two observations: (i) double bouquet cells generate $GABA_B$ receptor-mediated inhibition (Kang *et al.*, 1994), and (ii) long-range lateral connections fail to evoke $GABA_B$ inhibition (Hirsch and Gilbert, 1991). With double bouquet cells not engaged, associative inputs to pyramidal cells will have unimpeded access to their somata and will be maximally effective in triggering spike discharges. Thus, with double bouquet cells not active, the balance between associatively evoked excitation and inhibition will shift towards excitation.

We investigated the effectiveness of this mechanism in a modeling study in which two cell populations, representing different cortical columns, were linked by Hebbian associative connections and exposed to a statistically structured sensory environment. Specifically, the environment consisted of two sets of stimuli, one for each cell population, and these stimuli were presented to the two populations in pairs. Different pairs of stimuli were presented to the network more or less frequently. As expected, this network developed associative interactions between cell groups representing co-occurring stimuli; these interactions switched sign depending on whether one or both stimuli were presented. Most significantly, the model showed that stimuli negatively correlated in their occurrences also generated associative interactions, but those interactions were opposite to those between positively correlated stimuli (Fig. 12.9).

Based on the results of the modeling study shown in Figure 12.9, we propose the following scheme for associative interactions among cortical representations of sensory events:

1. If two sensory events are *positively* correlated in their occurrences, when they both take place they will partially suppress each other's neural representation. However, when only one of the events takes place, it will activate, in addition to its own representation, the representation of the other event instead of suppressing it.

2. If two sensory events are *negatively* correlated in their occurrences, when they both take place they will enhance each other's neural representation. However, when only one of the events takes place, it will inhibit cells representing the other event.

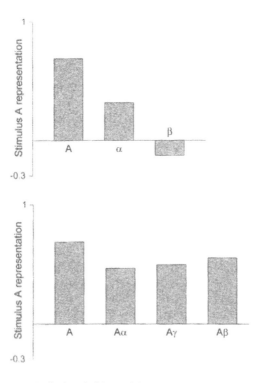

Figure 12.9 Stimulus-controlled switching of the sign of associative interactions. A model was set up consisting of two cell populations representing two cortical columns driven by different sensory events. Each population had 64 excitatory and 64 inhibitory cells, with Hebbian connections among them. Excitatory cells of one population also had Hebbian associative connections with excitatory and inhibitory cells of the other population. Approximating the proposed function of double bouquet cells, associative connections on excitatory cells were turned off when the target population was directly driven by a stimulus. Each population was stimulated with 5 different stimulus patterns. Stimuli to the two populations were paired in various combinations. After a period of repetitive stimulation, when intrinsic and associative connections settled, one population was tested for how it represented one of its stimuli (stimulus A). To identify the group of cells that represent stimulus A, correlation coefficients were computed between activities of the excitatory cells and the presence of stimulus A. The activity of the cell group that represents stimulus A was then computed as a sum of activities of all the population's excitatory cells, weighted by their correlation coefficients. The top plot shows the response of the A-representing cell group to stimulus A, and to two other stimuli, delivered to the other population (i.e. these stimuli activated this cell group via associative connections). One of these stimuli, stimulus α, was positively associated with stimulus A (they were frequently presented together, and correlated in appearances at $\rho = 0.5$). This stimulus activated the cell group A. The other stimulus, stimulus β, was negatively associated with stimulus A ($\rho = -0.25$). This stimulus evoked a negative representation of stimulus A. The bottom plot shows the response of cell group A to stimulus A alone and in combination with stimuli α, β, and γ, delivered to the other population (stimulus γ was not statistically associated with stimulus A; $\rho = 0$). Note that the presence of all these stimuli reduced the response of the cell group A to its own stimulus, but the greatest reduction was produced by the positively associated stimulus α, while the negatively associated stimulus β produced the smallest reduction.

The functional benefits of this proposed scheme of associative interactions are twofold. First, they might enable the cortical network to act predictively, as for example using the sensory context to interpret an ambiguous sensory event, to fill in a missing piece of information, or to interpret a large sensory pattern as containing distinct figures (figure/ ground separation, perceptual grouping). Second, associative interactions might enable the cortical network to emphasize unusual coincidences by representing unusual sensory events, which are anomalous with respect to their context, more strongly than sensory events that cohere with their context. These abilities, together, might underlie a number of perceptual phenomena, such as contour completion, weak signal enhancement, pop-out, fill-in, and enhancement of border between different textures.

CONCLUSIONS

This paper considered sensory cortical networks as a statistical predictive/inferential device. The core of this device is a massive system of lateral connections linking different regions of the network. By virtue of their Hebbian plasticity, lateral connections that link groups of neurons representing different sensory events come to embody statistical relations between those events. Consequently, such lateral connections enable groups of neurons to mimic in their interactions the statistical interactions between sensory events that these groups represent.

While lateral connections are the core of the cortical network in its role as a statistical device, what makes this device functionally useful and gives it its inferential power are two other putative properties of the cortical network. One of them is the network's self-organization under conditions of complex nonlinear dynamics, which is proposed to tune neurons to combinations of sensory events that are predictive of other such events. The other property is a proposed ability of the cortical network to switch associative interactions from excitatory to inhibitory and vice versa, depending on whether their sensory events are actually taking place or being inferred from the context. This property enables the network both to make inferences and to detect unusual coincidences. Without these properties, the network's inferential abilities would be trivial and of limited, if any, functional use.

The operation of sensory cortex as the proposed statistical device might underlie its central task of organizing patterns of peripheral receptor activation into neural representations of objects and their properties. Many information-processing operations that lead to perception make use of the statistical order in the sensory environment; these operations might essentially be associative interactions among cortical representations of sensory events, learned by the network exposed to its sensory environment. This possibility is appealing because of its algorithmic simplicity and universality: it requires minimal phylogenetically prespecified machinery, and it would adapt to any sensory environment that has an orderly structure.

The theoretical framework used in this paper to understand lateral connections predicts that these connections should be orderly at the level of groups of neurons that represent sensory events. Neurons belonging to such a representational group do not form a spatially easily identifiable local module, but are scattered among other neurons within local cortical populations. They are scattered because self-organization of the near-chaotic network diversifies functional properties of neighboring cortical cells, making them almost uncorrelated in their stimulus-evoked behaviors (Favorov and Kelly, 1996a; Gawne *et al.*, 1996).

At other levels — individual neurons, local cell clusters, or cortical columns — lateral connections and their physiological effects should appear much less orderly, because individual cells participate in multiple representational groups and therefore their lateral connections will reflect multiple associations, and because cell clusters isolated in experiments (and even more, cortical columns) contain multiple representational groups with different associative links. This expectation is consistent with the experimentally observed complexity of long-range lateral connections and their physiological effects, such as a degree of imprecision in patterns of termination of long-range lateral connections on orientation-specific cortical columns (Malach *et al.*, 1993; Yoshioka *et al.*, 1996), diverse effects of stimulation of these connections (Hirsch and Gilbert, 1991), and a rich repertoire of observed contextual effects on cells' responses to stimuli within their classical receptive fields (Kapadia *et al.*, 1995; Levitt and Lund, l997). The theoretical framework advanced in this paper suggests that behind the experimentally observed complexity of lateral interactions might lie a small set of basic principles.

ACKNOWLEDGMENTS

The authors thank Mr. Thomas Daniel Ryder for his helpful discussions and comments on the manuscript. The work was supported, in part, by ONR grant N00014-95-1-0113 to Oleg V. Favorov and NIH grant R29 NS-32358 to Mark Tommerdahl.

REFERENCES

Abeles, M. and Goldstein, M.H., Jr. (1970). Functional architecture in cat primary auditorycortex: columnar organization and organization according to depth. *Journal of Neurophysiology*, **33**, 172–187.

Aertsen, A., Gerstein, G.L., Habib, M.K. and Palm, G. (1989). Dynamics of neuronal fringe correlation: modulation of effective connectivity. *Journal of Neurophysiology*, **61**, 900–917.

Albus, K. (1975). A quantitative study of the projection area of the central and the paracentralvisual field in area 17 of the cat: I. The spatial organization of the orientation domain. *Experimental Brain Research*, **24**, 181–202.

Anderson J.A. (1995). *An Introduction to Neural Networks*. Cambridge, MA: MIT Press.

Bosking, W.H., Zhang, Y., Schofield, B. and Fitzpatrick, D. (1997). Orientation selectivity and the arrangement of horizontal connections in tree shrew striate cortex. *Journal of Neuroscience*, **6**, 2112–2127.

Brown, T.H., Kairiss, E.W. and Keenan, C.L. (1990). Hebbian synapses: biophysical mechanisms and algorithms. *Annual Review of Neuroscience*, **13**, 475–511.

Burton, H. and Fabri, M. (1995). Ipsilateral intracortical connections of physiologically defined cutaneous representations in areas 3b and 1 of macaque monkeys: projections in the vicinity of the central sulcus. *Journal of Comparative Neurology*, **355**, 508–538.

Chapman, B., Zahs, K.R. and Stryker, M. P. (1991). Relation of cortical cell orientation selectivity to alignment of receptive fields of the geniculocortical afferents that arborize within a single orientation column in ferret visual cortex. *Journal of Neuroscience*, **11**, 1347–1358.

DeFelipe, J., Conley, M. and Jones, E.G. (1986). Long-range focal collateralization of axonsarising from corticocortical cells in monkey sensory-motor cortex. *Journal of Neuroscience*, **6**, 3749–3766.

DeFelipe, J., Hendry, M.C. and Jones, E.G. (1989). Synapses of double bouquet cells in monkey cerebral cortex visualized by calbindin immunoreactivity. *Brain Research*, **503**, 49–54.

DeFelipe, J. and Farinas, I. (1992). The pyramidal neuron of the cerebral cortex: morphological and chemical characteristics of the synaptic inputs. *Progress in Neurobiology*, **39**, 563–607.

Edelman, G.M. (1978). Group selection and phasic re-entrant signalling: a theory of higher brain function. In *The Mindful Brain*, edited by G.M. Edelman and V.B. Mountcastle, pp. 51–100. Cambridge, MA: MIT Press.

Edelman, G.M. (1987). *Neural Darwinism*. New York: Basic Books.

Favorov, O.V. and Whitsel, B.L. (1988). Spatial organization of the peripheral input to area 1 cell columns: I. The detection of 'segregates'. *Brain Research Reviews*, **13**, 5–42.

Favorov, O.V. and Diamond, M.E. (1990). Demonstration of discrete place-defined columns — segregates — in the cat SI. *Journal of Comparative Neurology*, **298**, 7–112.

Favorov, O.V. and Kelly, D.G. (1994a). Minicolumnar organization within somatosensory cortical segregates: I. Development of afferent connections. *Cerebral Cortex* **4**, 408–427.

Favorov, O.V. and Kelly, D.G. (1994b). Minicolumnar organization within somatosensory cortical segregates: II. Emergent functional properties. *Cerebral Cortex*, **4**, 428–442.

Favorov, O.V., Kelly, D.G. and Lu, J. (1995). Complex nonlinear dynamics in a model of SI cortex. *Society of Neuroscience Abstracts*, **21**, 115.

Favorov, O.V. and Kelly, D.G. (1996a). Local receptive field diversity within cortical neuronal populations. In *Somesthesis and the Neurobiology of the Somatosensory Cortex*, edited by O. Franzen, R. Johansson and L. Terenius, pp. 395–408. Basel: Birkhauser Verlag.

Favorov, O.V. and Kelly, D.G. (1996b). Stimulus-response diversity in local neuronal populations of the cerebral cortex. *NeuroReport*, **7**, 2293–2301.

Gawne, T.J., Kjaer, T.W., Hertz, J.A. and Richmond, B.J. (1996). Adjacent visual cortical complex cells share about 20% of their stimulus-related information. *Cerebral Cortex*, **6**, 482–489.

Gilbert, C.G. and Wiesel, T.N. (1983). Clustered intrinsic connections in cat visual cortex. *Journal of Neuroscience*, **3**, 1116–1133.

Gilbert, C.G. and Wiesel, T.N. (1989). Columnar specificity of intrinsic horizontal and corticocortical connections in cat visual cortex. *Journal of Neuroscience*, **9**, 2432–2442.

Gilbert, C.D. (1992). Horizontal integration and cortical dynamics. *Neuron*, **9**, 1–13.

Grossberg, S., Mingolla, E. and Ross, W.D. (1997). Visual brain and visual perception: how does the cortex do perceptual grouping? *Trends in Neuroscience*, **20**, 106–111.

Hirsch, J.A. and Gilbert, C.D. (1991). Synaptic physiology of horizontal connections in the cat's visual cortex. *Journal of Neuroscience*, **11**, 1800–1809.

Hubel, D.H. and Wiesel, T.N. (1962). Receptive fields, binocular interaction, and functional architecture in the cat's visual cortex. *Journal of Physiology*, **160**, 106–154.

Hubel, D.H. and Wiesel, T.N. (1974). Sequence regularity and geometry of orientation columns in the monkey striate cortex. *Journal of Comparative Neurology*, **158**, 267–294.

Kang, Y., Kaneko, T., Ohishi, H., Endo, K. and Araki, T. (1994). Spatiotemporally differential inhibition of pyramidal cells in the cat motor cortex. *Journal of Neurophysiology*, **71**, 280–293.

Kapadia, M.K., Ito, M., Gilbert, C.D. and Westheimer, G. (1995). Improvement in visual sensitivity by changes in local context: parallel studies in human observers and in V1 of alert monkeys. *Neuron*, **15**, 843–856.

Kisvarday, Z.F., Martin, K.A.C., Freund, T.F., Magloczky, Z., Whitteridge, D. and Somogyi, (1986). Synaptic targets of HRP-filled layer III pyramidal cells in the cat striate cortex. *Experimental Brain Research*, **64**, 541–552.

Knierim, J.J. and Van Essen, D.C. (1992). Neuronal responses to static texture patterns in area V1 of the alert macaque monkey. *Journal of Neurophysiology*, **67**, 961–980.

Kruger, J. and Becker, J. D. (1990). Segregation of 'meaning' and 'importance' of neuronal messages. In *Parallel Processing in Neural Systems and Computers*, edited by R. Eckmiller, G. Hartmann, and G. Hauske, pp. 121–123. Amsterdam: Elsevier North-Holland.

Levitt, J.B. and Lund, J.S. (1997). Contrast dependence of contextual effects in primate visual cortex. *Nature*, **387**, 73–76.

Lund, J.S. (1984). Spiny stellate neurons. In *Cerebral Cortex*, edited by A. Peters and E.G. Jones, pp. 255–308. New York: Plenum Press.

Lund, J.S., Yoshioka, T. and Levitt, J.B. (1993). Comparison of intrinsic connectivity in different areas of macaque monkey cerebral cortex. *Cerebral Cortex*, **3**, 148–162.

Jones, E.G. (1975). Varieties and distribution of non-pyramidal cells in the somatic sensory cortex of the squirrel monkey. *Journal of Comparative Neurology*, **160**, 205–267.

Jones, E.G. (1981). Anatomy of cerebral cortex: columnar input-output organization. In *The Organization of the Cerebral Cortex*, edited by F.O. Schmitt, pp.199–235. Cambridge: MIT Press.

Malach, R., Amir, Y., Harel, M. and Grinvald, A. (1993). Relationship between intrinsic connections and functional architecture revealed by optical imaging and in vivo targeted biocytin injections in primate striate cortex. *Proceedings of National Academy of Science, USA*, **22**, 10469–10473.

McCasland, J.S. and Woolsey, T.A. (1988). High-resolution 2-deoxyglucose mapping of functional cortical columns in mouse barrel cortex. *Journal of Comparative Neurology*, **278**, 555–569.

McGuire, B., Gilbert, C.D., Wiesel, T.N. and Rivlin, P.K. (1991). Targets of horizontal connections in macaque primary visual cortex. *Journal of Comparative Neurology*, **305**, 370–392.

Merzenich, M.M., Sur, M., Nelson, R.J. and Kaas, J.H. (1981). Organization of the SI cortex: multiple cutaneous representations in areas 3b and 1 of the owl monkey. In *Cortical Sensory Organization*, edited by C.N. Woolsey, vol. 1, pp. 47–66. Clifton, N.J: Humana Press.

Middlebrooks, J., Clock, A.E. and Xu, L. (1994). A panoramic code for sound location by cortical neurons. *Science*, **264**, 842–844.

Mountcastle, V.B. (1957). Modality and topographic properties of single neurons of cat's somatic sensory cortex. *Journal of Neurophysiology*, **20**, 408–434.

Peters, A. and Yilmaz, E. (1993). Neuronal organization in area 17 of cat visual cortex. *Cerebral Cortex* **3**, 49–68.

Prince, D., Whitsel, B.L., Tommerdahl, M. and Favorov, O.V. (1994). Factors that influence neuronal response variability in somatosensory cortex: the role of NMDA receptors. *Society of Neuroscience Abstracts*, **20**, 1383.

Rakic, P. (1988). Specification of cerebral cortical areas. *Science*, **241**, 170–176.

Reid, R.C. and Alonso, J.M. (1995). Specificity of monosynaptic connections from thalamus to visual cortex. *Nature*, **378**, 281–284.

Richmond, B.J., Optican, L.M. and Poden, M. (1987). Temporal encoding of two-dimensional patterns by single units in primate inferior temporal cortex. I. Response characteristics. *Journal of Neurophysiology*, **57**, 132–146.

Richmond, B.J. and Optican, L.M. (1990). Temporal encoding of two-dimensional patterns by single units in primate primary visual cortex. II. Information transmission. *Journal of Neurophysiology*, **64**, 370–380.

Ruiz, S., Crespo, P. and Romo, R. (1995). Representation of moving tactile stimuli in the somatic sensory cortex of awake monkeys. *Journal of Neurophysiology*, **73**, 525–537.

Shaw, G.L., Kruger, J., Silverman, D.J., Aertsen, A.M., Aiple, F. and Liu, H.C. (1993). Rhythmic and patterned neuronal firing in visual cortex. *Neurological Research*, **15**, 46–50.

Singer, W. (1995). Development and plasticity of cortical processing architectures. *Science*, **270**, 758–764.

Somogyi, P. and Cowey, A. (1981). Combined Golgi and electron microscopic study on the synapses by double bouquet cells in the visual cortex of the cat and monkey. *Journal of Comparative Neurology*, **195**, 547–566.

Sorabji R (1969). *Aristotle on Memory*. Brown Univ. Press.

Stemmler, M., Usher, M. and Niebur, E. (1995). Lateral interactions in primary visual cortex: a model bridging physiology and psychophysics. *Science*, **269**, 1877–1882.

Tamas, G., Buhl, E.H. and Somogyi, P. (1997a). Fast IPSPs elicited via multiple synaptic release sites by different types of GABAergic neurone in the cat visual cortex. *Journal of. Physiology, (London)*, **500**, 715–738.

Tamas, G., Buhl, E.H., Jones, R.S.G. and Somogyi, P. (1997b). Interactions between distinct GABAergic neuron types eliciting fast IPSPs in cat visual cortex: number and placement of synapses. *Society Neuroscience Abstracts*, **23**, 1265.

Tolhurst, D.J., Movshon, J.A. and Thompson, I.D. (1981). The dependence of the response amplitude and variance of cat visual cortical neurones on stimulus contrast. *Experimental Brain Research*, **41**, 414–419.

Tommerdahl, M., Favorov, O.V., Whitsel, B.L., Nakhle, B. and Gonchar, Y.A. (1993). Minicolumnar activation patterns in cat and monkey SI cortex. *Cerebral Cortex*, **3**, 399–411.

Ts'o, D.Y., Gilbert, C.D. and Wiesel, T.N. (1986). Relationships between horizontal interactions and functional architecture in cat striate cortex as revealed by cross-correlation analysis. *Journal of Neuroscience*, **6**, 1160–1170.

Whitsel, B.L., Favorov, O.V., Delemos, K.A., Lee, C.J., Tommerdahl, M., Essick, G.K. and Nakhle, B. (1999). SI neuron response variability is stimulus-tuned and NMDA receptor-dependent. *Journal of Neurophysiology*, **81**, 2988–3006.

Van Vreeswijk, C. and Sompolinsky, H. (1996). Chaos in neural networks with balance dexcitatory and inhibitory activity. *Science*, **274**, 1724–1726.

Yoshioka, T., Blasdel, G.G., Levitt, J.B. and Lund, J.S. (1996). Relation between patterns of intrinsic lateral connectivity, ocular dominance, and cytochrome oxidase-reactive regions in macaque monkey striate cortex. *Cerebal Cortex*, **6**, 297–310.

CHAPTER 13

MODULATION OF SOMATOSENSORY CORTICAL RESPONSIVENESS FOLLOWING UNEXPECTED BEHAVIORAL OUTCOMES

Randall J. Nelson

Department of Anatomy and Neurobiology, College of Medicine, University of Tennessee, Memphis, TN, USA

INTRODUCTION

The somatosensory system conveys information to the central nervous system about the external and internal sensory environments. From many experiments using anaesthetised animals, much is known about this system's topographic organization and response properties (Mountcastle, 1997). Recently, however, it has been argued convincingly that only a portion of this system's role in perception and motor control can be gleaned from experiments involving animals that are not actively utilizing sensory inputs (Kalaska, 1994; Chapman *et al.*, 1996; Nelson, 1996).

This chapter will outline first some of the organizational features of sensorimotor cortices and the basis for response modulation of neurons within these areas. Second, it will discuss some aspects of when, where, and why neuronal responses are modulated during behavior. Next, details of one recent experiment will be presented. This experiment was designed to provide more information about sensory- and movement-related cortical activity that is modulated when 'expectation' is altered. Finally, these new data will be discussed in light of previous findings and new directions that are emerging in the study of somatosensory responsiveness. A somewhat extensive reference list is provided as a point of departure for investigation into the general topic of modulation of somatosensory responsiveness.

THE BASIS FOR MODULATION

Connections

Primary somatosensory cortex (SI) is composed of four cytoarchitectonically distinct regions — areas 3a, 3b, 1, and 2. It receives projections from the ventral thalamic nuclei that subserve exteroceptive, proprioceptive, thermal, and nociceptive inputs (e.g. Jones,

Address for correspondence: Randall J. Nelson, Department of Anatomy and Neurobiology, College of Medicine, University of Tennessee, Memphis, 855 Monroe Avenue, Memphis, TN 38163, USA. Tel: (901) 444-5979; Fax: (901) 448-7193; E-mail: rnelson@utmem1.utmem.edu

1986; Pons *et al.*, 1987; Pons *et al.*, 1992). In general, SI represents contralateral body inputs and preserves peripheral receptor adjacency within each of its four functional representations. While these representations have similar patterns of organization, the precise representational topography and proportional distribution of somatosensory submodalities differ from area to area. Moreover, these cortical areas have different subcortical and intracortical connections, most notably with the precentral cortices (Kaas, 1983; Mountcastle, 1984; Jones and Peters, 1986; Pons *et al.*, 1987; Kaas and Pons, 1988; Huerta and Pons, 1990; Pons *et al.*, 1992; Stepniewska *et al.*, 1993; Chmielowska and Pons, 1995). Regions within SI and the second somatosensory cortical area, by their interconnections, process somatosensory information both in serial and parallel (Pons *et al.*, 1987; Pons *et al.*, 1992; Turman *et al.*, 1992; Zhang *et al.*, 1996). Primary motor cortex (MI; area 4) and SI are interconnected; the interconnection between MI and area 3b varies across primate species (Stepniewska *et al.*, 1993). The more posteriorly situated region of SI, cortical area 2, and parietal cortical area 5 are interconnected with a portion of premotor cortex bordering on MI. The connections of premotor regions have recently received extensive review (Wise, 1985; Jones, 1986; Cavada and Goldman-Rakic, 1989; Dum and Strick, 1991; He *et al.*, 1995). The connectional scheme is presented schematically elsewhere (Nelson, 1996; Fig. 13.1).

SI and area 5 neurons have been classified by peripheral receptive field (RF) location (see Kaas, 1983), preferential responsiveness to peripheral stimuli, and adaptation properties exhibited during continuous stimulus presentation (e.g. Jones and Porter, 1980; Kaas, 1983; Nelson *et al.*, 1991a; Lebedev *et al.*, 1994 for review). Area 3a receives inputs from group Ia afferents, areas 3b and 1 receive predominantly cutaneous inputs, and areas 2 and 5 have neurons with increasing RF complexity (Hyvärinen and Poranen, 1978a; Hyvärinen and Poranen, 1978b; Iwamura and Tanaka, 1978; Iwamura *et al.*, 1980; Iwamura *et al.*, 1983). These areas are thought to be the sites where complex spatial analysis occurs (Mountcastle *et al.*, 1975; Georgopoulos *et al.*, 1983; Kalaska, 1988; Kalaska and Crammond, 1995). Areas 2 and 5 have been shown to participate in the initiation and execution of arm movements (Duffy and Burchfiel, 1971; Hyvärinen and Poranen, 1978a; Hyvärinen and Poranen, 1978b; Iwamura and Tanaka, 1978; Hyvärinen *et al.*, 1980; Iwamura *et al.*, 1980; Iwamura *et al.*, 1983; Kalaska, 1988; Crammond and Kalaska, 1989; Kalaska *et al.*, 1989; Burbaud *et al.*, 1991). The posterior parietal cortex, of which area 5 is a part, perhaps maintains maps of impending movements (Kalaska *et al.*, 1990; Assad and Maunsell, 1995; Colby *et al.*, 1995; Kalaska and Crammond, 1995). As such, the posterior portions of SI and the posterior parietal cortex not only provide sensory information to motor cortical regions (Jones, 1986; see discussion in Pons and Kaas, 1986), they also have been implicated in the transition from sensation to behavioral intention associated with active movement (Andersen, 1995).

Types of Inputs

SI may be unique among primary sensory cortices, because it is the recipient of two distinct classes of ascending information during sensorimotor behavior. Like other sensory cortices, SI receives information about sensory events arising in the external environment (exteroception). These inputs may arise from objects coming casually in contact with skin surfaces, active manipulation of the environment, and the feedback from muscles and joints

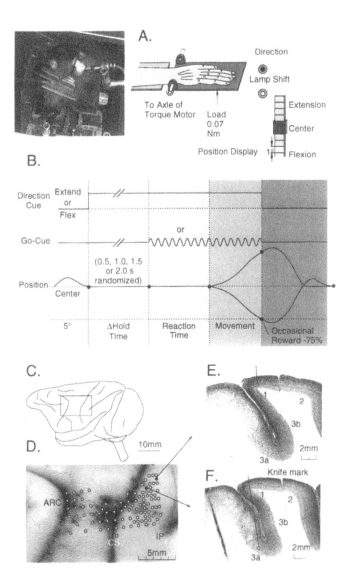

Figure 13.1 **A:** Two rhesus monkeys made wrist flexions or extensions in response to vibratory cues to obtain a fruit juice reward. Animals viewed a visual display of their wrist position and initiated trials by holding a centered position. **B:** Experimental paradigm. Hold times were 0.5, 1.0, 1.5, or 2.0 s and were varied pseudorandomly. The somatosensory go-cues were palmar vibratory stimulation of 57 Hz. This vibration remained on until the animal moved at least 5 from the previously held position. Monkeys were rewarded 75% of the time for correct performance. **C:** Drawing of the dorsolateral surface of the brain of the more extensively studied monkey. **D:** Recorded region as marked in box in C. Circles indicate surface location of penetrations. **E:** Location of the area 1 neuron whose recordings are illustrated in Figure 3. **F:** Location of the area 3a neuron whose responses are illustrated in Figure 4. ARC = arcuate, CS = central and IP = intraparietal sulci. Shown are calibration marks for reconstruction.

involved in active or passive movements. The somatosensory system also monitors the state of the internal sensory environment by contributing to information processing of body and limb position and muscular activity (proprioception). In addition, SI neurons are thought to receive centrally generated inputs related to impending movements and expected stimuli that, in turn, have the capability of further modifying SI neuronal responsiveness to peripheral inputs (Chapman *et al.*, 1996; Nelson, 1996).

SI can be characterized by the remarkable fidelity with which its neurons respond to peripheral sensory inputs. Sensations probably are controlled dynamically, because the same sensory receptors (muscle and joint afferents and, to a certain extent, afferents from skin receptors) convey feedback about muscle tension and limb position and information about external sensory events (Evarts *et al.*, 1984). This dual function is perhaps most important during voluntary or active movement.

Active Movement and Somatosensory Information

Active movement involves stimulus detection and classification, as well as response programming, selection, and production (Stelmach *et al.*, 1986; Stelmach and Nahom, 1992). Proper motor control also involves the use of information about external environmental conditions and the current 'internal state' of the parts of the body that will be moved (Matthews, 1988). The availability and utilization of sensory information may be altered by attention and expectation of successful behavioral outcome (Robinson and Petersen, 1986a; Robinson and Petersen, 1986b; Derryberry, 1989; Posner and Petersen, 1990; Aston-Jones *et al.*, 1994). Internal representations of current behavioral circumstances, as they relate to impending movements, also must be updated efficiently to account for any changes in conditions that could affect movements waiting to be planned or executed if previously programmed (Kalaska and Crammond, 1995).

ACTIVITY MODULATION AND BEHAVIOR

Utilization of Sensory Information

To be of use in the control of motor behavior, information about body parts and external sensory events must be conveyed unambiguously to regions of the brain involved in initiating and executing purposeful movements. Matthews (1988) has suggested that, to do so, the central nervous system should have a means of enhancing behaviorally important sensory inputs. In addition, he has suggested that inputs that are not as important, that compete for limited attentional resources, or that interfere with the fidelity of information about necessary sensations, should be suppressed. To date, there is no clear consensus about where behaviorally associated somatosensory modulation occurs. Changes in sensory responsiveness accompanying movement initiation and execution could occur at or before the primary cortical level or at higher cortical levels (see Mountcastle *et al.*, 1990; and Nelson, 1996 for review). Neither is there a clear indication under what behavioral conditions modulation of somatosensory responsiveness occurs.

Theories of the control of skilled movements involve the interaction between motor centers and sensory recipient zones and the modulation of sensory responsiveness. In humans, gating of somatosensory-evoked potentials before the onset of muscle activity (measured by electromyography (EMG)) is thought to result from cortically generated premotor events and is not due to corticofugal influences (Rushton *et al.*, 1981; Cohen and Starr, 1987; Schmidt *et al.*, 1990a,b; Papa *et al.*, 1991). This gating may represent 'corollary discharge' (Sperry, 1950) or 'efference copy' (von Holst and Mittelstaedt, 1950; von Holst, 1954) by which motor centers inform the sensory cortices that subsequent movements are centrally generated (Miles and Evarts, 1979; Seal and Commenges, 1985; Seal and Commenges, 1985; Nelson, 1987; Chapman *et al.*, 1988; Pertovaara *et al.*, 1992). It has been suggested that 'the cutaneous input may be inhibited [before and] during movement to facilitate transmission of information from other receptors more important for the control of movement' (Dyhre-Poulsen, 1978) (such as those from joint afferents and muscle receptors). This modulation '...has been interpreted as [resulting in] an improvement in signal-to-noise ratio' (Coquery, 1978) and may also be important in reducing both cutaneous and deep inputs that might interfere with motor control (Dyhre-Paulsen, 1975; Dyhre-Poulsen, 1978; Wiesendanger and Miles, 1982; Jiang *et al.*, 1990a,b). While theories about corollary discharge and efference copy have been in the literature for many years, it has not been possible to unequivocally prove when and where motor commands modify sensory responses and movement-related neuronal activity.

Modulation of Responsiveness

It has become increasingly clear in the last several years that SI representations are dynamically maintained and can be changed in adults (Pons *et al.*, 1991; Merzenich and Jenkins, 1993; Merzenich and Sameshima, 1993; Merzenich *et al.*, 1993 for review). Changes appear to be use dependent. Representations are strengthened functionally and expanded spatially by inputs resulting from repetitive behavior (Recanzone *et al.*, 1992; Merzenich and Sameshima, 1993). However, we still know very little about how somatosensory inputs are processed during behavior and what behaviors modify cortical responsiveness (Lee *et al.*, 1992; Lee and Whitsel, 1992; Essick and Whitsel, 1993; Lebedev *et al.*, 1994; Prud'homme *et al.*, 1994; Ruiz *et al.*, 1995; Lebedev and Nelson, 1996). It has been suggested that structured behavior in the form of retraining can result in recovery of function following central or peripheral injury (Merzenich *et al.*, 1993). To facilitate recovery of function through retraining, it is crucial to understand how somatosensory cortical activity is modulated during common behaviors.

Over the past two decades increasingly sophisticated methods have been used to determine when, where, and perhaps why sensory responsiveness in the primary cortical recipient zones for exteroceptive and proprioceptive information channels might be modulated during behavior. We know, not only from published accounts but also by introspection, that sensory responsiveness must be modulated. Taking the latter first, every day we verify work by Coquery (1978) who showed that before and after active movements, the ability of subjects to perceive peripheral stimuli is diminished. Important work by Chapin and Woodward (1982a,b) began to elucidate the neuronal substrates of this diminished responsiveness and began to suggest when it might occur, that being whenever sensory inputs are redundant or competitive with other more behaviorally important inputs. Thus, they built on the seminal work of others who suggested that there is central control over peripheral input (Sperry, 1950; von Holst and Mittelstadt, 1950; Teuber, 1966; Matthews, 1988; Prochazka, 1993).

In recent years, many studies, almost too numerous to mention here, have investigated how somatosensory information is processed. Investigators have realized that they must consider two very important concepts: perceptual and motor set (Evarts *et al.*, 1984). Using only slight modifications from the definitions of these terms as used by Evarts and co-workers, 'perceptual set' can be defined as the state of readiness to receive a stimulus that has not yet arrived. In the same way, 'motor set' can be defined as a state of readiness to make a response that will probably result in the attainment of a desired reward. The common denominator that these two concepts share is that of 'expectation'.

Attentional Effects

To determine how neuronal responsiveness varies as a function of perceptual and motor set, neuronal recording is conducted often while animals perform controlled behavioral tasks designed to dissociate desired effects while striving to control others. Selective attention paradigms as employed by Hyvärinen, Poranen, and colleagues (1980), and more recently Iwamura and colleagues (Iriki *et al.*, 1996) and Johnson and coworkers (Hsiao *et al.*, 1993), have shown what have been referred to as enhanced responses to attended stimuli. Several studies have demonstrated sensory stimulus response enhancement if stimuli were relevant (e.g. the go-cue for monkeys to make movements) as compared to irrelevant for movement, as in the case when stimuli are no longer linked to movement initiation ('no-go' tasks; Hyvärinen *et al.*, 1980; Rizzolatti *et al.*, 1981; Nelson, 1984; Nelson *et al.*, 1991b; Kalaska and Crammond, 1995). Responsiveness in the ascending somatosensory system has been studied during active movements compared with passive movements made through the same trajectory (Sinclair and Burton, 1991; Ageranioti-Belanger and Chapman, 1992; Chapman and Ageranoti-Belanger, 1992). Modulations of sensory and movement-related activity during the performance of some tasks may result from motor cortical outflow to sensory cortical and subcortical targets (e.g. Matthews, 1988; Nelson *et al.*, 1991a; Chapman and Ageranoti-Belanger, 1992; reviewed in Nelson, 1996). Studies involving picking appropriate stimuli (Whang *et al.*, 1991) and demonstrating that monkey somatosensory cortical neurons can be selectively activated during categorization (Romo *et al.*, 1996) have strengthened the contention that somatosensory cortical responsiveness is under dynamic control. If selective attention involves deciding what to pay attention to, then perhaps the 'go vs. no-go' paradigms that we and others have used are the motor equivalent (Hyvärinen *et al.*, 1980; Rizzolatti *et al.*, 1981; Nelson, 1984; Nelson *et al.*, 1991b; Kalaska and Crammond, 1995). Commonly, there are profound differences in the responses of SI neurons to sensory stimuli, depending upon when, how, and if monkeys or other subjects are prepared to respond to them (Chapman *et al.*, 1996; Nelson, 1996). Despite some disagreement about when cortical activity is modulated relative to movement onset, it is clear that active engagement in motor behavior has profound effects on sensory responsiveness when compared with passive stimulus presentation.

Most studies discussed above suggest that both facilitation and suppression of sensation and perception occur before expected sensory events and the motor responses to them. Missing from the literature are demonstrations from single-cell recordings that unwanted or redundant somatic sensations are suppressed while, at the same time, behaviorally significant ones are facilitated. Also missing is a demonstration that facilitation and suppression occur simultaneously but at spatially separate locations.

Expectation

There has been interest in what happens to SI neuronal sensory responsiveness and movement-related activity when an animal's expectation of being rewarded for successful performance of a required behavior is altered. An experiment will be described below that does not always 'go according to plan'. By occasionally and unpredictably withholding fruit juice reward despite proper behavior, it is probable that a condition is established in which the linkage between conditioned stimuli, conditioned responses, and desired goals is temporarily brought into question. It has been suggested that activity modulation must be shown to be independent of changes in intention to move (Robinson and Petersen, 1986a). Because animals receive the same stimuli and make the same responses to them immediately after this temporary disruption of the behavior-reward linkage, the paradigm creates a situation in which expectation is altered while motor set remains the same. In essence, the experiment tested the hypothesis that sensory responsiveness and movement-related activity are modified when behavioral outcome becomes suddenly less predictable. It also tested the hypothesis that these activities are more tightly coupled to sensory stimulus characteristics and movement kinematics. It is conceivable that there may be little change in perceptual set as well, because the same stimulus is used regardless of whether the animals were previously rewarded. These studies have led to the construction of three basic hypotheses, addressed in the experiment described below and in others to follow.

This first hypothesis is that, during motor behavior, the correlation of sensorimotor cortical neuronal activity with sensory cues and movement kinematics is diminished as long as behavioral outcome is predictable. As a corollary to this, one might predict that following unexpected behavioral outcomes, such as a withheld reward despite correct performance, the fidelity of sensory responses and the association of neuronal activity with movement kinematics improve. The second hypothesis is that the fidelity of sensory responses is best when sensory inputs are necessary for movement guidance. Under these conditions, movement-related activity may be more tightly coupled to movement kinematics. The final hypothesis is that, once motor responses are selected, sudden changes in sensory inputs can modify pre-programmed behaviors, but only to a certain extent. The SI activity changes that have been studied previously have been related to the processing of peripheral and centrally generated inputs during several phases of sensory-triggered wrist movements (Rushton *et al.*, 1981; Nelson 1984; Nelson, 1987; Nelson, 1988; Nelson *et al.*, 1990; Nelson *et al.*, 1991a,b; Lebedev *et al.*, 1994; Lebedev and Nelson, 1995; Lebedev and Nelson, 1996). Despite these studies and those of several other groups, changes in sensory responsiveness that accompany stimulus identification, response selection, and response programming (Rushton *et al.*, 1981; Salinas and Abbott, 1995) leading to appropriate motor output are still far from completely understood.

An Experiment

We recently conducted experiments that tested the hypothesis that variations in expectation alter SI neuronal sensory responsiveness and premovement activity (PMA). Using an unpredictable reward schedule for correct task performance, we created a condition where monkeys sometimes are not reinforced for seemingly appropriate movements. Several results were thought to be possible. In trials immediately following correct but unrewarded performance, both sensory responsiveness and premovement activity might have been either enhanced or suppressed. We sought to describe these changes quantitatively and qualitatively.

This experiment was fueled by our interest in sensorimotor cortex participation in initiating and executing arm movements (Duffy and Burchfiel, 1971; Hyvärinen and Poranen, 1978a,b; Iwamura and Tanaka, 1978; Hyvärinen *et al.*, 1980; Iwamura *et al.*, 1980; Iwamura *et al.*, 1983 Kalaska, 1988; Crammond and Kalaska, 1989; Kalaska *et al.*, 1989; Burbaud *et al.*, 1991) and the degree to which more posterior cortical areas interact with classically defined SI (Nelson, 1996). We previously observed that early movement-related activity occurs, on average, earlier in area 1 than in area 3b of SI (Gardiner and Nelson, 1992). This timing is opposite the hierarchical projection pattern of afferent inputs (Jones *et al.*, 1978; Jones and Porter, 1980; Pons and Kaas, 1986; Garraghty *et al.*, 1990; Pons *et al.*, 1992) and suggests that movement-related activity may occur first in parietal sensorimotor cortices and then is propagated forward. This serial propagation is similar to that proposed for the serial processing of motor programming thought to occur in the premotor (PMd, PMd/MI) and MI cortices (Wise and Kurata, 1989; Kurata, 1993).

Methods

Two rhesus monkeys (*Macaca mulatta*) were trained to make wrist flexion and extension movements in response to vibratory and/or visual go-cues. Each monkey first held a centered wrist position and awaited the trial's go-cue. Upon receipt of that cue, in blocks of ten, each made ballistic wrist flexion or extension movements. Using a pseudorandom reward schedule, we created a condition in which behavioral outcome could not be reliably predicted. About 75% of the trials in which the monkeys performed correctly were re-warded; the other 25% were not. The activity patterns of 288 task-related neurons were studied in detail. A total of 69/288 were vibratory responsive. Vibratory responsive neurons exhibited sustained or transient changes in neuronal activity associated with stimulus presentation. A total of 235/288 sensorimotor cortical neurons exhibited statistically significant changes in premovement activity. Locations of penetrations in the more extensively studied monkey are shown in Figure 13.1. Tables listing recording locations and RF characteristics are included in Figure 13.2.

Descriptions of the animal care and training, the basic behavioral paradigm, preparation, and methods for single-unit recording, data collection, and histological reconstruction of recording sites are detailed elsewhere (Nelson *et al.*, 1991a; Lebedev *et al.*, 1994; Lebedev and Nelson, 1995, 1996). Each monkey was cared for in accordance with the *NIH Guide for Care and Use of Laboratory Animals, revised 1985*.

Experimental Paradigm

The task consisted of three basic parts: maintaining an initial wrist position, detecting a vibratory cue delivered to the palm of the hand, or a visual cue, and ballistic wrist flexion or extension movements (Fig. 13.1). Each trial began when the monkey positioned the handle in the centered zone. Thus, he actively held the handle against a small upward force generated by a DC torque motor. The monkey had to hold this centered position for 0.5, 1.0, 1.5, or 2.0 s (pseudorandomly varied, based on a computer random number generator).

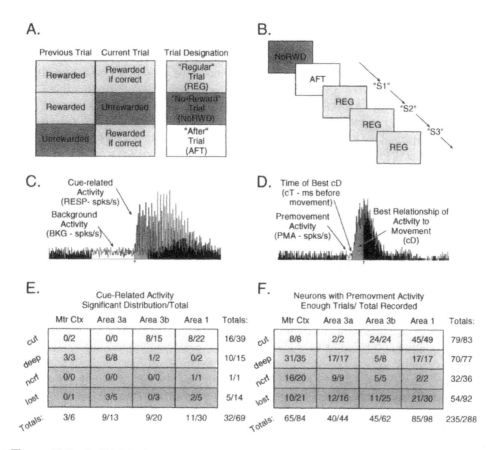

Figure 13.2   A: Trial designators. Rewards were given for 75% of the correctly performed trials using a pseudorandom schedule. Trial type was defined by whether the previous trial was rewarded and whether the current trial was scheduled to be rewarded if correctly performed. **B:** Sequences. Based on the 'reward history', sequences of trials occurred that could be defined based on the number of previously rewarded trials. **C:** Activity changes associated with vibratory go-cues. Abbreviations as in the text. **D:** Movement associated activity changes. **E** and **F:** Tables indicating the number of recorded neurons by cortical location, receptive field type, and whether there were significant activity changes in response to cues (E) or enough trials to determine relationships with movements.

If the animal moved prior to completing the hold period, the trial was cancelled. The monkey could initiate a new trial by returning the handle to the central zone.

If the monkey did not move during the hold period, he received a vibratory go-cue. The plate was vibrated, signaling that the movement could begin. The vibratory cue was delivered by adding a sine wave signal, generated by a signal generator, to the summing junction of the operational amplifier in the torque motor controller. The torque motor controller also produced a steady state torque. The vibratory stimulus amplitude was <0.057° or less than 100 μm peak-to-peak measured 10 cm distal to the coupling of the handle to the motor. This amplitude was sufficient to excite cutaneous receptors in the palm and fingers (Talbot *et al.*, 1968; Mountcastle *et al.*, 1969; LaMotte and Mountcastle, 1975). One stimulus frequency was used (57 Hz), which probably excited cutaneous rapid adaptors, Pacinian afferents, and muscle spindles (Talbot *et al.*, 1968; Wiesendanger, 1986). Fruit juice 'rewards' were given by a computer-controlled, solenoid-based system.

Given the random 75% reward schedule used in this paradigm, several types of trials occurred, represented schematically in Figure 13.2A. If the current trial was preceded by a rewarded trial and the monkey's performance was to be rewarded if correct, it was called a 'regular' trial (REG). If the current trial was preceded by a rewarded trial and the reward was withheld despite the monkey's correct performance, it was called a 'no-reward' trial (NoRWD). On the trial immediately following a NoRWD, the monkey's correct performance could once again elicit a reward. This was called an 'AFTER' trial (AFT). In no instance were AFT trials imposed sequentially. Because there was a 75% reward schedule, it was possible that there could be sequences of several REG trials (Fig. 13.2B). Each trial in a sequence was designated by the number of consecutive rewarded trials that immediately preceded the current trial.

Analyses

Data analysis was conducted in several stages. The first trial in each block was excluded, because it was not influenced by a previous trial. Graphic and numerical displays of the neuronal activity and wrist position were reconstructed by an off-line data analysis routine. Peri-event histograms, raster displays of the neuronal activity, and analog displays of the animal's behavioral performance were examined. These displays were oriented in time with either the onset of the vibratory go-cues or the onset of the sensory triggered movements.

Background activity (BKG; Fig. 13.2C) for each trial was the mean discharge rate (measured in spikes/s) occurring while the monkey actively held against the load of the handle. Neuronal activity associated with sensory stimulus onset (RESP) was measured by determining the first monotonic change in activity after stimulus onset in which the magnitude of the activity was statistically different from background (Fig. 13.2C). Premovement activity was measured from displays centered on movement onset using the same temporal and magnitude criteria (Fig. 13.2D). PMA was designated as the first significant activity change following the return to background after any (if present) go-cue related response and continued until movement onset. We used the cumulative sum method (CUSUM; Ellaway, 1977; Jiang *et al.*, 1991) to determine the first change in mean firing rate for neurons which was ± 3 standard deviations from the background level and which occurred between 20–250 ms before movement onset. This occurrence served as the PMA onset (Lebedev and Nelson, 1995). Cue-related and premovement activity magnitude changes were compared by subtracting the background activity from each. Times of activity changes (onsets) were temporally referenced to important behavioral events. Stimulus-related events such as onsets and offsets of instructions and stimuli were entered into the data stream. Movement-related events such as movement onset from center and the time of 5° handle deflection were recorded. Thus, we were able to determine if neuronal activity was temporally correlated with, and preceded or followed, stimulus and movement-related events.

We used two methods to determine the tightness of activity coupling to stimuli and movements. We used the length of the mean vector (VEC) to determine how well stimulus responses were related to stimulus characteristics (Lebedev and Nelson, 1996). Our implementation varied little from circular-statistical parameters described previously (Batschelet, 1981; Zar, 1974):

$$r = (1/n) \cdot [\sum_{i=1}^{n} \sin(2\pi f t_i))^2 + (\sum_{i=1}^{n} \cos(2\pi f t_i))^2]^{1/2}; \qquad (1)$$

where r is the mean vector length, f is the frequency of vibration, t_i represents the time of spike occurrences, and n is the number of spikes. The value of r can change in the range of 0.0 to 1.0, where 0.0 corresponds to the uniform cycle distribution, and 1.0 corresponds to discharges at a constant phase relative to the stimulus cycle. We measured stimulus-related activity occurring in the period from stimulus onset to 125 ms after its occurrence, which was well before activity related to the initiation of movement commonly occurred (Lebedev and Nelson, 1996).

A multiple regression analysis was used to determine how well movement-related activity covaried with movement parameters such as position, velocity, and acceleration (Ashe and Georgopoulos, 1994; Taira *et al.*, 1996). RTs and MTs were measured by conventional means (Welford, 1980). Handle position was continuously sampled. The onset of neuronal activity relative to movement onset was used to determine, on a trial-by-trial basis, if activity changes were best correlated with movement onset or stimulus presentation. By plotting single-trial lag times (activity onset time after go-cue; R1) or lead times (activity onset time before movement; R2) against the reaction times and examining the linear regressions for significance, the neurons could be defined as sensory-related, movement-related, related to both events, or related to neither. The tightness of coupling to vibratory stimuli was done by methods previously described (Lebedev *et al.*, 1994; Lebedev and Nelson, 1996). REG- and AFT-trial stimulus responsiveness was compared. We used multivariant analysis procedures to describe the relationship between single-unit activity and movement kinematics in our ballistic wrist movement paradigm (adapted from Ashe and Georgopoulos, 1994; Taira *et al.*, 1996). In brief, after appropriate data conditioning, using procedures identical to those previously described, we compared the changes in wrist position, velocity, and acceleration with the time varying frequency of single-unit discharge $f_{(t+\tau)}$ on a trial-by-trial basis using the equation

$$f_{(t+\tau)} = \text{const} + b_1 * \text{position}_{(t)} + b_2 * \text{velocity}_{(t)} + b_3 * \text{acceleration}_{(t)} + \varepsilon \qquad (2)$$

where const is the intercept, b_{1-3} are partial regression coefficients for the variables, t is an array of points the length of the movement time divided by the bin width of the filtered spike function, and ε is an error term. The coefficient of determination (cD = R^2) was calculated and served as a description of the goodness of fit of the predicted function to the low-pass filtered spike record (Taira *et al.*, 1996). This equation was applied to a time-shifted spike function to determine at which time, relative to movement, the best correspondence between neuronal activity and the movement kinematics occurred. The cD and the best time (cT) then served as comparative measures of the degree of co-variance of single-unit activity with movement. Examples of this analysis are seen in Figures 13.3–5.

Once onsets, offsets, and magnitudes of activity, mean vectors, and movement-activity relationships were determined, the results were compared quantitatively and qualitatively. Repeated-measures analyses of variance (ANOVA) were conducted on all measures, grouped by recording location, movement direction, and trial type (REG vs. AFT). Dependent variables included mean background activity, cue response magnitude, movement-associated activities, reaction times, and movement times. The data were split only by those independent paradigm and activity variables having significant influence over the dependent variables. Groups were compared by cortical location. Group statistics were conducted using either parametric or, when appropriate, non-parametric procedures, because there was no *a priori* reason to assume that the distributions were evenly distributed or of equal number. Thus, we determined which experiment manipulations had influences over the firing patterns of sensorimotor cortical neurons.

Observations

Results from previous experiments indicate that SI neurons commonly show different responses, depending on the behavioral contexts under which peripheral stimuli are delivered and movements are made (Chapman *et al.*, 1996; Nelson, 1996). Changes in neuronal activity are related to whether monkeys make movements in response to sensory inputs that are delivered to the same body part that is subsequently moved (Hyvärinen *et al.*, 1980; Nelson *et al.*, 1991b). We have now been able to show that the activity of SI neurons changes depending on whether a reward for requested movements is predictable or not.

Results from the recordings from 288/327 task-related sensorimotor cortical neurons indicate that when behavioral outcome suddenly becomes unpredictable, responsiveness to peripheral sensory and centrally generated inputs increases. Moreover, the correlation between neuronal activity and movements that are subsequently made improves for a short time. We view both occurrences as indicative of a release from the tonic attenuation that probably occurs during the performance of stereotypic behaviors. We compared neuronal activity in trials following rewarded trials (REG) with that occurring in trials following withheld rewards despite correct performance (AFT; Figure 13.2). In both instances, monkeys made virtually indistinguishable wrist movements. As noted earlier, we calculated the onsets and magnitudes of stimulus-related activity (RESP) and premovement activity (PMA) changes for all neurons.

Neuronal Activity Types

Examining vibratory stimulus-related activity can indicate the fidelity with which sensorimotor cortical neurons respond to task-related peripheral sensory events. Qualitative differences were readily noticeable in the activity patterns of about 5% of the 69 stimulus-responsive sensorimotor cortical neurons recorded during REG and AFT trials. For example, Figure 13.3 (A. and A') compares the PMA changes that occurred at about 100 ms before movement onset. Following withheld rewards (AFT trials), there was a cessation in the ongoing vibratory stimulus-related activity that was not present in trials that followed predictable reward presentation. In many ways, however, these records are typical of most that came from sensorimotor cortical neurons with stimulus-related activity. In general, when the activity associated with AFT trials was compared with that associated with REG trials, several common features emerged. BKG tended to be elevated and RESP tended to be greater (Fig. 13.6). There was often a reduction in the length of VEC, indicating a reduction in the fidelity with which the temporal characteristics of vibratory stimuli were represented in the activity patterns of these neurons.

The vast majority of sensorimotor cortical neurons having both vibratory stimulus-related activity changes and PMA had activity changes that necessitated quantitative assessment. Moreover, as previously demonstrated, vibratory stimulus-related changes were not limited to neurons located in cortical regions traditionally thought to receive predominantly from exteroceptors, but were evident for neurons located in areas 3a and 4 (Lebedev and Nelson, 1995). Figure 13.4 illustrates the records of a neuron that was located in area 3a, the cortical regions receiving input predominantly from proprioceptors associated with muscles (see Wiesendanger, 1986). Changes in BKG, RESP, and VEC paralleled those illustrated in Figure 13.3, in that during trials immediately following withheld rewards, BKG and RESP were elevated while there was a reduction in the length of the mean VEC.

PMA commonly occurs early enough before the onset of muscle activity and the onset of movement to suggest that this activity may be related to the initiation as well as the execution of active movements (Nelson *et al.*, 1991a). This early activity in 235/288 sensorimotor cortical neurons, for which complete recordings were obtained, was probably subject to modulation by motor commands, because the generation of these commands often coincided with changes in cortical neuronal activity, sensations, and perceptions (see Chapman *et al.*, 1996, and Nelson, 1996, for review). Figure 13.5 illustrates an example of phasic activity changes that occurred at approximately 170 ms prior to the onset of vibratory stimulus-triggered wrist flexion movements, regardless of whether or not the previous behavioral trial resulted in the monkey receiving a reward. In panels B and B', BKG, PMA magnitudes, and onsets are given for REG and AFT trials. For this neuron, as was the case for many of those with PMA, there were only modest changes in BKG and PMA magnitudes. Often, as well, the onsets of PMA in the REG and the AFT trials occurred at about the same time.

Several things characterized the modulation of sensorimotor cortical neuronal activity that occurred prior to movement. Figure 13.5C illustrates an example of the low-pass filtered spike function, represented in the form of a histogram, that was the result of applying the fractional interspike interval algorithm (Ashe and Georgopoulos, 1994; Taira *et al.*, 1996). Four consistent findings are illustrated in this panel. First, the best coefficients of determination (cDs) for the correlation between the PMA and the kinematic of the initial movement made in response to the sensory go-cue were greater in the AFT trials than they were in the regular trials. Thus, the sensorimotor neuronal activity was more tightly coupled to the subsequently occurring movement after rewards were unpredictably withheld. Second, the times of occurrences of the best correlation (cT) between neuronal activity and movement kinematics were earlier, relative to movement onset, in the trials following withheld rewards when compared with trials following rewarded trials. Third, the time between the onset of significant changes in PMA and cT, which was termed the 'rise time' (Rise T), was shorter in this case by about 23 ms. Finally, in this figure, as in the two previous figures, the mean RTs are listed. It was common for mean RTs to be significantly shorter in trials immediately following withheld rewards than in trials that followed predictable reward delivery.

As a means of comparing the modulation of somatosensory cortical responses during unpredictable rewarding conditions, the data were subjected to a factorial ANOVA and then

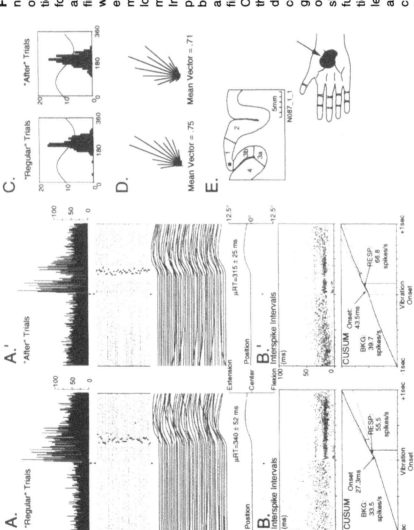

Figure 13.3 A–B': Displays of neuronal activity and behavior as a function of 'behavioral time', centered on vibration onset. Records are from trials that followed a rewarded trial (regular trials). **A–A':** Histograms showing mean firing rates (bin width 5 ms). Rasters in which each dot represents a spike and each row represents a single trial. Dark marks indicate movement onset. Below, average position traces. **B–B'** and mean reaction times (μRT±1 SD). Interspike intervals (ISI) constructed by plotting the ISI of the *n*th spike in behavioral time. Rhythmic firing appears as horizontal bands. Calculations of firing rates and their onsets using the CUSUM method whereby the slope of the CUSUM is the mean firing rate and deviations of ±3 SDs for the mean constitute the onset. **C:** Cycle histograms for spikes during the first 125 ms of stimulus presentation. **D:** Vector plots showing a significant distribution as a function of stimulus phase (see equation 1). Below are listed mean vector lengths. **E:** Recording location. This area 1 neuron (see Figure 1C) had a cutaneous receptive field on the thenar pad extending to the hypothenar eminence (D). 'Regular' and 'After' trials are defined in the text, as are abbreviations.

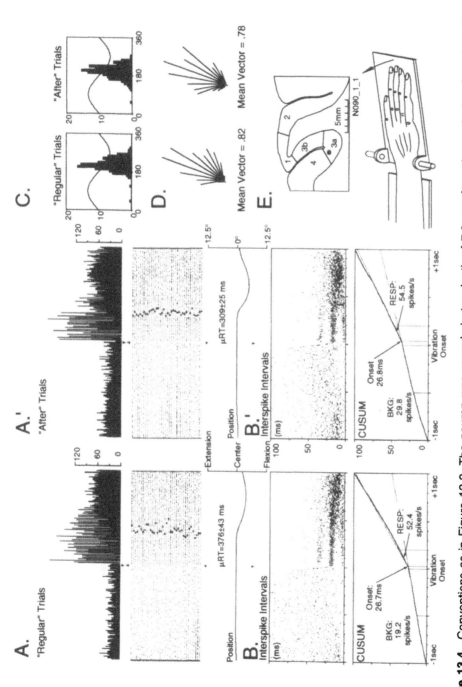

Figure 13.4 Conventions as in Figure 13.3. The neuron was recorded at a depth of 7.8 mm from the cortical surface (see Fig. 1F). It responded to passive wrist extension and palpation of the flexor carpi ulnaris muscle.

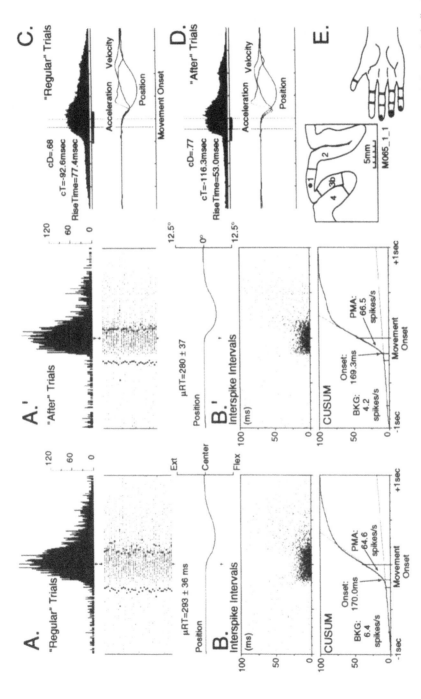

Figure 13.5 A–B': General conventions as in Figure 3, except that displays are centered on movement onset, left marks indicate go-cue onset and right marks indicate movement of 5° and reward delivery. **C–D:** Low-pass filtered spike function as a histogram and the average position, velocity, and acceleration traces centered on movement onset. The times (cTs) of best activity movement-relations (cDs; see equation 2) are shown along with the times from activity onset until cTs (rise times). The neuron was recorded at a depth of 1.4 mm from the cortical surface within area 1. It had a strong response to stimulation of the tip of the fourth digit and weaker response when the tips of the third and fifth digits were stimulated.

spilt by those variables that had significant influences on these values. A schematic depiction of the comparative results is presented in Figure 13.6. Also included for comparison in Figure 6 are entries for the population of area 4 motor cortex neurons that were recorded during the performance of this task. In general, the most profound changes in both sensory and movement-related activities were seen in area 3a neurons. The least profound were in area 3b neurons.

Activity Changes Related to Vibratory Cues

In general, when significant changes in sensory stimulus-related activity occurred, they consisted of increases in BKG and RESP but decreases in the fidelity with which peripheral stimuli were represented by cortical neuronal activity (VEC). BKG increased significantly (10–22%) during AFT trials for neurons located in each of the four cortical areas. Only in area 3a were these increases observed regardless of movement direction. For area 1 neurons, BKG increased significantly in flexion trials following withheld rewards. Flexions in this paradigm were made toward the stimulated surface of the hand and against the modest load that assisted extension movements. For areas 4 and 3b, increases in BKG were seen during extension trials, which required movements to be made away from the stimulated surface of the hand and with the handle's load. Both increases (6–11%) and decreases (one instance of 17%) in RESP were observed when REG and AFT trial magnitudes were compared as a function of recording location and movement direction. Notably, the predominantly cutaneous input recipient zones (areas 3b and 1) exhibited opposite changes in RESP, and these changes were significant only for flexion trials, when the animals made movements toward the stimulated surface of the hand and opposing the handle's load. Area 3a RESP in AFT trials was elevated by an almost equal amount independent of movement direction. Significant changes in the fidelity of vibratory-related responses during AFT trials occurred only for area 3a and the few area 4 neurons that exhibited cue responses. Even then, these changes were significant only when extension trials were compared.

Activity Changes Related to Impending Movements

Movement-related activity changes were characterized by modest increases in the correlation between cortical neuronal activity and movement kinematics without substantial changes in the magnitudes of PMA. In addition, the times of best activity-movement correlations, in some instances, shifted toward the onset of the sensory go-cues. Although PMA was generally greater in AFT trials, significant differences occurred only for those neurons located in area 3a. As with cue-related changes, PMA changes for area 3a neurons did not vary as a function of movement direction. The period between the onset of PMA and the time at which the activity was best related to movement kinematics, Rise T, was significantly shorter for area 1 neurons during flexion trials. The activity-movement correlations, as measured by cDs, were consistently, albeit only slightly, better during AFT trials. Increases in mean cDs of 2–4% were seen during flexion AFT trials for neurons in each somatosensory cortical area studied as well as in area 4 motor cortex. In addition, cDs were larger for area 3a neurons during extension AFT trials. Figure 13.6B depicts the distribution of the shifts in cDs for area 1 neurons. It can be seen that most AFT cDs were greater than the cDs for the corresponding REG trials when the data are viewed on a case-

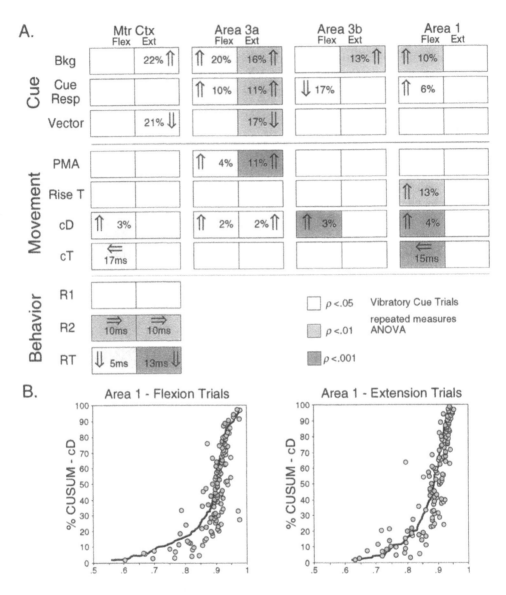

Figure 13.6 **A:** Table of significant changes in stimulus- (cue) and movement-related measures. Shown are the mean percentage increases and decreases in AFT trials as compared with REG trials. Shading depicts significance level of repeated measures ANOVA. **B:** Two-variable cumulative sum (CUSUM) plots of cDs during REG (line) and AFT (circles) trials. The cDs for REG trials were ordered by increasing value and plotted as a CUSUM. At each 'level' on the y-axis is plotted the cD for the AFT trials that corresponds to the cD for the REG trial. As such, dots to the right of the line indicate that the cD for that case was greater for the AFT trials. Distributions for area 1 flexion (left) and extension (right) movement are shown.

by-case basis. Although the mean percentage increases were small for data grouped by cortical area and movement direction, the overall consistency of the effect within individual groups probably led to the findings of statistical significance using repeated measures ANOVA. There were, for each area and movement direction, individual cases for which there were substantial differences in cDs in favour of AFT trials. This actually may be an indication that the neuronal population within each cortical area is heterogeneous with respect to the goodness of fit between neuronal activity profiles and movement parameters. However, to date, the characteristics by which the populations might be split have not been determined and there was no *a priori* reason to do so at the time of the original data analyses.

Two of three behaviorally related measures were different in AFT trials when compared with REG trials. We divided the time from go-cue onset to movement onset (reaction time; RT) into two intervals. The first is the interval between go-cue onset and the onset of PMA (R1); the second interval is from PMA onset to movement onset (R2). R1s did not change in trials following withheld rewards. However, R2s were shorter by a mean value of 10 ms; thus PMA onset occurred nearer movement onset by a mean value of 10 ms. This is depicted in Figure 6B as a shift to the right. Overall, RTs were decreased by a mean of 5–13 ms depending upon the direction of movement. Thus, the reduction in mean RTs was of approximately the same order as the rightward shift of the mean onset of PMA.

From these initial PMA analyses, we developed working hypotheses to guide subsequent analyses. When movements are triggered by vibratory cues, PMA onsets occur earlier in REG as compared with AFT trials (Fig. 13.6). Thus, PMA changes occur nearer movement onset when the behavioral conditions are less predictable. Second, PMA magnitudes are not consistently greater during AFT trials than during REG trials, although the activity-movement correlations are better as evidenced by slight increases in cDs. Our findings indicate that these sensorimotor cortical neurons tend to be more responsive to peripheral stimuli in AFT trials as compared with REG trials. Statistically significant increases in BKG and increases and decreases in RESP as well as the changes in movement related activities, as recounted above, suggest that both sensorimotor cortical responses to both central and peripheral inputs are dynamically regulated (Nelson *et al.*, 1991a; Kalaska, 1994). We also observed that RTs in AFT trials were shorter and less variable than in REG trials. This final observation is consistent with the hypothesis that additional selective attention is directed toward the task following unpredictable behavioral outcome (Derryberry, 1989).

Duration of Effects — Sequences

One of several remaining unanswered questions is whether sensorimotor neuronal response modulation in AFT trials continues for long periods of time, without change, after a single episode of withheld rewards. It is possible that within a block of trials, one or more measures of activity may systematically vary due to adaptation to the partial reward schedule. This does not appear to be the case. Figures 13.3–5 are presented such that REG and AFT trials are ordered chronologically from top to bottom in the raster displays. No significant serially dependent variations in background, vibratory-related activity, nor premovement activity are evident upon visual inspection in these nor in more than 98% of the 288 cases examined. Over 10% of these have been examined using the Wald-Wolfowitz Runs Test to determine if there is any serial order to the magnitude of changes in activity, as a function of trial type, during background, cue, and premovement epochs. No significant differences were found. Second, it is possible that the magnitude of the effect(s) may vary as a function of recording session, thus indicating a long-term adaptation to the paradigm conditions. We examined changes in cDs as a function of the order of observation for instances of area 1 neuronal PMA recorded from a single animal. No serial order was found. These observations suggest that the modulations of activity and the resultant alterations in activity-movement coupling occur repeatedly despite the fact that the animals are over-trained prior to the beginning of neuronal recording. Yet, for there to be differences in REG and AFT trials, there must be some transience in the increases and decreases in the measured parameters.

To test the hypothesis that modulations of responsiveness were related to the predictability of reward receipt for correct behavioral performance, we recoded our data to reflect the number of sequentially rewarded trials that preceded each trial. As depicted in Figure 13.2B, AFT trials were immediately preceded by a withheld reward. Subsequent trials may have been preceded by one to several rewarded trials until the occurrence of the next withheld reward. Two possible patterns of modulation were thought possible. In the first, it is possible that modulation of sensorimotor cortical neuronal responsiveness might occur immediately after a withheld reward and the effects might gradually return, over several trials, to levels exhibited in blocks of REG trials, when reward delivery was predictable. The second possibility is that a single rewarded trial is sufficient to reset mechanisms that are influenced by reward predictability and are responsible for responsiveness modulations that occur after a single withheld reward.

Figure 13.7A–D illustrates that the latter is probably true. This figure shows the records of an area 1 neuron where the data have been grouped by movement direction and the number of rewarded trials preceding each trial in the group. While the general features of the PMA are similar regardless of the number of rewards that preceded each trial, the data show that the mean of the cDs for the AFT trials is greater than for each other group. Moreover, the cDs for groups of trials that followed rewarded trials were not different as a function of the number of previously rewarded trials. Figure 13.7E presents the results for the entire population of area 1 neurons. For flexion trials, cDs were slightly, but significantly, greater (103.4%, $\rho < .005$; repeated measures ANOVA) in AFT trials compared with groups of trials where there had been one, two, or three previously rewarded trials in a sequence. The cDs for extension trials were not significantly different as a function of the number of previous trials that were rewarded.

The modulations that resulted in changes in the responsiveness of sensorimotor cortical neurons to peripheral and central inputs appeared to affect only the AFT trials. For all the measures listed in Figure 13.6, significant changes occurred only in the trials immediately following withheld rewards. Once there had been at least one previously rewarded trial, all subsequent trials had cue-related and movement-related activity patterns and magnitudes that were not significantly different from one another.

DISCUSSION

When behavioral outcome is predictable (i.e. when animals are consistently rewarded for performing correctly), both sensory responsiveness and PMA in SI neurons appear to become stabilized at some baseline level. In trials that follow withheld rewards despite correct performance, outcome predictability may be thought of as having been altered.

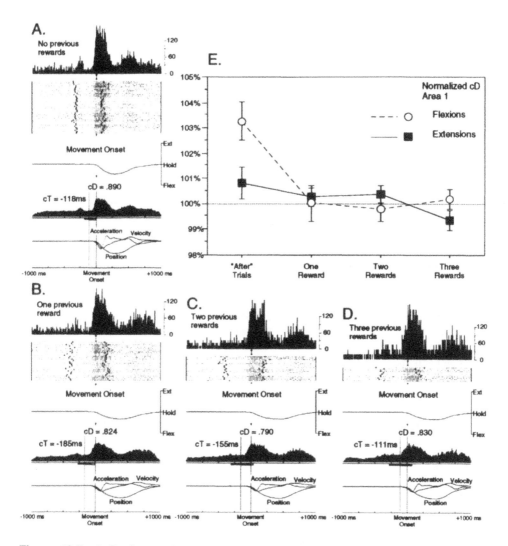

Figure 13.7 A–D: Conventions as in Figures 3–5 for trials preceded by 0 through 3 rewarded trials, respectively. **E:** Mean percentage changes (± 1 standard error of the mean) in cDs as a function of the number of consecutively rewarded trials in a sequence (see Fig. 2). The value for the 'after' trial cDs during flexion trials was significantly different from the other three values ($p < .01$; repeated measures ANOVA). Values for extension trials were not significantly different from one another.

When behavioral conditions become less predictable, the overall activity occurring both while animals await vibratory go-cues and in response to those cues is elevated. The single notable exception to this generalization is for area 3b neurons. These neurons do not respond as well to vibratory cues in AFT trials involving flexion movements. In addition, the coupling between movements and the neuronal activity that precedes them becomes better. However, decreased predictability also appears to be associated with later PMA onsets, perhaps because of behavioral uncertainty. Movements are made more quickly and there is less variation in both the RTs and the movement times (MTs).

Intepretations

These data are open to at least two types of interpretations. We have previously argued that PMA may gate the response of other SI neurons that do not play an important part in the upcoming movement (Nelson *et al.*, 1991a; Lebedev *et al.*, 1994; Lebedev and Nelson, 1995, 1996). We have also argued that PMA may reflect a corollary discharge because neurons in SI areas receiving direct MI projections tend to show PMA, while those in areas without direct MI projections do not (Nelson 1996). MI stimulation is known to alter SI neuronal responsiveness to cutaneous and proprioceptive inputs (see Chapman *et al.*, 1996, for review). Finally, we have observed in records of the 32/69 sensory-responsive neurons that the fidelity with which vibratory stimuli are presented by SI neuronal activity is at times reduced in AFT trials. This last observation may suggest that during conditions of predictable behavioral outcome, the responsiveness of sensorimotor cortical neurons is attenuated — when performance outcome becomes less predictable this attenuation appears to be removed, but that the removal of attenuation is accompanied by a reduction in the fidelity of stimulus representation.

The fact that the correlation between neuronal activity and movement kinematics, as indicated by cDs, is in general greater in flexion movement AFT trials fits the hypothesis that a general mechanism is responsible for decreasing the activity of SI neurons in response to both peripheral and centrally generated inputs associated with predictable behavioral outcomes. One possible role for an increase in activity-movement coupling might then be to suppress the activity of other neurons that do not convey important information about the subsequent movements or that might disrupt the fidelity with which information is conveyed. This possibility could be examined in future experiments by simultaneously recording from several single-units at sites both within and next to sensory fields and 'movement fields' involved in the behavioral task.

Another interpretation is that the neurons with increased activity are themselves the behaviorally important ones and that changes in activity-movement coupling actually reflect increased information transfer, perhaps related to selective attention directed toward behaviorally important events (Whang *et al.*, 1991; Hsiao *et al.*, 1993; Iriki *et al.*, 1996). This possibility could be evaluated by examining the correlation of sensory and movement-related activity with movements kinematics in paradigms using movement with more degrees of freedom or using sequences of novel movements rather than stereotyped ones. The two roles alluded to above probably are not mutually exclusive and probably could occur simultaneously. Regardless of which of many interpretations is favored for these and similar observations, it appears that sensory gating depends upon whether there is a reasonable expectation that attenuating some inputs and strengthening others may actually improve performance by removing potentially competitive sources of information coming from the periphery.

Models of Sensorimotor Integrations

Recent models that have attempted to organize concepts of sensorimotor integration have several things in common. They usually incorporate some sort of comparison between predicted environmental and behavioral conditions and actual outcomes (Bullock *et al.*, 1993; Prochazka, 1993; Salinas and Abbott, 1995; Schultz *et al.*, 1995a,b; 1997; Johansson, 1996; Schultz *et al.*, 1997). Mismatches between these are thought to generate error signals (Fig. 13.8; Nelson, 1996). Other model features are switching between reactive and predictive modes and attentional control of sensory input gain. For new behaviors, the 'actual' outcome signal results from comparing peripheral feedback and motor commands; this may be indicative of reactive mode. Once behavioral outcome is predictable, peripheral feedback may be replaced by internal signals or signals from other modalities such as vision (predictive mode). Most models include a differential reliance on sensations and their suppression or facilitation at cortical levels as behavioral constraints allow, as well as neuronal activity representing parameters of intended movements that exists before movements are begun (Bullock *et al.*, 1993).

Model makers suggest that there are basic principles that should be considered when demonstrating the influences of behavioral context upon sensory responsiveness, movement-related activity, and ultimately upon sensorimotor integration (Schmidt, 1988; Bullock *et al.*, 1996). There is a behavioral 'steady state' achieved during performance of necessarily over-trained and stereotypic behaviors. Changes in sensory responsiveness, movement-related activity, and motor performance are referenced to this baseline. Regions of the CNS involved in sensorimotor integration are more responsive to inputs when animals or subjects are actively engaged in the task. More complex behavioral requirements, such as discrimination as compared with detection of inputs, engage the sensorimotor integration system even more. This appears to result in responsiveness modulation prior to response onset to readjust certain parts of the system for their participation in response programming, selection, and production. Nearly identical behavioral responses can be made over a number of behavioral trials, and these motor responses can be elicited by several types of go-cues. Thus, subtle effects of continuous stimulus presentation and possible effects of stimulus interference with response programming, selection, and production can be determined. Finally, while the source of the central influences that maintain response integrity and regulate sensory responses and movement-related activity is not known, these influences may result from interactions at the sensorimotor cortical level.

Internal Feedback

Part of our working hypothesis is that stereotypic movements are made under conditions in which peripheral inputs are attenuated, because their outcome is predictable and 'internal feedback' is used. Shifts to reliance on peripheral feedback would then occur when outcome becomes unpredictable. Bioulac *et al.* (1995) have suggested that MI normally suppresses early PMA in area 5 because MI ablation results in an increase in the number of area 5 neurons with very early discharges locked to the onset of arm movements. In addition, neurons in MI, dorsal premotor (PMd), and parietal cortices recorded during behavioral tasks similar to ours are activated at times (> 100 ms before movement) that precede PMA onset in SI (Seal and Commenges, 1985; Kurata and Tanji, 1986; Vaadia *et al.*, 1988; Riehle and Requin 1989; Wise and Kurata, 1989; Crammond and Kalaska, 1994) and thus could be involved in the modulation of SI activity. These cortical regions have preparatory neurons that encode the goal or target and the intended direction of movement independent

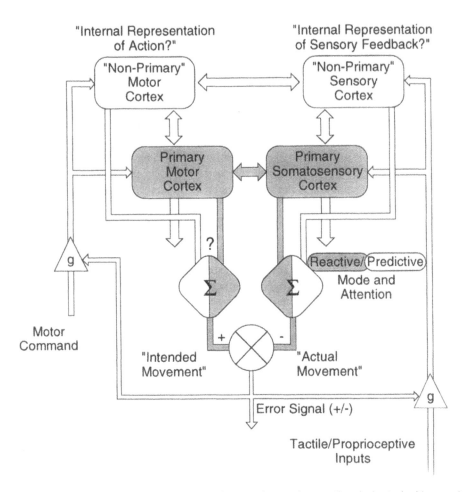

Figure 13.8 A: Model of some aspects of sensorimotor integration (adapted with permission from R.J. Nelson (1996) (Interactions between motor commands and somatic perceptions in sensorimotor cortex. *Curr. Opinion Neurobiol.*, **6**, 801–810). The sigma (Σ) represents a summing junction, the 'g' depicts and variable gain amplifier, and the circle with a cross (⊗) represents a comparator. Represented are primary and higher order ('non-primary') sensory and motor cortices receiving inputs from the sensory periphery or from command generating apparatus, respectively. As suggested (see reference above), when behavioral conditions are predictable, the central nervous system may operate in a 'predictive mode', utilizing internal representations of actions and the 'feedback' that would result from them to generate an 'error signal'. This error signal is thought to provide an evaluation of outcome. Internal representations of actions and feedback may be located in 'non-primary' cortices. As such, the primary cortices may have their responsiveness attenuated because, provided that conditions remain predictable, some inputs from these primary areas may not actually be needed for successful initiation and execution. By this model, when behavioral conditions become unpredictable, there is a switch to 'reactive' mode, which involves utilization of input from the periphery for 'updating' and error signal generation. Output to effectors from primary motor cortex probably occurs at all times, as does intracortical activity involving SI. However, under predictable behaviour conditions, the activity coming from these areas may be attenuated. (See Chapman *et al.*, 1996 and Nelson, 1996 for review).

of their movement-related discharge (Alexander and Crutcher, 1990). Additional experiments are needed in which recordings are made from neurons in 'nonprimary' motor and somatosensory cortices for comparison with previously collected data from SI.

It has been suggested that positive expectation leads to the use of internal representations of the sensory information that normally accompanies motor execution (Kalaska and Crammond, 1995; Johansson, 1996; Nelson 1996). Johansson suggests that '...disturbances in task execution due to erroneous anticipatory settings of motor commands are reflected by discrete mechanical events' (p. 413). It is tempting, therefore, to suggest that during expected behavior, sensory inputs are actually suppressed and premovement activities are less correlated with subsequent movements as the result of using these internal maps by the parietal cortex rather than relying on peripheral inputs. The improvement of cD with failure to receive a reward may reflect a tighter coupling between afferent inputs and subsequent movements. A rightward shift in the activation might suggest a reduction of the reliance on predictive signals. However, activation occurs 10 ms closer to movement onset only for area 4 and area 1 neurons. In general, RTs decrease by about the same amount. Shifts in cT to the left and cD increases may mean that PMA (which may represent corollary discharge from the motor systems) is better related to movements. This might be the result of a release from tonic inhibition. However, these initial experiments suggest that tighter coupling between cortical activity and movements may last only as long as 'uncertainty' exists because the activity-movement coupling seems to return to the 'steady-state' once behaviors are again rewarded (Fig. 13.7).

SUMMARY

These experiments, by examining short-term changes in neuronal responsiveness resulting from internal sources as well as those from external sensory cues, further our understanding of neuronal activities that help to maintain cortical organization when consistent with, and change it when different from, some internally maintained templates. This template matching, we and others believe, is influenced by an animal's expectation. Additional experiments are needed to better understand the role the sensorimotor cortices play in motor control during stimulus detection and classification, response programming and selection (initiation), and response production (execution).

ACKNOWLEDGMENTS

The author wishes to thank Dr Michael Lebedev for his contributions to the experiments described and Dr David Armbruster for editorial assistance in the preparation of this manuscript. Supported by USPHS Grant NS 36860-01 and USAF Grant AFOSR 91-0333.

REFERENCES

Ageranioti-Belanger, S.A. and Chapman, C.E. (1992). Discharge properties of neurons in the hand area of primary somatosensory cortex in monkeys in relation to the performance of an active tactile discrimination task. II. Area 2 as compared to areas 3b and 1. *Experimental Brain Research*, **91**, 207–228.

Alexander G.E. and Crutcher, M.D. (1990). Preparation for movement: neural representations of intended direction in three motor areas of the monkey. *Journal of Neurophysiology*, **64**, 133–149.

Andersen, R.A. (1995). Encoding of intention and spatial location in the posterior parietal cortex. *Cerbral Cortex*, **5**, 457–469.

Ashe, J. and Georgopoulos, A.P. (1994). Movement parameters and neural activity in motor cortex and area 5. *Cerebral Cortex*, **6**, 590–600.

Assad, J.A. and Maunsell, J.H.R. (1995) Neuronal correlates of inferred motion in primate posterior parietal cortex. *Nature*, **373**, 518–521.

Aston-Jones, G., Rajkowski, J., Kubiak, P. and Alexinsky, T. (1994). Locus coeruleus neurons in monkey are selectively activated by attended cues in a vigilance task. *Journal of Neuroscience*, **14**, 4467–4480.

Batschelet, E. (1981). *Circular Statistics in Biology*. New York, Academic Press.

Bioulac, B., Burband P. and Varoqueaux, D. (1995). Activity of area 5 neurons in monkey during arm movements: effects of dentate nucleus lesion and motor cortex ablation. *Neuroscience Letters*, **192**, 189–192.

Bullock, D., Grossberg, S. and Guenther, F.H. (1993). A self-organizing neural model of motor equivalent reaching and tool use by a multijoint arm. *Journal of Cognitive Neuroscience*, **5**, 408–435.

Bullock, D., Grossberg. S. and Guenther, F. (1996). Neural network modeling of sensory-motor control in animals. In *Advances in Motor Learning and Control*, edited by H.N. Zelaznik. Champaign, IL: Human Kinematic.

Burbaud, P., Doegle, C., Gross, C. and Bioulac, B. (1991). A quantitative study of neuronal discharge in areas 5, 2, and 4 of the monkey during fast arm movements. *Journal of Neurophysiology*, **66**, 429–443.

Cavada C. and Goldman-Rakic, P.S. (1989). Posterior parietal cortex in Rhesus monkey: I. Parcellation of areas based on distinctive limbic and sensory corticocortical connections. *Journal Comparative Neurology*, **287**, 393– 421.

Chapman, C.E. and Ageranioti-Belanger, S.A (1992). Discharge properties of neurons in the hand area of primary somatosensory cortex in monkeys in relation to the performance of an active tactile discrimination task. I. Areas 3b and 1. *Experimental Brain Research*, **87**, 319–339.

Chapman, C.E., Jiang,.W. and Lamarre, Y. (1988). Modulation of lemniscal input during conditioned arm movements in the monkey. *Experimental Brain Research*, **72**, 316–334.

Chapman, C.E., Tremblay, F. and Ageranioti-Belanger, S.A. (1996). Role of primary somatosensory cortex in active and passive touch. In *Hand and Brain: The Neurophysiology and Psychology of Hand Movements*, edited by A.M. Wing, P. Haggard, and J.R. Flanagan, pp. 329–347. San Diego: Academic Press.

Chapin, J.K. and Woodward, D.J. (1982a). Somatic sensory transmission to the cortex during movement: gating of single cell responses to touch. *Experimental Neurology*, **78**, 654–669.

Chapin, J.K. and Woodward, D.J. (1982b). Somatic sensory transmission to the cortex during movement: phasic modulation over the locomotor step cycle. *Experimental Neurology*, **78**, 670–684.

Chmielowska, J. and Pons, T.P. (1995). Patterns of thalamocortical degeneration after ablation of somatosensory cortex in monkeys. *Journal of Comparative Neurology*, **360**, 377–392.

Cohen, L.G. and Starr, A. (1987). Localization, timing and specificity of gating of somatosensory evoked potentials during active movement in man. *Brain*, **110**, 451–467.

Colby, L.C., Duhamel, J.R. and Goldberg M.E. (1995). Oculocentric spatial representation in parietal cortex. *Cerebral Cortex*, **5**, 470–481.

Coquery, J.-M. (1978). Role of active movement in control of afferent input from skin in cat and man. In *Active Touch: The Mechanisms of Object Manipulation: A Multidisciplinary Approach*, edited by G. Gordon, pp. 161–169. Oxford: Pergamon.

Crammond, D.J. (1997). Motor imagery: never in your wildest dream. *TINS*, **20**, 54–57.

Crammond, D.J. and Kalaska, J.F. (1989). Neuronal activity in primate parietal cortex area 5 varies with intended movement direction during an instructed-delay period. *Experimental Brain Research*, **76**, 458–462.

Crammond, D.J. and Kalaska J.F. (1994). Modulation of preparatory neuronal activity in dorsal premotor cortex due to stimulus-response compatibility. *Journal of Neurophysiology*, **71**, 1281–1284.

Derryberry, D. (1989). Effects of goal-related motivational states on the orienting of spatial attention. *Acta Psychologica*, **72**, 199–220.

Duffy, F.H. and Burchfiel, J.L. (1971). Somatosensory system: Organizational hierarchy from single units in monkey area 5. *Science*, **172**, 273–275.

Dum, R.P. and Strick, P.L. (1991). The origin of corticospinal projections from premotor areas in the frontal lobe. *Journal of Neuroscience*, **11**, 667–689.

Dyhre-Poulsen, P. (1975). Increased vibration threshold before movements in human subjects. *Experimental Neurology*, **47**, 516–522.

Dyhre-Poulsen, P. (1978). Perception of tactile stimuli before ballistic and during tracking movements. In *Active Touch: The Mechanisms of Object Manipulation: A Multidisciplinary Approach*, edited by G. Gordon, pp. 171–176. Oxford: Pergamon.

Ellaway, P.H. (1977). An application of cumulative sum technique (cusums) to neurophysiology. *Journal of Physiology (London)*, **265**, 1–2P.

Essick G.K. and Whitsel, B.L. (1993). The response of SI directionally selective neurons to stimulus motion occurring at two sites within the receptive field. *Somatosensory and Motor Research*, **10**, 97–113.

Evarts, E.V., Shinoda, Y. and Wise, S.P. (1984). *Neurophysiological Approaches to Higher Brain Functions*. New York: Wiley.

Gardiner, T.W. and Nelson, R.J. (1992). Striatal neuronal activity during the initiation and execution of hand movements made in response to visual and vibratory cues. *Experimental Brain Research*, **92**, 15–26.

Garraghty, P.E., Florence, S.L. and Kaas, J.H. (1990). Ablations of areas 3a and 3b of monkey somatosensory cortex abolish cutaneous responsivity in area 1. *Brain Research*, **528**, 165–169.

Georgopoulos, A.P., Kalaska, J.F. and Caminiti, R. (1983). Relations between two-dimensional arm movements and single-cell discharge in motor cortex and area 5, Movement direction versus movement end point. In *Hand Function and the Neocortex: Experimental Brain Research, Suppl 10.*, edited by A. Goodwin and I. Darian-Smith, pp 175–183. Heidelberg: Springer-Verlag.

He, S.Q., Dum, R.P. and Strick, P.L. (1995). Topographic organization of corticospinal projections from the frontal lobe: motor areas on the medial surface of the hemisphere *Journal of Neuroscience*, **15**, 3284–3306.

Hsiao, S.S., O'Shaughnesy, D.M. and Johnson, K.O. (1993). Effects of selective attention on spatial form processing in monkey primary and secondary somatosensory cortex. *Journal of Neurophysiology*, **70**, 444–447.

Huerta, M.F. and Pons, T.P. (1990). Primary motor cortex receives input from area 3a in macaques. *Brain Research*, **537**, 367–371.

Hyvärinen, J. and Poranen, A. (1978a). Movement-sensitive and direction and orientation-sensitive cutaneous receptive fields in the hand area of postcentral gyrus in monkeys. *Journal of Physiology (London)*, **283**, 523–537.

Hyvärinen, J. and Poranen, A. (1978b). Receptive field integration and submodality convergence in the hand area of the postcentral gyrus of alert monkeys. *Journal of Physiology (London)*, **283**, 539–556.

Hyvärinen, J., Poranen, A. and Jokinen, Y. (1980). Influence of attentive behavior on neuronal responses to vibration in primary somatosensory cortex in the monkey. *Journal of Neurophysiology*, **43**, 870–882.

Iriki, A., Tanaka, M. and Iwamura, Y. (1996). Attention-induced neuronal activity in the monkey somatosensory cortex revealed by pupillometrics. *Neuroscience Research*, **25**, 173–181.

Iwamura, Y. and Tanaka, M. (1978). Postcentral neurons in hand region of area 2: Their possible role in the form discrimination of tactile objects. *Brain Research*, **150**, 662–666.

Iwamura, Y., Tanaka, M. and Hikosaka, O. (1980). Overlapping representation of the fingers in somatosensory cortex (area 2) of the conscious monkey. *Brain Research*, **197**, 516–520.

Iwamura, Y., Tanaka, M., Sakamoto, M. and Hikosaka, O. (1983). Converging patterns of finger representation and complex response properties of neurons in area 1 of the first somatosensory cortex of the conscious monkey. *Experimental Brain Research*, **51**, 327–337.

Jiang, W., Chapman, C.E. and Lamarre, Y. (1990a). Modulation of cutaneous cortical evoked potentials during isometric and isotonic contractions in the monkey. *Brain Research*, **536**, 69–78.

Jiang, W., Chapman, C.E. and Lamarre, Y. (1990b). Modulation of somatosensory evoked responses in the primary somatosensory cortex produced by intracortical microstimulation of the motor cortex in the monkey. *Experimental Brain Research*, **80**, 333–344.

Jiang, W., Lamarre, Y. and Chapman, C.E. (1990b). Modulation of cutaneous cortical evoked potentials during isometric and isotonic contractions in the monkey. *Brain Research*, **536**, 69–78.

Jiang, W., Chapman, C.E. and Lamarre, Y. (1991) Modulation of the cutaneous responsiveness of neurones in the primary somatosensory cortex during conditioned arm movements in the monkey. *Experimental Brain Research*, **84**, 342–354.

Johansson, R.S. (1996). Sensory control of dexterous manipulations in humans. In *Hand and Brain: The Neurophysiology and Psychology of Hand Movements*, edited by A.M. Wing, P. Haggard, and J.R. Flanagan, pp. 381–414. San Diego: Academic Press.

Jones, E.G. (1986). Connectivity of the primate sensory-motor cortex. In *Cerebral Cortex: Sensory-Motor Areas and Aspects of Cortical Connectivity*, edited by E.G. Jones and A. Peters, Vol. 5, pp. 113–184. New York: Plenum Press.

Jones, E.G., Coulter, J.D. and Hendry, S.H.C. (1978). Intracortical connectivity of architectonic fields in the somatic sensory, motor and parietal cortex of monkeys. *Journal of Comparative Neurology*, **181**, 291–348.

Jones, E.G. and Peters, A. (1986). *Cerebral Cortex: Sensory-Motor Areas and Aspects of Cortical Connectivity*, Vol. 5. New York: Plenum Press.

Jones, E.G. and Porter, R. (1980). What is area 3a? *Brain Research Reviews*, **2**, 1–43.

Kaas, J.H. (1983). What, if anything, is SI? Organization of first somatosensory area of cortex. *Physiological Reviews*, **63**, 206–231.

Kaas, J.H. and Pons, T.P. (1988). The somatosensory system of primates. In *Comparative Primate Biology*, Vol. 4, pp. 421–468. New York: Alan Liss, Inc.

Kalaska, J.F. (1988). The representation of arm movements in postcentral and parietal cortex. *Canadian Journal of Physiology and Pharmacology*, **66**, 455–63.

Kalaska, J.F. (1994). Central neural mechanisms of touch and proprioception. *Canadian Journal of Physiology and Pharmacology*, **72**, 542–545.

Kalaska, J.F., Cohen, D.A., Hyde, M.L. and Prud'homme, M. (1989). A comparison of movement direction-related versus load direction-related activity in primate motor cortex, using a two-dimensional reaching task. *Journal of Neuroscience*, **9**, 2080–2102.

Kalaska, J.F., Cohen, D.A., Prud'homme, M. and Hyde, M. (1990). Parietal area 5 neuronal activity encodes movement kinematics, not movement dynamics. *Experimental Brain Research*, **80**, 351–364.

Kalaska, J.F. and Crammond, D.J. (1995). Deciding not to go: neuronal correlates of response selection in a go/nogo task in primate premotor and parietal cortex. *Cerebral Cortex*, **5**, 410–428.

Kurata, K. (1993). Premotor cortex of monkeys: set- and movement-related activity reflecting amplitude and direction of wrist movements. *Journal of Neurophysiology*, **69**, 187–200.

Kurata, K. and Tanji, J. (1986). Premotor cortex neurons in macaques: activity before distal and proximal forelimb movements. *Journal of Neuroscience*, **6**, 403–411.

LaMotte, R.H. and Mountcastle, V.B. (1975). The capacities of humans and monkeys to discriminate between vibratory stimuli of different frequency and amplitude: A correlation between neural events and psychophysical measurements. *Journal of Neurophysiology*, **38**, 539–559.

Lebedev, M.A., Denton, J.M. and Nelson, R.J. (1994). Vibration-entrained and premovement activity in monkey primary somatosensory cortex. *Journal of Neurophysiology*, **72**, 1654–1673.

Lebedev, M.A. and Nelson, R.J. (1995). Rhythmically firing (20–50 Hz) neurons in monkey primary somatosensory cortex: Activity patterns during initiation of vibratory-cued hand movements. *Journal of Computational Neuroscience*, **2**, 313–334.

Lebedev, M.A. and Nelson, R.J. (1996). High-frequency vibratory sensitive neurons in monkey primary somatosensory cortex: Entrained and nonentrained responses to vibration during the performance of vibratory-cued hand movements. *Experimental Brain Research*, **111**, 313–325.

Lee, C.J. and Whitsel, B.L. (1992). Mechanisms underlying somatosensory cortical dynamics: I. In vivo studies. *Cerebral Cortex*, **2**, 81–106.

Lee, C.J., Whitsel, B.L. and Tommerdahl, M. (1992). Mechanisms underlying somatosensory cortical dynamics: II. In vitro studies. *Cerebral Cortex*, **2**, 107–133

Matthews, P.B.C. (1988). Proprioceptors and their contribution to somatosensory mapping: Complex messages require complex processing. *Canadian Journal of Physiology and Pharmacology*, **66**, 430–438.

Merzenich, M.M. and Jenkins, W. (1993). Reorganization of cortical representations of the hand following alterations of skin inputs induced by nerve injury, skin island transfers and experience. *Journal of Hand Therapy*, **6**, 89–104.

Merzenich, M.M. and Sameshima, K. (1993). Cortical plasticity and memory. *Current Opinion in Neurobiology*, **3**, 187–196.

Merzenich, M.M., Schreiner, C., Jenkins, W. and Wang, X. (1993). Neural mechanisms underlying temporal integration, segmentation, and input sequence representation: Some implication for the origin of learning disabilities. *Annuals of the New York Academy of Science*, **682**, 1–22.

Miles, F.A. and Evarts, E.V. (1979). Concepts of motor organization. *Annual Review of Psychology*, **30**, 327–362.

Mountcastle, V.B. (1984). Central nervous mechanisms in mechanoreceptive sensibility. In *Handbook of Physiology – The Nervous System III*, edited by I. Darian-Smith, pp. 789–878. London: Oxford Univ. Press.

Mountcastle, V.B. (1997). The coulmnar organization of the neocortex. *Brain*, **120**, 701–722.

Mountcastle, V.B., Lynch, J. C., Georgopoulous, A., Sakata, H. and Acuna, C. (1975). Posterior parietal association cortex of the monkey: Command functions for operations in extrapersonal space. *Journal of Neurophysiology*, **38**, 871–908.

Mountcastle, V.B., Steinmetz, M.A. and Romo, R. (1990). Frequency discrimination in the sense of flutter: Psychophysical measurements correlated with postcentral events in behaving monkeys. *Journal of Neuroscience*, **10**, 3032–3044.

Mountcastle, V.B., Talbot, W.H., Sakata, H. and Hyvärinen, J. (1969). Cortical neuronal mechanisms in flutter-vibration studies in unanesthetized monkeys. Neuronal periodicity and frequency discrimination. *Journal of Neurophysiology*, **32**, 452–484.

Nelson, R.J. (1984). Responsiveness of monkey primary somatosensory cortical neurons to peripheral stimulation depends on 'motor-set'. *Brain Research*, **304**, 143–148.

Nelson, R.J. (1987). Activity of monkey primary somatosensory cortical neurons changes prior to active movement. *Brain Research*, **406**, 402–407.

Nelson, R.J. (1988). Set related and pre-movement related activity in primate primary somatosensory cortical neurons depends upon stimulus modality and subsequent movement. *Brain Research Bulletin*, **21**, 411–424.

Nelson, R.J. (1996). Interactions between motor commands and somatic perceptions in sensorimotor cortex. *Current Opinion in Neurobiology*, **6**, 801–810.

Nelson, R.J., Smith, B.N. and Douglas, V.D. (1991a). Relationships between sensory responsiveness and premovement activity of quickly adapting neurons in areas 3b and 1 of monkey primary somatosensory cortex. *Experimental Brain Research*, **84**, 75–90.

Nelson, R.J., Li, B. and Douglas, V.D. (1991b). Sensory response enhancement and suppression of monkey primary somatosensory cortical neurons. *Brain Research Bulletin*, **27**, 751–757.

Nelson, R.J, McCandlish, C.A. and Douglas, V.D. (1990). Reaction times for hand movements made in response to visual versus vibratory cues. *Somatosensory and Motor Research*, **7**, 337–349.

Papa, S.M., Artieda, J. and Obeso, J.A. (1991). Cortical activity preceding self-initiated and externally triggered voluntary movement. *Movement Disorders*, **6**, 217–224.

Pertovaara, A., Kemppainen, P. and Leppanen, H. (1992). Lowered cutaneous sensitivity to nonpainful electrical stimulation during isometric exercise in humans. *Experimental Brain Research*, **89**, 447–452.

Pons, T.P., Garraghty, P.E., Friedman, D.P. and Mishkin, M. (1987). Physiological evidence for serial processing in somatosensory cortex. *Science*, **237**, 417–420.

Pons, T.P., Garraghty, P.E. and Mishkin, M. (1992). Serial and parallel processing of tactual information in somatosensory cortex of rhesus monkeys. *Journal of Neurophysiology*, **68**, 518–527.

Pons, T.P., Garraghty, P.E., Ommaya, A.K., Kaas, J.H., Taub, E. and Mishkin, M. (1991). Massive cortical reorganization after sensory deafferentation in adult macaques. *Science*, **252**, 1857–1860.

Pons, T.P. and Kaas, J.H. (1986). Corticocortical connections of area 2 of somatosensory cortex in macaque monkeys: A correlative anatomical and electrophysiological study. *Journal of Comparative Neurology*, **248**, 313–335.

Pons, T.P., Wall, J.T., Garraghty, P.E., Cusick, C.G. and Kaas, J.H. (1987). Consistent features of the representation of the hand in area 3b of macaque monkeys. *Somatosensory Research*, **4**, 309–331.

Posner, M.I. and Petersen, S.E. (1990). The attention system of the human brain. *Annual Review of Neuroscience*, **13**, 25–42.

Prochazka, A. (1993). Comparison of natural and artificial control of movement. *IEEE Transactions in Rehabilitation Engineering*, **1**, 7–16.

Prud'homme, M.J., Cohen, D.A. and Kalaska, J.F. (1994). Tactile activity in primate somatosensory cortex during active arm movements: Cytoarchitectonic distribution. *Journal of Neurophysiology*, **71**, 173–181.

Recanzone, G.H., Merzenich, M.M. and Schneiner, C.E. (1992). Changes in the distributed temporal response properties of SI cortical neurons reflect improvements in performance on a temporally based tactile discrimination task. *Journal of Neurophysiology*, **67**, 1071–1091.

Riehle, A. and Requin, J. (1989). Monkey primary motor and premotor cortex: Single-cell activity related to prior information about direction and extent of an intended movement. *Journal of Neurophysiology*, **61**, 534–549.

Rizzolatti, G., Scandolara, C., Matelli, M., Gentilucci, M. and Camarda, R. (1981). Response properties and behavioral modulation of 'mouth' neurons of the postarcuate cortex (area 6) in macaque monkeys. *Brain Research*, **225**, 421–424.

Robinson, D.L. and Petersen, S.E. (1986a). The neurobiology of attention. In *Mind and Brain: Dialogues in Cognitive Neuroscience*, edited by J.E. LeDoux and W. Hirst, pp. 142–171. Cambridge, UK: Cambridge Univ. Press.

Robinson, D.L. and Petersen, S.E. (1986b). A neurobiological view of the psychology of attention. In *Mind and Brain: Dialogues in Cognitive Neuroscience*, edited by J.E. LeDoux and W. Hirst, pp. 172–178. Cambridge, UK: Cambridge Univ. Press.

Romo, R., Merchant, H., Zainos, A. and Hernandez, A. (1996). Catagorization of somaesthetic stimuli: Sensorimotor performance and neuronal activity in primary somatic sensory cortex of awake monkeys. *Neuroreport*, **7**, 1273–1279.

Ruiz, S., Crespo, P. and Romo, R. (1995). Representation of moving tactile stimuli in the somatic sensory cortex of awake monkeys. *Journal of Neurophysiology*, **73**, 525–537.

Rushton, D.N., Rothwell, J. C. and Craggs, M.D. (1981). Gating of somatosensory evoked potentials during different kinds of movement in man. *Brain*, **104**, 465–491.

Salinas, E. and Abbott, L.F. (1995). Transfer of coded information from sensory to motor networks. *Journal of Neuroscience*, **15**, 6461–6474.

Schmidt, R.A. (1988). *Motor Control and Learning*. Champaign, IL: Human Kinetics.

Schmidt, R.F., Schady, W.J.L. and Torebjork, H.E. (1990a). Gating of tactile input from the hand. I. Effect of finger movement. *Experimental Brain Research*, **79**, 97–102.

Schmidt, R.F., Torebjork. H.E. and Schady, W.J.L. (1990b). Gating of tactile input from the hand. II. Effect of remote movements and anaesthesia. *Experimental Brain Research*, **79**, 103–108.

Schultz, W., Apicella, P., Romo, R. and Scarnati, E. (1995a). Context-dependent activity in primate striatum reflecting past and future behavioral events. In: *Models of Information Processing in the Basal Ganglia*, edited by J.C. Houk, J.L. Davis and D.G. Beiser, pp. 11–28. Boston: M.I.T. Press.

Schultz, W., Romo, R., Ljungberg, T., Mirenowicz, J., Hollerman, J.R. and Dickinson, A. (1995b). Reward-related signals carried by dopamine neurons. In: *Models of Information Processing in the Basal Ganglia*, edited by J.C. Houk, J.L. Davis and D.G. Beiser, pp. 233–248. Boston: M.I.T. Press.

Schultz, W. (1997). Dopamine neurons and their role in reward mechanisms. *Current Opinions of Neurobiology*, **7**, 191–197.

Schultz, W., Dayan, P., and Read-Montague, P. (1997). A neural substrate of prediction and reward. *Science*, **275**, 1593–99.

Seal, J. and Commenges, D. (1985). A quantitative analysis of stimulus- and movement-related responses in the posterior parietal cortex of monkey. *Experimental Brain Research*, **58**, 144–153.

Sinclair, R.J. and Burton, H. (1991). Neuronal activity in the primary somatosensory cortex in monkeys (Macaca mulatta) during active touch of textured surface gratings: Responses to groove width, applied force, and velocity of motion. *Journal of Neurophysiology*, **66**, 153–169.

Soso, M.J. and Fetz, E.E. (1980). Responses of identified cells in postcentral cortex of awake monkeys during comparable active and passive joint movements. *Journal of Neurophysiology*, **43**, 1090–1110.

Sperry, R.W. (1950). Neural basis of the spontaneous optokinetic response produced by visual inversion. *Experimental Zoology*, **92**, 263–279.

Stelmach, G.E. and Nahom, A. (1992). Cognitive-motor abilities of the elderly driver. *Human Factors (United States)*, **34**, 53–65.

Stelmach, G.E., Worringham, C.J. and Strand, E.A. (1986). Movement preparation in Parkinson's disease. *Brain*, **109**, 1179–1194.

Stepniewska, I., Preuss, T.M. and Kaas, J.H. (1993). Architectonics, somatotopic organization, and ipsilateral cortical connections of the primary motor area (M1) of owl monkeys. *Journal of Comparative Neurology*, **330**, 238–271.

Taira, M., Boline, J., Smyrnis, N., Georgopoulos, A.P. and Ashe, J. (1996). On the relations between single cell activity in the motor cortex and the direction and magnitude of three-dimensional static isometric force. *Experimental Brain Research*, **109**, 367–376.

Talbot, W.H., Darian-Smith, I., Kornhuber, H.H. and Mountcastle, V.B. (1968). The sense of flutter-vibration: Comparison of the human capacity with response patterns of mechanoreceptive afferents from the monkey hand. *Journal Neurophysiology*, **31**, 301–334.

Teuber, H-L. (1966). Alterations of perception after brain injury. In *Brain and Conscious Experience*, edited by J.C. Eccles, pp. 182–216. New York: Springer Verlag.

Turman, A.B., Ferrington, D.G., Ghosh, S., Morley, J.W. and Rowe, M.J. (1992). Parallel processing of tactile information in the cerebral cortex of the cat: effect of reversible inactivation of SI on responsiveness of SII neurons. *Journal of Neurophysiology*, **67**, 411–429.

Vaadia, E., Kurata, K. and Wise, S.P. (1988). Neuronal activity preceding directional and nondirectional cues in the premotor cortex of Rhesus monkey. *Somatosensory and Motor Research,* **6,** 207–230.

von Holst, E. (1954). Relations between the central nervous system and the peripheral organs. *The British Journal of Animal Behaviour,* **2,** 89–94.

von Holst, E. and Mittelstaedt, H. *(1973).* Das Reafferenzprinzip. Wechselwirkungen zwischen zentralnervensystem und peripherie. Naturwissenschaften, 150, 37, 464–476 (English translation in *The Behavioral Physiology of Animals and Man,* pp. 139–173. London: Methuen,

Welford, A.T. (1980). *Reaction Times.* New York: Academic Press.

Whang, K.C., Burton, H. and Shulman, G.L. (1991). Selective attention in vibrotactile tasks: Detecting the presence and absence of amplitude change. *Perception and Psychophysics,* **50,** 157–165.

Wiesendanger, M. (1986). Some concluding remarks about general concepts in studies of the skeletomotor system. *Progress in Brain Research,* **64,** 419–423.

Wiesendanger, M. and Miles, T.S. (1982). Ascending pathway of low-threshold muscle afferents to the cerebral cortex and its possible role in motor control. *Physiological Reviews,* **62,** 1234–1270.

Wise, S.P. (1985). The primate premotor cortex fifty years after Fulton. *Behavioural Brain Research,* **18,** 79–88.

Wise, S.P. and Kurata, K. (1989). Set-related activity in the premotor cortex of rhesus monkeys: effects of triggering cues and relatively long delay intervals. *Somatosensory and Motor Research,* **6,** 455–476.

Zar, J.H. (1974). *Biostatistic Analysis.* Englewood Cliffs, NJ: Prentice-Hall.

Zhang, H.Q., Murray, G.M., Turman, A.B., Mackie, P.E., Coleman, G.T. and Rowe, M.J. (1996). Parallel processing of tactile information in the cerebral cortex of the marmoset monkey: effect of reversible inactivation of SI on the responsiveness of SII neurons. *Journal of Neurophysiology,* **76,** 3633–3655.

CHAPTER 14
SOMATOSENSORY EVOKED MAGNETIC FIELDS IN HUMANS

Ryusuke Kakigi

Department of Integrative Physiology, National Institute for Physiological Sciences, Myodaiji, Okazaki, Japan

INTRODUCTION

The evaluation of averaged electroencephalography (EEG) following somatosensory stimulation, i.e. somatosensory evoked potential (SEP), is one of the most useful methods for investigating the human somatosensory system. A large number of studies have reported the computerized bit-mapped images of scalp topography of SEP in attempts to elucidate the mechanisms of each identifiable component. However, scalp-recorded EEG could not provide enough resolution to estimate the precise electrical source location in the brain. This is because the influence of volume currents on scalp-recorded EEG severely affects precise source identification. Furthermore, large interindividual variability of intervening tissues, including scalp, skull and cerebrospinal fluid, makes it difficult to estimate the dipole location.

Another non-invasive technique for investigating the bioelectrical functions of the brain, magnetoencephalography (MEG), has been developed during the past 25 years. MEG has several theoretical advantages over EEG in localizing cortical sources (brain dipoles), because the magnetic fields recorded on the scalp are less affected by volume currents and anatomical inhomogeneities.

MEG has excellent spatial and temporal resolution in the order of mm and msec. The spatial resolution of MEG is almost the same as that of functional magnetic resonance imaging (fMRI) and positron emission tomography (PET), but its temporal resolution is much better than fMRI and PET. Therefore, we can analyze MEG responses to somatosensory stimulation for not only detecting cortical sources but also measuring periods for signal transfer, or activity, in the brain in the order of msec. However, there are four main disadvantages of MEG. The first disadvantage is that it is difficult for MEG to detect brain dipoles radial to the skull – in particular, those in a gyrus immediately beneath the skull generated in the gyrus. However, dipoles tangential to the skull generated in the wall of the sulcus, i.e. area 3b or 4 along the central sulcus, are easily detected by MEG. Second, activities in the white matter are not detected by MEG, since generators of MEG are apical dendrites of the pyramidal cells in the cortex. The third disadvantage is that it is difficult for MEG

Address for correspondence: Ryusuke Kakigi, M.D., Ph.D., Professor and Chairman, Department of Integrative Physiology, National Institute for Physiological Sciences, Myodaiji, Okazaki, 444, Japan. Tel: +81 564 55 7765, 7766; Fax: +81 564 52 7913; E-mail: kakigi@nips.ac.jp

to detect dipoles generated in the deep areas, since magnetic fields recorded from outside the scalp decline rapidly with increasing depth of the generators. Fourth, it is difficult for MEG to detect multiple generators (dipoles). Therefore, a new algorithm is necessary for calculating multiple sources. Researchers using MEG must be aware of the disadvantages.

Many studies of averaged MEG values following somatosensory stimulation, i.e. somatosensory evoked magnetic fields (SEF), have been reported over the past twenty years (Brenner *et al.*, 1978; Hari *et al.*, 1990; Wood *et al.*, 1985; Suk *et al.*, 1991; Gallen *et al.*, 1993; Huttunen *et al.*, 1986, Forss *et al.*, 1994; Mauguiere *et al.*, 1997a, b). In this chapter I will introduce SEF findings in humans, in particular the recent studies of SEF in our department. At first, our MEG system is explained briefly, and then the general findings, mainly topography of the primary and secondary somatosensory cortex, SI and SII, respectively, are described. I will focus on: SEF to passive movement, findings on the effects of sensori-motor interactions on SEF, and pain-related SEF.

MEG SYSTEM

The magnetic fields recorded from the human brain are very small, approximately ten thousand to a million times smaller than the Earth's steady magnetic field and environmental fields (caused by a train, for example). A superconducting quantum interference device (SQUID) is necessary to detect such weak brain fields. We use dual 37-channel axial-type first-order biomagnetometers (Magnes, Biomagnetic Technologies inc. (BTi), San Diego, CA) (Fig. 14.1). The waveforms from 74 channels are simultaneously recorded. The detection coils of the biomagnetometers are arranged in a uniformly distributed array in concentric circles over a spherically concave surface. Each device is 144 mm in diameter and 122 mm in a radius of curvature. The outer coils are 72.5° apart. The coils are 20 mm in diameter and the spacing between centers of coils is 22 mm.

With our MEG system, dual probes centered at the subject's C3 and C4 positions. The International 10/20 system was used in all instances, which cover the left and right hemisphere, respectively. When the lower limb area of the primary sensory cortex (SI) was examined, one probe was centered at the Cz position, around the vertex. The responses were usually recorded with a 0.1–100 or 200 Hz bandpass filter, and digitized at a sampling rate of 2048 or 4096 Hz.

A spherical model (Sarvas, 1987) was fitted to the digitized shape of the head of each subject, and the location (x, y and z location), orientation and amplitude of a best-fitted single equivalent current dipole (ECD) were estimated at each time point. The correlation between the recorded measurements and the values expected from the ECD estimate was calculated as a measure of how closely the measured values corresponded to the theoretical field generated by the model and the observed field. The MEG source locations were superimposed on Magnetic resonance imaging (MRI) with a contiguous 1.5 mm slice thickness.

SEF FOLLOWING STIMULATION APPLIED TO VARIOUS PARTS OF THE BODY

Somatosensory Homunculus in SI

Since the landmark studies of Penfield and Boldrey (1937) on the sensory representations in the human cerebral cortex based on direct electrical stimulation of the cortical surface,

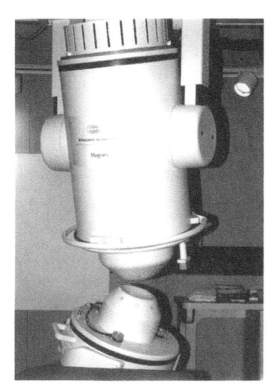
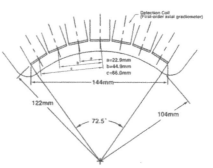
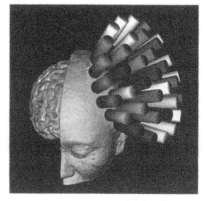

Figure 14.1 Scheme of our MEG system. Thirty-seven coils are placed on the head as shown in the right figure.

it has been well known that SI is organized in an orderly somatotopic way which has been termed the 'homunculus' representation of the cutaneous body surface. We are now studying SEF following mechanical stimulation applied to various parts of the body in normal subjects to examine the homunculus by this non-invasive method. As described in the Introduction, MEG detects only a specific orientation of brain current tangential to the skull. Therefore, dipoles generated in area 3b (which is located on the posterior bank of the central sulcus) are easily detected, but dipoles in area 1, 2 or 3a (which are located on the bottom or the crown of the central sulcus) are not. Therefore responses to cutaneous tactile stimulation arriving at area 3b are easily detected by MEG, so that this form of stimulation is the preferred method for SEF. Electrical stimulation is also used, since its stimulus onset is very sharp, and it generates much larger magnetic field than cutaneous stimulation. Since findings of both SEP and SEF to electrical stimulation are very similar to those to cutaneous stimulation in terms of waveforms and generators' location, it is generally considered that the signals ascend through the same pathways as those traversed by inputs generated by cutaneous mechanical stimulation (Kakigi and Shibasaki, 1984).

In order to clarify the homunculus in normal subjects in detail, we studied SEF findings following stimulation applied to the lower limb (Kakigi *et al.*, 1995b, Shimojo *et al.*, 1996b, Hoshiyama *et al.*, 1997), upper limb (Kakigi, 1994, Xiang *et al.*, 1997a, Hoshiyama *et al.*, 1997), lip (Hoshiyama *et al.*, 1996, 1997) and scalp (Hoshiyama *et al.*, 1995). Although

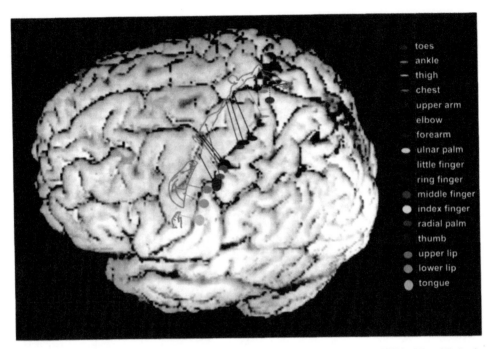

toes
ankle
thigh
chest
upper arm
elbow
forearm
ulnar palm
little finger
ring finger
middle finger
index finger
radial palm
thumb
upper lip
lower lip
tongue

Figure 14.2 Detailed somatosensory receptive map represented by MEG. The 3D brain image was reconstructed using MRI of this subject. Each receptive area, which was estimated to be located in the posterior bank of the central sulcus, was projected onto the cortical surface. The size of each ellipse reflects the presumed size of activated cortical area. Note that the receptive area for the toes is in the medial side of the left hemisphere. (Adapted from Nakamura *et al.*, 1998).

interindividual differences were identified, the topography of SI was generally consistent with the plan of the homunculus reported by Penfield and Boldrey (1937). One of the most interesting findings is that SI of the bilateral hemispheres were activated following stimulation of the unilateral side of face (lip and scalp) (Hoshiyama *et al.*, 1995, 1996). The peak latency of the primary component is approximately 20 and 40 msec to stimulation of the upper and lower limb, respectively, and 2 or 3 sequential components are identified within 100 msec following stimulation.

For summarizing these previous studies, we mapped a complete homunculus in 5 normal subjects (Nakamura *et al.*, 1998). We stimulated 19 sites, tongue, lower lip, upper lip, thumb, index finger, middle finger, ring finger, little finger, radial palm, ulnar palm, forearm, elbow, upper arm, chest, thigh, ankle, big toe, second toe and five toe. After measuring the location of single ECD of the primary component for each stimulus site, we put them on the MRI of each subject (Fig. 14.2). Then, we could map a large part of the somatosensory receptive fields (topography) on SI, including tongue, lips, finger, palm, arm and toes. These representation areas were generally arranged in the above order from inferior to superior, lateral to medial, and anterior to posterior. These coordinates were compatible with the anatomy of the central sulcus and the homunculus. ECD location to the upper lip could be distinguished from that to the lower lip, the former located more superior than the latter in all subjects. Each finger representation area of the thumb, index finger, middle finger, ring finger and little finger was also distinguishable from the others and was represented sequentially from thumb to little finger, ascending the postcentral sulcus.

Topography of SII

One of the major advantages of SEF is that it easily records activity in SII, where it is difficult for SEP to detect activity due to the location and direction of dipole sources. We analyzed the topography of SII to somatosensory stimulation applied to various parts of the body of normal subjects using SEF (Maeda *et al.*, 1999). SII components were found about 80–100 msec after stimulation as the middle-latency components. SII areas in both hemispheres were activated by unilateral body stimulation; that is, SII in humans has 'bilateral function'. Although there were large inter-individual differences, the following orders of receptive fields were found (Fig. 14.3): (i) Anterior-posterior direction: lower lip-upper lip-thumb-middle finger-foot, (ii) Medial-lateral direction: foot-middle finger-thumb-upper lip-lower lip, and (iii) Lower-upper direction: lower lip-upper-lip-thumb-middle finger-foot. In general, these findings are similar to those obtained in studies of animals (Whitsel *et al.*, 1969). However, the differentiation was not as clear as that seen in the SI homunculus, the SII area is located at a site more anterior, medial and above the auditory cortex.

SEF Following Passive Movement

As already described, inputs generating SEP or SEF following electrical or cutaneous stimulation of the skin probably ascend through cutaneous fibers and reach area 3b (Kakigi and Shibasaki, 1984). In contrast, after passive movements of the fingers, signals mainly ascend through proprioceptive and muscle afferents and reach area 2 and/or 3a in SI. To our knowledge, ours were the first reports on SEF in response to passive movement (Xiang *et al.*, 1997a,b). We made a new device for measuring SEF in response to passive movement of the middle finger by approximately 20°, and at an angular velocity of approximately 525–530°/s.

SEF waveforms to conventional electrical stimulation applied to the middle finger are shown first to compare results with those to passive movement (Fig. 14.4). Five components, 1M(E), 2M(E), 3M(E), 4M(E) and 4M(EI), were identified following electrical stimulation. 1, 2, 3, and 4 mean the order of appearance of each component, and 'M' and 'E' indicate magnetic component and electrical stimulation, respectively. Therefore, 1M(E) indicates the primary magnetic component to electrical stimulation recorded from the hemisphere contralateral to the stimulation. 'I' indicates the hemisphere ipsilateral to the stimulation. Only one component, termed 4M(E), was identified from the hemisphere ipsilateral to the stimulation and it appeared to correspond to the 4M(E). Therefore it was termed 4M(EI).

Four main components were identified within 100 msec in latency from the hemisphere contralateral to the moved finger in all 10 subjects: 1M(P), 2M(P), 3M(P) and 4M(P). 'P' indicates passive movement. Therefore, 1M(P) indicates the primary magnetic component

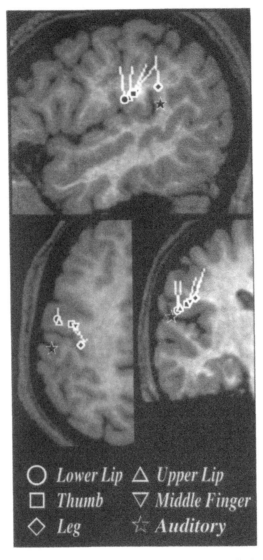

Figure 14.3 ECDs location to somatosensory stimulation applied to various parts of the body and auditory stimulation overlapped on MRI in a representative subject. All ECDs are projected to a slice in which ECDs to the thumb stimulation were found, since it is easily understood by this procedure, and since it is impossible to show all slices in which each ECD is located. The relationship of each ECD was easily found by these figures. There was a large inter-individual topographic difference in SII, and there was no clear topographic order in SII, unlike the homunculus in SI. However, there was a tendency in the topographic order as follows:

Anterior-posterior direction: Lower lip-Upper lip-Thumb-Middle finger-tibial nerve.
Medial-lateral direction: Tibial nerve-Middle finger-Thumb-Upper lip-Lower lip.
Lower-upper direction: Lower lip-Upper-lip-Thumb-Middle finger-Tibial nerve.
The auditory cortex is located at a site more posterior, lateral and lower compared to SII.

(Adapted from Maeda *et al.*, 1999).

Subject 1

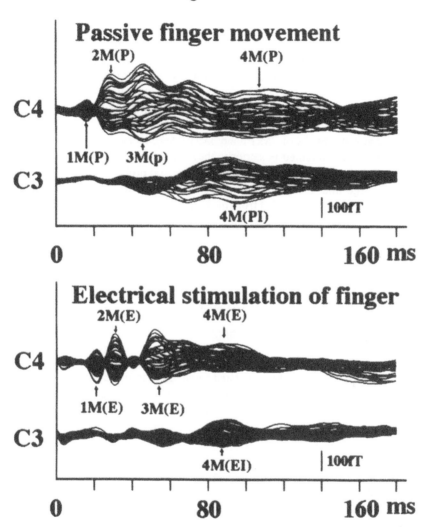

Figure 14.4 The SEFs following passive movement and electrical stimulation of the left middle finger recorded at the C4 (contralateral hemisphere to stimulation) and C3 (ipsilateral hemisphere to stimulation) position simultaneously in a representative subject showing the nomenclature of each recognizable component. Waveforms recorded at the 37 channels are superimposed. 1, 2, 3, and 4 mean the order of appearance of each component, and 'M' and 'E' indicate magnetic component and electrical stimulation, respectively. Therefore, 1M(E) indicates the primary magnetic component to electrical stimulation recorded from the hemisphere contralateral to the stimulation. 1M(P), the first response following passive finger movement, is small in amplitude or absent. However, 2M(P) and 3M(P) are clearly larger than 2M(E) and 3M(E). 2M(P) and 3M(P) appear to be combined as one component with two peaks. Only one component, 4M(PI) and 4M(EI), was recorded from the hemisphere ipsilateral to the moved or electrically stimulated finger. 'I' indicates the hemisphere ipsilateral to the stimulation. (Adapted from Xiang *et al.*, 1997a).

to passive movement. The 1M(P) was clearly identified only in three subjects and was smaller than other components, although 1M(E) was clearly recorded in all subjects (Fig. 14.4). The ECDs of 1M(P) were located around the finger area of the SI and oriented either posteriorly or anteriorly. Since area 3a is situated at the bottom of the central sulcus, the orientation of the ECD generated in area 3a is mainly radial. As described in the Introduction, it is difficult for MEG to detect radial dipoles. However, since every generator dipole is separated into radial and tangential vectors, we suspect that the tangential vector of the dipole generated in area 3a could be detected in the three subjects, but was too small to be detected in the seven other subjects.

The 2M(P) and 3M(P) were usually combined as one large deflection with two peaks (Fig. 14.4). Because the ECDs of 2M(P) and 3M(P) were located around the finger area of the SI and both were oriented posteriorly, they were thought to be generated in area 3b and/or 4, and their activities have temporal overlap and probably spread from area 3a. The 4M(P) has large inter-individual difference in terms of amplitude and latency. The 4M(PI), the main component recorded from the hemisphere ipsilateral to the moved finger, was located in the upper bank of the Sylvian fissure, probably in SII. The SEF following passive movement was clearly different from the SEF following electrical stimulation, in terms of waveforms and source locations, probably due to differences in the ascending fibers and receptive fields.

SENSORI-MOTOR INTERACTION ON SEF

Effects of Movement Interference (gating)

SEP and SEF following electrical stimulation are markedly modified by voluntary (active) movements of the body part near the stimulated nerve; this particular phenomenon is known as 'gating' (Kakigi *et al.*, 1995a, 1997). We therefore examined the 'gating' effects of active and passive movements of the toes and 'movement imagery' (mental moving of the toe without actual movements) on SEF following stimulation of the posterior tibial nerve in normal subjects (Kakigi *et al.*, 1997). The SEF was triggered not to the continuous movements but to the time-locked electrical stimulation to the posterior tibial nerve, to determine the effects of continuous and concurrent movements on time-locked electrically stimulated SEF. Active and passive movements significantly attenuated the short- and long-latency cortical components ($P < 0.001$) with no latency change, and the effects of the active movements were larger than those of the passive movements (Fig. 14.5). Therefore, both centrifugal and centripetal mechanisms should be considered. The probable mechanism is based on the effects of the continuous excitations of neurons in area 3a and/or 4 by active movements on neurons in area 3b concerned with the processing of input from the cutaneous mechanoreceptors. Active movements of the toes contralateral to the stimulated nerve caused no significant gating effect.

The short-latency components were not consistently changed by 'movement imagery', but the middle- and long-latency components were enhanced (Fig. 14.5). Their ECDs were located in SI contralateral to the stimulated nerve and in SII in both hemispheres. The reason why the movement imagery enhanced the components generated in SI remains to be elucidated, but neurons in the motor cortex might be activated by continuous movement imagery and affect the somatosensory evoked brain responses. Enhancement of the SEF

Posterior tibial nerve stimulation (Cz)

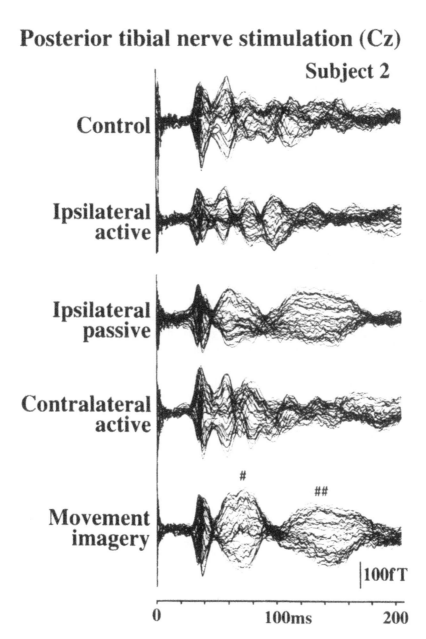

Figure 14.5 The SEF recorded at the Cz position around the vertex following stimulation of the left posterior tibial nerve in the control and each interference session in one subject. The waveforms of all 37 channels are superimposed. The short-latency components are attenuated in the ipsilateral active and ipsilateral passive sessions. There is no significant change by contralateral active interference. In the movement imagery session, the short-latency components showed no significant change. However, the middle-latency component indicated by # was enhanced, and the long-latency component indicated by ## was found. (Adapted from Kakigi *et al.*, 1997).

components generated in SII was observed when subjects paid close attention to the infrequent (target) stimulation when the so-called 'oddball' paradigm was used (Hari *et al.*, 1990). In contrast, activities in SII following median nerve stimulation were markedly reduced or disappeared during sleep (Kitamura *et al.*, 1996). Therefore, we speculated that brain responses to somatosensory stimulation, particularly components generated in SII, were affected by volitional changes. Similar findings were identified following median nerve stimulation (Kakigi *et al.*, 1995a).

EFFECTS OF TACTILE INTERFERENCE STIMULATION

A change of SEF waveforms by tactile interference was clearly different from that due to movement interference as described above, probably due to differences in the ascending fibers of the peripheral nerve and receptive sites in the sensory cortex.

We examined the changes of SEF following electrical stimulation of the right median nerve (Kakigi *et al.*, 1996). Continuous light tactile stimulation was delivered to the right (ipsilateral) and left (contralateral) palm by the experimenter concurrently and continuously with the right median nerve stimulus, using a soft wad of tissue paper. When stimulation was applied to the ipsilateral hand, all components except for one component indicated by **, whose peak latency was about 25–40 msec, were attenuated in all subjects (P < 0.0001) (Fig. 14.6). The SEF is thought to have been generated in area 3b which responds mainly to tactile stimulation. The attenuation of the components generated in SI by tactile interference of the ipsilateral hand was therefore considered to be a result of partial 'saturation' of a cortical area concerned with the processing of input from the tactile mechanoreceptors, such that it was unable to respond to the coherent volley evoked by the median nerve impulses.

When tactile interference was applied to the contralateral hand, only one component indicated by *, whose peak latency was about 20–30 msec, was significantly enhanced (P < 0.05); the other components showed no significant change (Fig. 14.6). Iwamura *et al.* (1994) reported the presence of neurons that receive somatosensory signals from both hands in the postcentral somatosensory cortex (areas 2 and 5) in the monkey. Therefore, the enhancement of this component may be due to the effects of the ipsilateral response activity on current dipoles generated in area 3b, probably through the corpus callosum. Similar results were obtained when tactile interference was applied to SEF following posterior tibial nerve stimulation (Naka *et al.*, 1998).

Effects of Visual and Auditory Stimulation on SEF

Some areas of the brain such as Brodmann's areas 5 and 7 in the parietal lobe or superior part of the temporal lobe have been considered polysensory (Hyvärinen and Poranen, 1974). We therefore suspect that the integrative processing of various sensory inputs, such as somatosensory, visual and auditory also takes place in these areas. However, to our knowledge, no study has reported the effects of visual or auditory stimulation on SEF. We therefore investigated the effects of continuous visual (cartoon and random dots motion) and auditory (piano playing) stimulation on SEF following electrical stimulation of the median nerve (Lam *et al.*, 1999). In the hemisphere contralateral to the stimulated nerve, the middle-latency components (35–60 msec in latency) were significantly enhanced by

Control

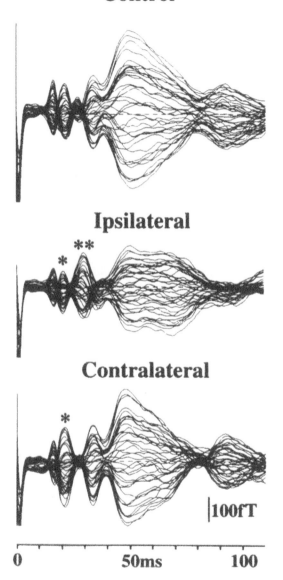

Ipsilateral

Contralateral

100fT

0   50ms   100

Figure 14.6 SEF following the right median nerve stimulation recorded at the C3 position (left hemisphere) in one subject. Waveforms recorded at all 37 channels are superimposed. Control: SEF with no interference; Ipsilateral and Contralateral: SEF with tactile interference applied to the hand ipsilateral and contralateral to the stimulated nerve, respectively. After interference was applied to the ipsilateral hand, one component (indicated by **) was clearly increased, but the other components including the second component (indicated by *) were attenuated. After interference was applied to the contralateral hand, the component indicated by * was clearly enhanced, but other components were not changed. (Adapted from Kakigi *et al.,* 1996).

Somatosensory Processing

Left median nerve stimulation (C4)
Subject 12

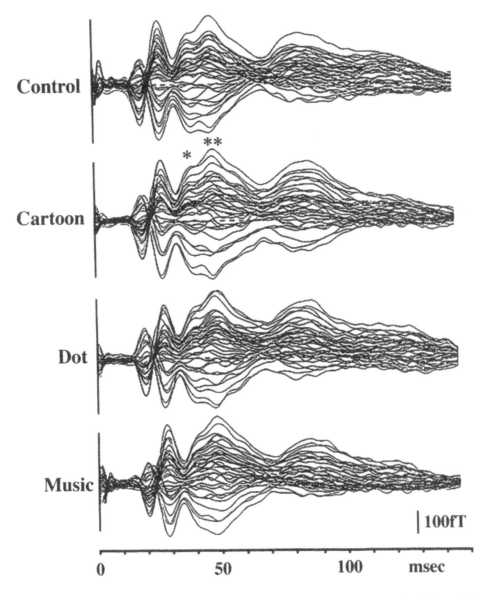

Figure 14.7 SEF recorded at the C4 position (contralateral) following stimulation of the left median nerve in the control and each multi-modal stimulus condition in one subject. The waveforms of all 37 channels are superimposed. The middle-latency components indicated by * and ** were enhanced in the cartoon and dots sessions. (Adapted from Lam *et al.*, 1999).

visual, but were not enhanced by auditory stimulation (Fig. 14.7). The dipoles of all components within 60–70 msec following stimulation were estimated to be very close to each other, around the hand area of the SI. In the ipsilateral hemisphere, the middle-latency components (70–100 msec in latency), the dipoles of which were estimated to be in SII, were markedly decreased in amplitude by both the visual and auditory stimulation. These changes of waveforms by visual and auditory stimulation are thought to be due to the effects of the activation of polymodal neurons, which receive not only somatosensory but also visual and/or auditory inputs, in areas 5 and/or 7 as well as in the median superior temporal region (MST) and superior temporal sulcus (STS), although a change of attention might also be one of the factors causing such findings to some degree.

Intracerebral Interaction Produced by Bilateral Stimulation

To investigate the intracerebral interactions in response to somatosensory stimulation, we recorded SEF following the simultaneous stimulation of median nerves of both sides ('bilateral' waveform) (Shimojo *et al.*, 1996a). The SEF following right median nerve stimulation and that following left median nerve stimulation were summated ('summated' waveform). A 'difference' waveform was induced by subtraction of the bilateral waveform from the summated waveform. Short-latency deflections showed no consistent differences between the summated and bilateral waveforms. The long-latency deflection indicated by the arrow, whose peak latency was about 80–100 msec, in the bilateral waveform was markedly ($P < 0.001$) reduced in amplitude compared with the summated waveform (Fig. 14.8). The differences were clearly identified in the difference waveform in all subjects. The ECDs of the short- and middle-latency deflections were located in SI, but the ECDs of the components indicated by the arrow was located in the bilateral SII. Similar findings were found in SEF following bilateral posterior tibial nerve stimulation (Shimojo *et al.*, 1997).

Neurons in SII differ from those in SI in that their receptive fields are larger, encompassing ipsilateral as well as contralateral areas of the body surface for 90% of the units in unanesthetized monkeys (Whitsel *et al.*, 1969). It therefore seems appropriate that some interactions take place in SII following bilateral stimulation.

SEF FOLLOWING PAINFUL CO$_2$ LASER STIMULATION

General Findings

The cerebral representation of pain perception in humans is poorly understood compared with other modalities of sensation such as touch or vibration. This is mainly because of a lack of the appropriate instrumentation for stimulation and recording (Kakigi and Shibasaki, 1984). Few systematic studies of SEF following painful stimulation have been reported. Hari *et al.* (1983) used painful electrical stimulation of tooth pulp. Huttunen *et al.* (1986) used CO$_2$ gas stimulation of the nasal mucosa. Howland *et al.* (1995) applied high-intensity painful electrical stimulation. They reported that the current sources were located at or near SII or the frontal operculum. We have also used high-intensity electrical stimulation to the upper and lower limbs, and found ECDs in not only SII but also the cingulate cortex (Kitamura *et al.*, 1995, 1997).

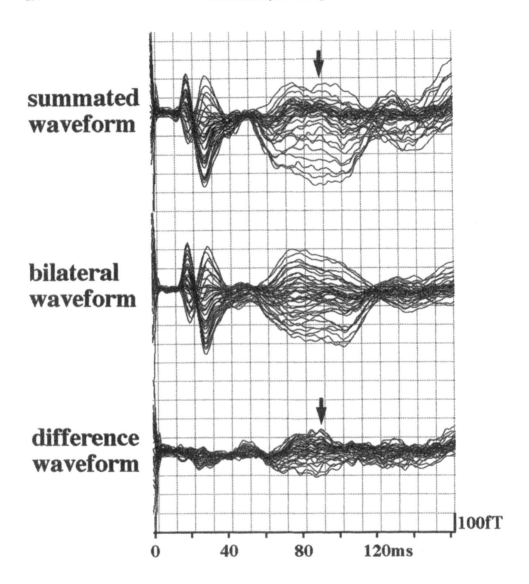

Figure 14.8 The 'bilateral', 'summated' and 'difference' waveforms recorded at the C3 position in one subject. Each waveform was obtained by the superimposition of all 37 channels. Waveforms recorded at all 37 channels are superimposed. The middle-latency component indicated by the arrow in the bilateral waveform is smaller than those in the summated waveform, and the difference is more clearly found in the difference waveform. (Adapted from Shimojo *et al.*, 1996a).

A low power and long wavelength CO_2 laser beam induces sensations of pain or heat when applied to the skin. From studies in normal subjects and in patients with various types of sensory impairment, it has been established that CO_2 laser stimuli excite skin nociceptors and send their signals, first over small peripheral myelinated fibers (Aδ) and then probably via the spinothalamic tract with an approximate conduction velocity of 10–15 m/sec. We have, therefore, studied SEF following painful CO_2 laser stimulation to elucidate the mechanisms of pain processing in the human brain (Kakigi *et al.*, 1995c; Watanabe *et al.*, 1998; Yamasaki *et al.*, 1999).

A consistent and clear magnetic field (termed 1M) was identified bilaterally in the cerebral hemispheres following stimulation of each arm in all 11 subjects examined (Fig. 14.9). Its onset and peak latencies varied among the subjects, being about 120–180 and 180–220 msec, respectively. When a conventional single ECD analysis was used, the ECDs were estimated to be located around SII or the insula in both hemispheres (Kakigi *et al.*, 1995c; Watanabe *et al.*, 1998; Yamasaki *et al.*, 1999) (Fig. 14.10). Peak latency of the SII response recorded from the hemisphere contralateral to the stimulation was significantly ($P < 0.01$) shorter (approximately 20–25 msec) than that recorded from the ipsilateral hemisphere (Yamasaki *et al.*, 1999).

Analysis Using Multi-dipole Model

Since various sites are thought to be responsible for pain perception, we adopted a multi-dipole model, brain electric source analysis (BESA) (Scherg, 1995), for the study (Watanabe *et al.*, 1998). The residual variance (%RV) indicated the percentage of data which cannot be explained by the model. The goodness of fit (GOF) was expressed in % as (100–%RV). Results showed that the 4-dipole model was the most appropriate (Fig. 14.11). Sources 1 and 2 were located in SII in the hemisphere contralateral and ipsilateral to the stimulation, respectively, and sources 3 and 4 were located at the anterior medial temporal area, around the amygdalar nuclei or hippocampal formation, in the hemisphere contralateral and ipsilateral to the stimulation, respectively. The GOF of this model was over 90%. When we placed source 5 around the arm area of SI in the hemisphere contralateral to the stimulated arm, its activity was very small or absent and the change of GOF was small (Fig. 14.11). When we placed source 6 in the cingulate cortex, its waveform was unusual and a change of GOF was also small (Fig. 14.11). The contributions of sources 5 and 6 were thus very small or absent. In addition, although we tested to calculate the other dipole (source 7) in all subjects, source 7 accounted only for drift and not for any significant activity.

Effects of Distraction on Pain SEF

Pain perception was changed by attention or distraction effects. Therefore, we aimed to compare the effects of distraction on pain SEF and SEP (Yamasaki *et al.*, 1999). Painful CO_2 laser stimuli were applied to the right forearm of ten healthy subjects. A table with 25 random two-digit numbers was shown to the subjects, who were asked to add 5 numbers of each line in their mind (calculation task) or to memorize the numbers (memorization task) during the recording. Subjects used a visual analogue scale (VAS) for rating subjective pain during each condition. In the SEF recording, the 1M was clearly identified in both hemispheres in each distraction task in all subjects. Because the sequential activities were

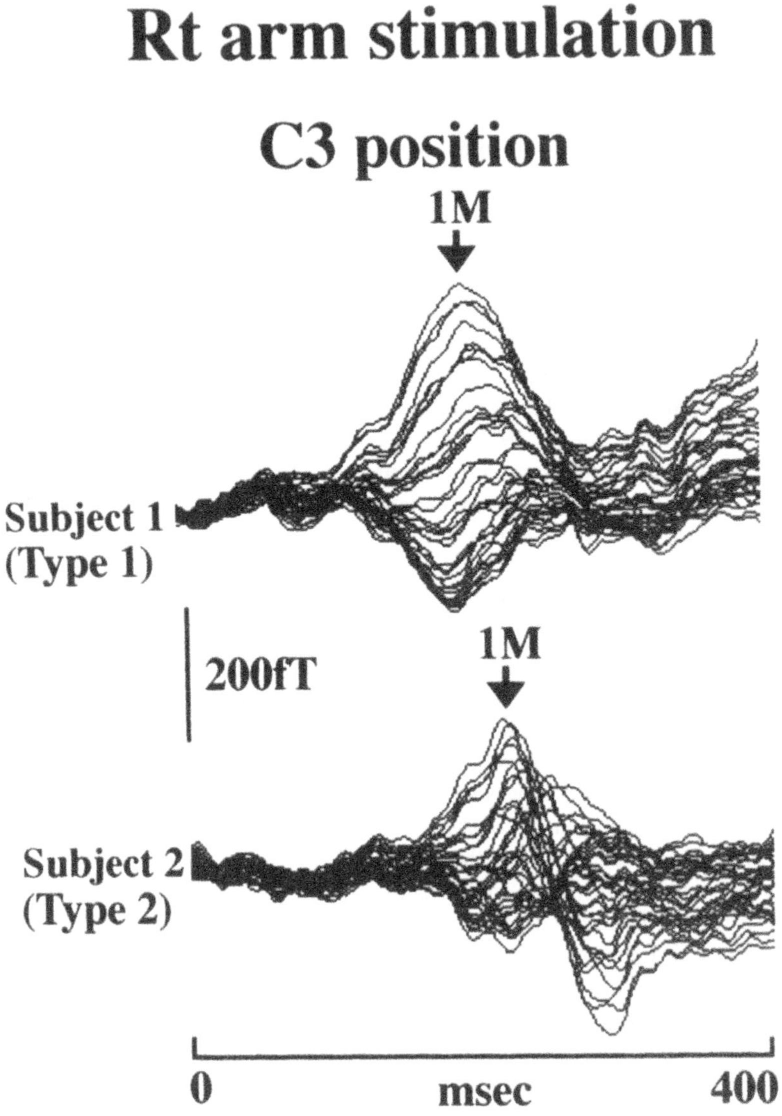

Rt arm stimulation

C3 position

Figure 14.9 Superimposed waveforms recorded from all 37 channels from the hemisphere contralateral to the painful CO_2 laser stimulation in two subjects. The first component (1M) appeared to be a simple polarity-reversal deflection in subject 1, but it appeared to be a hybrid of some components in subject 2. (Adapted from Watanabe *et al.*, 1998).

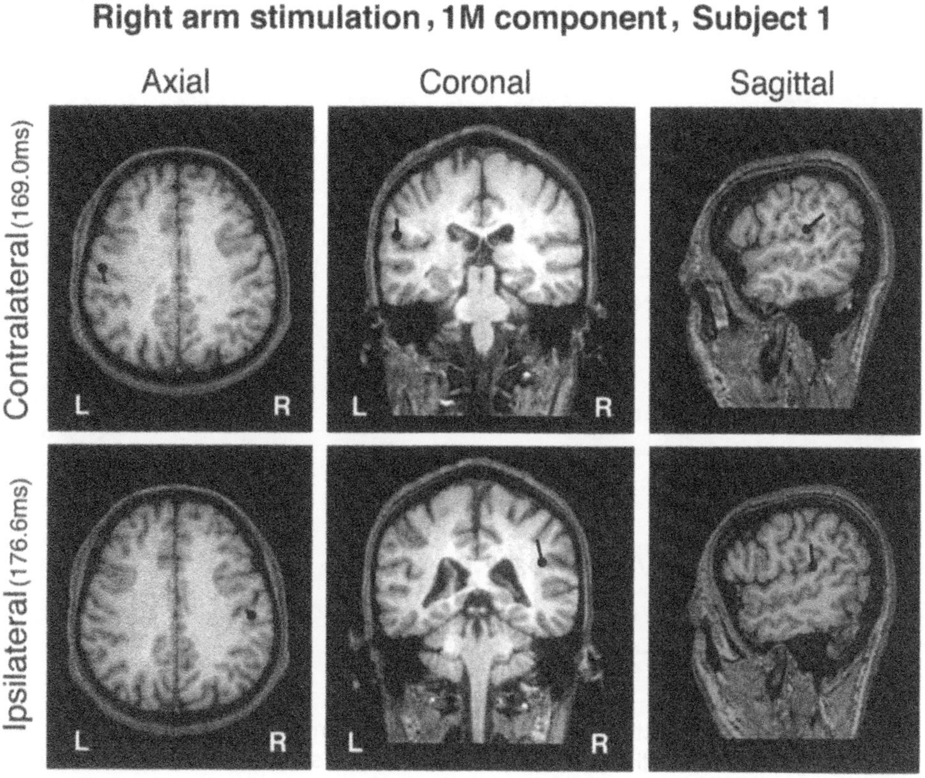

Right arm stimulation, 1M component, Subject 1

Figure 14.10 The location of ECD of the 1M overlapped on the axial, coronal and sagittal images of the MRI. The ECD was estimated to be located in the upper bank of the Sylvian fissure, around the SII, in bilateral hemispheres. (Adapted from Yamasaki *et al.*, 1999).

not clearly identified in all subjects, we did not analyze them in this study. In the SEP recording, the middle-latency components whose peak latencies were approximately 240 msec and 340 msec, N240 and P340, were identified in all subjects. N and P meant negative and positive components, respectively. The distraction tasks did not affect 1M component, but significantly attenuated EEG N240–P340 components (Fig. 14.12). A change of VAS score showed a positive correlation with a change of amplitude of N240–P340 complex, but not with 1M. Accounting for the differences of latency and distraction effects between the 1M and N240, the 1M and N240–P340 complex probably reflect different brain activities. We suspect that the 1M generated in SII-insula reflects the initial cortical activation to painful stimulation. The N240–P340 is considered to represent the activity of multiple areas including the insula and the limbic system, and may be affected by attentional changes.

General Discussion for Pain SEF

In our studies of SEF following painful CO_2 laser stimulation (Kakigi *et al.*, 1995c, Watanabe *et al.*, 1998; Yamasaki *et al.*, 1999), activity in SI was not clearly identified,

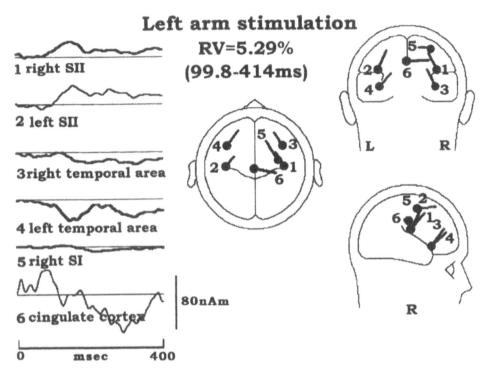

Figure 14.11 Temporal activation pattern and localization of each 6 source of pain SEFs following right arm stimulation calculated by BESA in one subject. **Left:** Temporal activation pattern of each source obtained by spatio-temporal source analysis. The horizontal bar indicates the time axis (extending from 0–400 msec). **Right:** The head diagrams indicate the locations of the dipole sources of the subject. The line and its length from each point indicate the direction and the magnitude of the dipole current, respectively. The four-source model, in which sources 1 and 2 are located in the SII in the hemispheres contralateral and ipsilateral to the stimulated arm, and sources 3 and 4 are located in the anterior medial temporal area (probably in the amygdalar nuclei or hippocampal formation) in the hemispheres contralateral and ipsilateral to the stimulated arm, was considered to be the most informative model, since the GOF was over 90%. When we added source 5 in the SI contralateral to the stimulation, its activity was found to be small or absent. When we added source 6 in the cingulate cortex, its activity was unusual and did not increase the GOF. This is probably due to its deep location, where it is difficult for MEG to detect its activity. (Adapted from Watanabe *et al.*, 1998).

perhaps for three reasons. First, the number of neurons activated by nociceptive stimulation may be very small; second, a relatively large difference of latency in each trial (jittering) may make this component small when responses are averaged; third, if principal activation is in the neurons in area 1, as reported in monkeys by Kenshalo and Willis (1991), the dipoles will usually be radially oriented, and the activity not clearly detected by MEGs. However, we do not consider the third possibility to be a main factor, because the activity thought to be generated in SI was absent or recorded as only a small notch, even by EEG (SEP) recording.

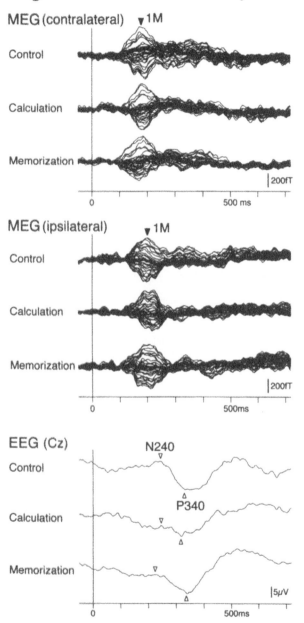

Figure 14.12 Waveforms of pain SEF (MEG) and pain SEP (EEG). MEG waveforms recorded from 37 channels in the contralateral and ipsilateral hemisphere to the stimulated arm are superimposed. EEG recorded at the Cz (vertex) is shown. The amplitude of the 1M component of the MEG in the bilateral hemispheres did not show a significant change in the calculation or memorization task. The N240–P340 component of EEG showed a significant reduction in amplitude, particularly in the calculation task. (Adapted from Yamasaki *et al.*, 1999).

The cingulate cortex may be activated by CO_2 laser stimulation, but its waveform as analyzed by BESA in the present study was unusual and did not increase, perhaps due to the deep location of the cingulate cortex, where it was very difficult for the MEG to detect activity. Compared with the cingulate cortex, the roles of the amygdalar nuclei and hippocampal formation in pain perception are not clearly elucidated. Inputs responsible for these activities were thought to ascend through the spino (trigemino)-ponto-amygdaloid pathway. It is therefore, possible that nociceptive inputs arrive at the contralateral SII, then the ipsilateral SII, the cingulate cortex, and then reach the amygdalar nuclei or the hippocampal formation through this pathway.

ACKNOWLEDGMENTS

I am very grateful to colleagues Dr Y. Kaneoke, Dr M. Hoshiyama, Dr S. Koyama, Dr Y. Kitamura, Dr D. Naka, Dr M. Shimojo, Dr S. Watanabe, Dr J. Xiang, Dr K. Lam, Dr K. Maeda, Dr H. Yamasaki, Mr Y. Takeshima and Mr O. Nagata for their support.

REFERENCES

Brenner, D., Lipton, J., Kaufman, L. and Williamson, S.J. (1978). Somatically evoked fields of the human brain. *Science*, **189**, 81–83.

Forss, N., Hari, R., Salmelin, R., Ahonen, A., Hamalainen, M., Kajola, M., Knuutila, J. and Simola, J. (1994). Activation of the human posterior parietal cortex by median nerve stimulation. *Experimental Brain Research*, **99**, 309–315.

Gallen, C.C., Sobel, D.F., Waltz, T., Aung, M., Copeland, B., Schwartz, B.J., Hirschkoff, E.C. and Bloom, F.E. (1993). Noninvasive presurgical neuromagnetic mapping of somatosensory cortex. *Neurosurgery*, **33**, 260–268.

Hari, R., Kaukoranta, E., Reinikainen, K., Huopaniemie, T. and Mauno, J. (1983). Neuromagnetic localization of cortical activity evoked by painful dental stimulation in man. *Neuroscience Letters*, **42**, 77–82.

Hari, R., Hamalainen, H., Hamalainen, M., Kekoni, J., Sams, M. and Tiihonen, J. (1990). Separate finger representations at the human second somatosensory cortex. *Neuroscience*, **37**, 245–249.

Hoshiyama, M., Kakigi, R., Koyama, S., Kitamura Y., Shimojo, M and Watanabe, S. (1995). Somatosensory evoked magnetic fields after mechanical stimulation of the scalp in humans. *Neuroscience Letters*, **195**, 29–32.

Hoshiyama, M., Kakigi, R., Koyama, S., Kitamura, Y., Shimojo, M and Watanabe, S. (1996). Somatosensory evoked magnetic fields following stimulation of the lip in humans. *Electroencephalography and Clinical Neurophysiology*, **100**, 96–104.

Hoshiyama, M., Koyama, S., Kitamura, Y., Watanabe, S., Shimojo, M. and Kakigi, R. (1997). Activity in parietal cortex following somatosensory stimulation in man: magnetoencephalographic study using spatio-temporal source analysis. *Brain Topography*, **10**, 23–30.

Howland, E.W., Wakai, R.T., Mjaanes, B.A., Balog, J.P. and Cleeland, C.S. (1995). Whole head mapping of magnetic fields following painful electric finger shock. *Brain Research Cognitive Brain Research*, **2**, 165–172.

Huttunen, J., Kobal, G., Kaukoranta, E. and Hari, R. (1986). Cortical responses to painful CO_2 stimulation of nasal mucosa; A magnetoencephalographic study in man. *Electroencephalography and Clinical Neurophysiology*, **64**, 347–349.

Hyvärinen, J. and Poranen, A. (1974). Function of the parietal associative area 7 as revealed from cellular discharges in alert monkeys. *Brain*, **97**, 673–692.

Iwamura, Y., Iriki, A. and Tanaka, M. (1994). Bilateral hand representation in the postcentral cortex. *Nature*, **369**, 554–556.

Kakigi, R. (1994). Somatosensory evoked magnetic fields following median nerve stimulation. *Neuroscience Research*, **20**, 165–174.

Kakigi, R., Koyama, S., Hoshiyama, M., Watanabe, S., Shimojo, M. and Kitamura, Y. (1995a). Gating of somatosensory evoked responses during active finger movement in man: magnetoencephalographic studies. *Journal of the Neurological Sciences*, **128**, 195–204.

Kakigi, R., Koyama, S., Hoshiyama, M., Shimojo, M., Kitamura, Y. and Watanabe, S. (1995b). Topography of somatosensory evoked magnetic fields following posterior tibial nerve stimulation. *Electroencephalography and Clinical Neurophysiology*, **95**, 127–134.

Kakigi, R., Koyama, S., Hoshiyama, M., Kitamura, Y., Shimojo, M. and Watanabe, S. (1995c). Pain-related magnetic fields following CO_2 laser stimulation in man. *Neuroscience Letters*, **192**, 45–48.

Kakigi, R., Koyama, S., Hoshiyama, M., Kitamura Y., Shimojo, M., Watanabe, S. and Nakamura, A. (1996). Effects of cutaneous interference stimulation on somatosensory evoked responses following median nerve stimulation in man; A magentoencephalographic study. *NeuroReport*, **7**, 405–408.

Kakigi, R. and Shibasaki, H. (1984). Scalp topography of mechanically and electrically evoked somatosensory potentials in man. *Electroencephalography and Clinical Neurophysiology*, **59**, 44–56.

Kakigi, R., Shimojo, M, Hoshiyama, M., Koyama, S., Watanabe, S., Naka D., Suzuki H. and Nakamura A. (1997). Effects of movement and movement imagery on somatosensory evoked magnetic fields following posterior tibial nerve stimulation. *Brain Research Cognitive Brain Research*, **5**, 241–253.

Kenshalo, D.R. Jr. and Willis, W.D. Jr (1991). The role of the cerebral cortex in pain perception. In *The Cerebral Cortex*. Vol 9, edited by A. Peters and E.G. Jones, pp. 153–211. New York, Plenum Press.

Kitamura, Y., Kakigi, R., Hoshiyama, M., Koyama, S., Shimojo, M. and Watanabe, S. (1995). Pain-related somatosensory evoked magnetic fields. *Electroencephalography and Clinical Neurophysiology*, **95**, 463–474.

Kitamura, Y., Kakigi, R., Hoshiyama, M., Koyama, S. and Nakamura, A. (1996). Effects of sleep on somato-sensory evoked responses in human: a magnetoencephalographic study. *Brain Research Cognitive Brain Research*, **4**, 275–279.

Kitamura, Y., Kakigi, R., Hoshiyama, M., Koyama, S., Shimojo, M and Watanabe, S. (1997). Pain-related somatosensory evoked magnetic fields following lower limb stimulation. *Journal of the Neurological Sciences*, **145**, 187–194.

Lam, K., Kakigi, R., Kaneoke, Y., Naka, D., Maeda, K. and Suzuki, H. (1999). Effects of visual and auditory stimulation on somatosensory evoked magnetic fields. *Electroencephalography and Clinical Neurophysiology*, **110**, 295–304.

Maeda, K., Kakigi, R., Hoshiyama, M. and Koyama, S. (1999). Topography of the secondary somatosensory cortex in humans: a magnetoencephalographic study. *NeuroReport*, **10**, 301–306.

Mauguiere, F., Merlet, I., Forss, N., Vanni, S., Jousmaki, V., Adeleine, P. and Hari, R. (1997). Activation of a distributed somatosensory cortical network in the human brain. A dipole modelling study of magnetic fields by median nerve stimulation. Part I: location and activation of SEF sources. *Electroencephalography and Clinical Neurophysiology*, **104**, 281–289.

Naka, D., Kakigi, R., Koyama, S.. Watanabe, S. and Kaneoke, Y. (1998). Effects of tactile interference stimulation on somatosensory evoked magnetic fields following tibial nerve stimulation. *Electroencephalography and Clinical Neurophysiology*, **109**, 168–177.

Nakamura, A., Yamada, T., Goto, A., Kato, T., Ito, K., Abe, Y., Kachi, T. and Kakigi, R. (1998). Somatosensory homunculus as drawn by MEG. *Neuroimage*, **7**, 377–386.

Penfield, W. and Boldrey, E. (1937). Somatic motor and sensory representation in the cerebral cortex of man as studied by electrical stimulation. *Brain*, **60**, 389–443.

Sarvas, J. (1987). Basic mathematical and electromagnetic concepts of the biomagnetic inverse problem. *Physics and Medical Biolology*, **32**, 11–22.

Scherg, M. (1995). BESA-M (Version 2.1). MEGIS Software GmbH, Munich, FRG.

Shimojo, M., Kakigi, R., Hoshiyama, M., Koyama, S., Kitamura, Y. and Watanabe, S. (1996a). Intracerebral interactions caused by bilateral median nerve stimulation in man: a magnetoencephalographic study. *Neuroscience Research*, **24**, 175–181.

Shimojo, M., Kakigi, R., Hoshiyama, M., Koyama, S., Kitamura, Y. and Watanabe, S. (1996b). Differentiation of receptive fields in the sensory cortex following stimulation of various nerves of the lower limb in humans: a magnetoencephalographic study. *Journal of Neurosurgery*, **85**, 255–262.

Shimojo, M., Kakigi, R., Hoshiyama, M., Koyama, S. and Watanabe, S. (1997). Magnetoencephalographic study of intracerebral interactions caused by bialteral posterior tibial nerve stimulation in man. *Neuroscience Research*, **28**, 41–47.

Suk, J., Ribary, U., Cappell, J., Yamamoto, T. and Llinas, R. (1991). Anatomical localization revealed by MEG recordings of the human somatosensory system. *Electroencephalography and Clinical Neurophysiology*, **78**, 185–196.

Watanabe, S., Kakigi, R., Koyama, S., Hoshiyama, M and Kaneoke, Y. (1998). Pain processing traced by magnetoencephalography in the human brain. *Brain Topography*, **10**, 255–264.

Whitsel, B.L., Petrucelli, L. and Werner, G. (1969). Symmetry and connectivity in the body surface in somatosensory cortex: identification by combined magnetic and potential recordings. *Journal of Physiology*, **32**, 218–223.

Wood, C.C., Cohen, D., Cuffin, B.N., Yarita, M. and Allison, T. (1985). Electrical sources in human somatosensory cortex: Identification by combined magnetic and potential recordings. *Science*, **227**, 1051–1053.

Xiang, J., Hoshiyama, M., Koyama, S., Kaneoke, Y., Suzuki, H., Watanabe, S., Naka, D. and Kakigi, R. (1997a). Somatosensory evoked magnetic fields following passive finger movement. *Brain Research Cognitive Brain Research*, **6**, 73–82.

Xiang, J., Kakigi, R., Hoshiyama, M., Kaneoke, Y., Naka, D., Takeshima, Y. and Koyama, S. (1997b). Somatosensory evoked magnetic fields and potentials following passive toe movement in humans. *Electroencephalography and Clinical Neurophysiology*, **104**, 393–401.

Yamasaki, H., Kakigi, R., Watanabe, S. and Naka, D. (1999) Effects of distraction on pain perception: magneto- and electro-encephalographic studies. *Brain Research Cognitive Brain Research*, **8**, 73–76.

INDEX

Printed and bound by CPI Group (UK) Ltd, Croydon, CR0 4YY

23/10/2024

01778226-0005